AS-Level Mathematics

The Revision Guide

Contents

Section One — The Riveting Basics of Algebra

A Few Definitions and Stuff 1
Multiplying Out Brackets 2
Taking Out Common Factors 3
Algebraic Fractions 4
Simplifying Expressions 5
Laws of Indices 6
Surds 7
Exam Questions 8

Section Two — Quadratics and the Factor Theorem

Sketching Quadratic Graphs 9
Factorising a Quadratic 10
Completing the Square 12
The Quadratic Formula 14
Linear Inequalities 17
Quadratic Inequalities 18
'Almost' Quadratic Equations 19
The Remainder and Factor Theorems 20
The Fantastic Factor Theorem 21
Factorising Cubics 22
Exam Questions 23

Section Three — Simultaneous Eqns and Geometry

Simultaneous Equations 24
Simultaneous Equations with Quadratics 25
Geometric Interpretation 26
Coordinate Geometry 27
Circles 30
Curve Sketching 32
Graph Transformations 33
Exam Questions 34

Section Four — Differentiation

Differentiation 35
Curve Sketching 38
Finding Tangents and Normals 39
Exam Questions 40

Section Five — Sequences and Series

Sequences 41
Arithmetic Progressions 43
Geometric Progressions 46
Binomial Expansions 49

Section Six — Trigonometry

The Trig Formulas You Need to Know 51
Using the Sine and Cosine Rules 52
Arc Lengths and Sector Areas 53
Graphs of Trig Functions 54
Transformed Trig Graphs 55
Solving Trig Equations in a Given Interval 56
Exam Questions 60

Section Seven — Log_A

Logs 61
Exponentials 63

Section Eight— Integration

Integration 64
Areas Between Curves 67
Numerical Integration of Functions 68
Exam Questions 70

Contents

Section Nine — Data
Histograms ... 71
Presenting Data ... 72
Stem and Leaf Diagrams ... 73
Mean, Median, Mode and Range ... 74
Cumulative Frequency Diagrams ... 76
Variance and Standard Deviation ... 78
Coding ... 80
Skewness and Outliers ... 82
Populations and Sampling ... 84

Section Ten — Probability
Random Events and Their Probability ... 86
Probability ... 88
Conditional Probability ... 90
Permutations and Combinations ... 92

Section Eleven — Probability Distributions
Probability Distributions ... 94
The Distribution Function ... 96
Discrete Uniform Distributions ... 97
Expected Values, Mean and Variance ... 98
The Binomial Distribution ... 100
The Geometric Distribution ... 102
The Normal Distribution ... 104
Confidence Intervals ... 108

Section Twelve — Correlation and Regression
Correlation ... 110
Linear Regression ... 114

Section Thirteen — Kinematics
Constant Acceleration Equations ... 116
Motion Graphs ... 118
Variable Acceleration ... 120

Section Fourteen — Vectors
Vectors ... 122
Vector Motion ... 124

Section Fifteen — Statics
Mathematical Modelling ... 126
Forces are Vectors ... 128
Types of Forces ... 130
Friction ... 132
Moments ... 134

Section Sixteen — Dynamics
Newton's Laws ... 136
Friction and Inclined Planes ... 138
Constant Acceleration and Path Equations ... 140
Connected Particles ... 142
Momentum ... 146
Impulse ... 148

Section Seventeen — Projectiles
Projectiles ... 150
Equations of Path ... 152
Cows ... 154

Answers ... 155
Index ... 186

For each module there are some formulas you've really got to know.
We've put them on the Online Revision bit of our website — www.cgpbooks.co.uk
That way you can print them out, and stick them above your revision desk.

Contributors:
Andy Ballard, Paul Conner, Charley Darbishire, Janet Dickinson, Bill Dolling, Dominic Hall, Dave Harding, Claire Jackson, Simon Little, Tim Major, Mark Moody, Iain Nash, Sam Norman, Ali Palin, Andy Park, Glenn Rogers, Garry Rowlands, Rob Savage, Emma Singleton, Mike Smith, Claire Thompson, Julie Wakeling, James Paul Wallis, Kieran Wardell, Sharon Watson, Chris Worth

Published by Coordination Group Publications Ltd.

ISBN: 978 1 84146 988 1

Groovy website: www.cgpbooks.co.uk
Jolly bits of clipart from CorelDRAW®
Printed by Elanders Hindson Ltd, Newcastle upon Tyne.

A Few Definitions and Stuff

This page is for AQA Core 1, AQA Core 2, OCR Core 1, OCR Core 2, Edexcel Core 1, Edexcel Core 2

Yep, this is a pretty dull way to start a book. A list of definitions. But in the words of that annoying one-hit wonder bloke in the tartan suit, things can only get better. Which is nice.

Polynomials

POLYNOMIALS are expressions of the form $a + bx + cx^2 + dx^3 + \dots$

$5y^3 + 2y + 23$ ← Polynomial in the variable y.

$1 + x^2$

$z^{42} + 3z - z^2 - 1$ ← Polynomial in the variable z.

The bits separated by the +/– signs are terms.

x, y and z are VARIABLES. They're usually what you solve equations to find. They often have more than one possible value.

Letters a, b and c are CONSTANTS. Constants never change. They're fixed numbers — but can be represented by letters. π is a good example. You use the symbol π, but it's just a number = 3.1415...

Functions

FUNCTIONS take a value, do something to it, and output another value.

$f(x) = x^2 + 1$ ← function f takes a value, squares it and adds 1.

$g(x) = 2 - \sin 2x$ ← function g takes a value (in degrees), doubles it, takes the sine of it, then takes the value away from 2.

You can plug values into a function — just replace the variable with a certain number.

$f(-2) = (-2)^2 + 1 = 5$

$f(0) = (0)^2 + 1 = 1$

$f(252) = (252)^2 + 1 = 63505$

$g(-90) = 2 - \sin(-180°) = 2 - 0 = 2$

$g(0) = 2 - \sin 0° = 2 - 0 = 2$

$g(45) = 2 - \sin 90° = 2 - 1 = 1$

Exam questions use functions all the time. They generally don't have that much to do with the actual question. It's just a bit of terminology to get comfortable with.

Multiplication and Division

There's three different ways of showing MULTIPLICATION:

1) with good old-fashioned "times" signs (×):

$$f(x) = (2x \times 6y) + (2x \times \sin x) + (z \times y)$$

← The multiplication signs and the variable *x* are easily confused.

2) or sometimes just use a little dot:

$$f(x) = 2x.6y + 2x.\sin x + z.y$$

← Dots are better for long expressions — they're less confusing and easier to read.

3) but you often don't need anything at all:

$$f(x) = 12xy + 2x\sin x + zy$$

And there's three different ways of showing DIVISION:

1) $\frac{x+2}{3}$

2) $(x+2) \div 3$

3) $(x+2)/3$

Equations and Identities

This is an IDENTITY:

$$x^2 - y^2 \equiv (x+y)(x-y)$$

Make up any values you like for x and y, and it's always true. The left-hand side always equals the right-hand side.

But this is an EQUATION:

$$y = x^2 + x$$

This has at most two possible solutions for each value of y. e.g. if y=0, x can only be 0 or -1.

The difference is that the identity's true for all values of x and y, but the equation's only true for certain values.

NB: If it's an identity, use the ≡ sign instead of =.

Multiplying Out Brackets

This page is for AQA Core 1, AQA Core 2, OCR Core 1, OCR Core 2, Edexcel Core 1, Edexcel Core 2

In this horrific nightmare that is AS-level maths, you need to manipulate and simplify expressions all the time.

Remove brackets by **Multiplying** them out

Here's the basic types you have to deal with. You'll have seen them before. But there's no harm in reminding you, eh?

Multiply Your Brackets Here — we do all shapes and sizes

Single Brackets

$a(b+c+d) \equiv ab+ac+ad$

Double Brackets

$(a+b)(c+d) \equiv ac+ad+bc+bd$

Squared Brackets

$(a+b)^2 \equiv (a+b)(a+b) \equiv a^2+2ab+b^2$

Use the middle stage until you're comfortable with it. Just never make this mistake: $(a+b)^2 \equiv a^2+b^2$

Long Brackets

Write it out again with each term from one bracket separately multiplied by the other bracket.

$(x+y+z)(a+b+c+d)$

$\equiv x(a+b+c+d)+y(a+b+c+d)+z(a+b+c+d)$

Then multiply out each of these brackets, one at a time.

Single Brackets

$3xy(x^2+2x-8)$ ← Multiply all the terms inside the brackets by the bit outside — separately.

$(3xy \times x^2)+(3xy \times 2x)+(3xy \times (-8))$ ← I've put brackets round each bit to make it easier to read.

All the stuff in the brackets now needs sorting out. Work on each bracket separately.

$(3x^3y)+(6x^2y)+(-24xy)$ ← Multiply the numbers first, then put the x's and other letters together.

$3x^3y+6x^2y-24xy$

Squared Brackets

$(2y^2+3x)^2$

Either write it as two brackets and multiply it out...

$(2y^2+3x)(2y^2+3x)$

$2y^2.2y^2+2y^2.3x+3x.2y^2+3x.3x$ ← The dot just means 'multiplied by' — the same as the × sign.

$4y^4+6xy^2+6xy^2+9x^2$ ← From here on it's simplification — nothing more, nothing less.

$4y^4+12xy^2+9x^2$

...or do it in one go.

$(2y^2)^2+2(2y^2)(3x)+(3x)^2$

a^2 $\quad 2ab \quad$ b^2

$4y^4+12xy^2+9x^2$

Long Brackets

$(2x^2+3x+6)(4x^3+6x^2+3)$

$2x^2(4x^3+6x^2+3)+3x(4x^3+6x^2+3)+6(4x^3+6x^2+3)$ ← Each term in the first bracket has been multiplied by the second bracket.

Multiply out each of these brackets now.

$(8x^5+12x^4+6x^2)+(12x^4+18x^3+9x)+(24x^3+36x^2+18)$

Then simplify it all...

$\underline{8x^5+24x^4+42x^3+42x^2+9x+18}$

Go forth and multiply out brackets...

OK, so this is obvious, but I'll say it anyway — if you've got 3 or more brackets together, multiply them out 2 at a time. Then you'll be turning a really hard problem into two easy ones. You can do that loads in maths. In fact, writing the same thing in different ways is what maths is about. That and sitting in classrooms with tacky 'maths can be fun' posters...

Taking Out Common Factors

This page is for AQA Core 1, AQA Core 2, OCR Core 1, OCR Core 2, Edexcel Core 1, Edexcel Core 2

Common factors need to be hunted down, and taken outside the brackets. They are a danger to your exam mark.

Spot those ***Common Factors***

A bit which is in each term of an expression is a common factor.

Spot Those Common Factors $2x^3z + 4x^2yz + 14x^2y^2z$

Look for any bits that are in each term.

Numbers: there's a common factor of 2 here because 2 divides into 2, 4 and 14.

Variables: there's at least an x^2 in each term and there's a z in each term.

So there's a common factor of $2x^2z$ in this expression.

And Take Them Outside a Bracket

If you spot a common factor you can "take it out":

$2x^2z(x + 2y + 7y^2)$

Write the common factor outside a bracket.

and put what's left of each term inside the bracket:

Afterwards, always multiply back out to check you did it right:

Check by Multiplying Out Again

$2x^2z(x+2y+7y^2) \equiv 2x^3z + 4x^2yz + 14x^2y^2z$

But it's not just numbers and variables you need to look for...

Trig Functions: $\sin x \sin y + \cos x \sin y$

This has a common factor of sin y. So take it out to get...

$\sin y(\sin x + \cos x)$

Brackets: $(y+a)^2(x-a)^3 + (x-a)^2$

$(x - a)^2$ is a common factor — it comes out to give:

$(x-a)^2((y+a)^2(x-a)+1)$

Look for ***Common Factors*** *when* ***Simplifying Expressions***

Example: Simplify... $(x + 1)(x - 2) + (x + 1)^2 - x(x + 1)$

There's an (x+1) factor in each term, so we can take this out as a common factor (hurrah).

$(x+1)\{(x-2)+(x+1)-x\}$

The terms inside the big bracket are the old terms with an (x+1) removed.

At this point you should check that this multiplies out to give the original expression. (You can just do this in your head, if you trust it.)

Then simplify the big bracket's innards:

$(x+1)(\cancel{x}-2+x+1-\cancel{x})$

$\equiv (x+1)(x-1)$

$\equiv x^2 - 1$

Get this answer by multiplying out the two brackets (or by using the "difference of two squares").

Bored of spotting trains or birds? Try common factors...

You'll be doing this business of taking out common factors a lot — so get your head round this. It's just a case of looking for things that are in all the different terms of an expression, i.e. bits they have in common. And if something's in all the different terms, save yourself some time and ink, and write it once — instead of two, three or more times.

Algebraic Fractions

This page is for AQA Core 1, AQA Core 2, OCR Core 1, OCR Core 2, Edexcel Core 1, Edexcel Core 2

No one likes fractions. But just like Mondays, you can't put them off forever. Face those fears. Here goes...

The first thing you've got to know about fractions:

$$\frac{a}{x}+\frac{b}{x}+\frac{c}{x}\equiv\frac{a+b+c}{x}$$

You can just add the stuff on the top lines because the bottom lines are all the same.

x is called a common denominator — a fancy way of saying 'the bottom line of all the fractions is x'.

Add fractions by putting them over a **Common Denominator**...

Finding a common denominator just means 'rewriting some fractions so all their bottom lines are the same'.

Example: Simplify $\frac{1}{2x}-\frac{1}{3x}+\frac{1}{5x}$

You need to rewrite these so that all the bottom lines are equal. What you want is something that all these bottom lines divide into.

Put It over a Common Denominator

30 is the lowest number that 2, 3, and 5 go into.
So the common denominator is 30x.

$$\frac{15}{30x}-\frac{10}{30x}+\frac{6}{30x}$$

Always check that these divide out to give what you started with.

$$\equiv\frac{15-10+6}{30x}\equiv\frac{11}{30x}$$

...even **horrible** looking ones

Yep, finding a common denominator even works for those fraction nasties — like these:

Example: Find $\frac{2y}{x(x+3)}+\frac{1}{y^2(x+3)}-\frac{x}{y}$

Find the Common Denominator

Take all the individual 'bits' from the bottom lines and multiply them together. Only use each bit once unless something on the bottom line is squared.

The individual 'bits' here are x, (x+3) and y...

$$xy^2(x+3)$$

...but you need to use y^2 because there's a y^2 in the second fraction's denominator.

Put Each Fraction over the Common Denominator

Make the denominator of each fraction into the common denominator.

$$\frac{y^2\times 2y}{y^2x(x+3)}+\frac{x\times 1}{xy^2(x+3)}-\frac{xy(x+3)\times x}{xy(x+3)y}$$

Multiply the top and bottom lines of each fraction by whatever makes the bottom line the same as the common denominator.

Combine into One Fraction

Once everything's over the common denominator — you can just add the top lines together.

$$\equiv\frac{2y^3+x-x^2y(x+3)}{xy^2(x+3)}$$

As always — if you see a minus sign, look out for possible problems.

All the bottom lines are the same — so you can just add the top lines.

$$\equiv\frac{2y^3+x-x^3y-3x^2y}{xy^2(x+3)}$$

All you need to do now is tidy up the top.

Not the nicest of answers. But it is the answer, so it'll have to do.

Well put me over a common denominator and pickle my walrus...

Adding fractions — turning lots of fractions into one fraction. Sounds pretty good to me, since it means you don't have to write as much. Better do it carefully, though — otherwise you can watch those marks shoot straight down the toilet.

Simplifying Expressions

This page is for AQA Core 1, AQA Core 2, OCR Core 1, OCR Core 2, Edexcel Core 1, Edexcel Core 2

I know this is basic stuff but if you don't get really comfortable with it you will make silly mistakes. You will.

Cancelling stuff on the top and bottom lines

Cancelling stuff is good — because it means you've got rid of something, and you don't have to write as much.

Example: Simplify $\frac{ax+ay}{az}$

You can do this in two ways. Use whichever you prefer — but make sure you understand the ideas behind both.

Factorise — then Cancel

$$\frac{ax+ay}{az} \equiv \frac{a(x+y)}{az}$$

Factorise the top line.

Cancel the 'a'.

$$\equiv \frac{\not{a}(x+y)}{\not{a}z} \equiv \frac{x+y}{z}$$

Split into Two Fractions — then Cancel

$$\frac{ax+ay}{az} \equiv \frac{ax}{az} + \frac{ay}{az}$$

This is an okay thing to do — just think what you'd get if you added these.

$$\equiv \frac{\not{a}x}{\not{a}z} + \frac{\not{a}y}{\not{a}z} \equiv \frac{x}{z} + \frac{y}{z}$$

This answer's the same as the one from the first box — honest. Check it yourself by adding the fractions.

*Simplifying complicated-looking **Brackets***

Example: Simplify the expression $(x-y)(x^2+xy+y^2)$

There's only one thing to do here.... Multiply out those brackets!

$$(x-y)(x^2+xy+y^2) \equiv x(x^2+xy+y^2) - y(x^2+xy+y^2)$$

Multiplying each term in the first bracket by the second bracket.

$$\equiv (x^3+x^2y+xy^2) - (x^2y+xy^2+y^3)$$

Multiplying out each of these two brackets.

$$\equiv x^3+x^2y+xy^2-x^2y-xy^2-y^3$$

Don't forget these become minus signs because of the minus sign in front of the bracket.

And then the x^2y and the xy^2 terms disappear...

$$\equiv x^3-y^3$$

*Sometimes you just have to do **Anything** you can think of and **Hope...***

Sometimes it's not easy to see what you're supposed to do to simplify something.
When this happens — just do anything you can think of and see what 'comes out in the wash'.

Example: Simplify $4x + \frac{4x}{x+1} - 4(x+1)$

There's nothing obvious to do — so do what you can. Try adding them as fractions...

$$4x + \frac{4x}{x+1} - 4(x+1) \equiv \frac{(x+1)\times 4x}{x+1} + \frac{4x}{x+1} - \frac{(x+1)\times 4(x+1)}{x+1}$$

The common denominator is $(x+1)$.

$$\equiv \frac{4x^2+4x+4x-4(x+1)^2}{x+1}$$

Still looks horrible. So work out the brackets — but don't forget the minus signs.

$$\equiv \frac{4x^2+4x+4x-4x^2-8x-4}{x+1}$$

$$\equiv -\frac{4}{x+1}$$

Aha — everything disappears to leave you with this. And this is definitely simpler than it looked at the start.

Don't look at me like that...

Choose a word, any word at all. Like "Simple". Now stare at it. Keep staring at it. Does it look weird? No? Stare a bit longer. Now does it look weird? Yes? Why is that? I don't understand.

Laws of Indices

This page is for AQA Core 1, AQA Core 2, OCR Core 1, OCR Core 2, Edexcel Core 1, Edexcel Core 2

You use the laws of indices a helluva lot in maths — when you're integrating, differentiating and ...er... well loads of other places. So take the time to get them sorted now.

Three mega-important *Laws of Indices*

The Laws of Indices are the same thing as The Power Laws

You must know these three rules. I can't make it any clearer than that.

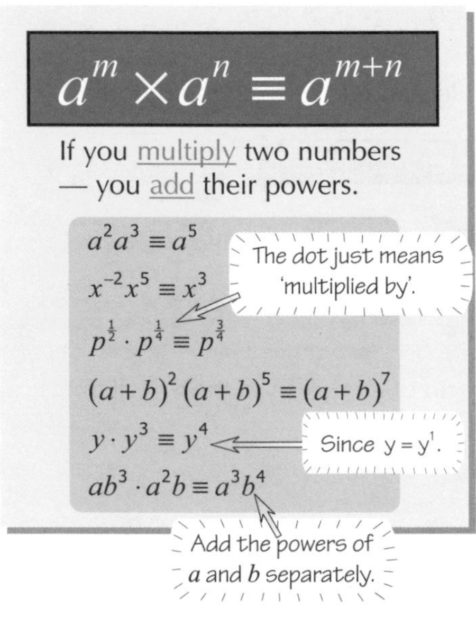

$a^m \times a^n \equiv a^{m+n}$
If you multiply two numbers — you add their powers.
$a^2a^3 \equiv a^5$
$x^{-2}x^5 \equiv x^3$
$p^{\frac{1}{2}} \cdot p^{\frac{1}{4}} \equiv p^{\frac{3}{4}}$
$(a+b)^2(a+b)^5 \equiv (a+b)^7$
$y \cdot y^3 \equiv y^4$
$ab^3 \cdot a^2b \equiv a^3b^4$
The dot just means 'multiplied by'.
Since $y = y^1$.
Add the powers of a and b separately.

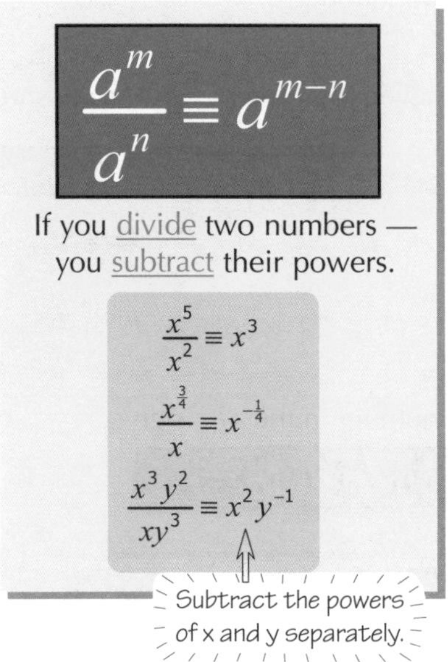

$\frac{a^m}{a^n} \equiv a^{m-n}$
If you divide two numbers — you subtract their powers.
$\frac{x^5}{x^2} \equiv x^3$
$\frac{x^{\frac{3}{4}}}{x} \equiv x^{-\frac{1}{4}}$
$\frac{x^3y^2}{xy^3} \equiv x^2y^{-1}$
Subtract the powers of x and y separately.

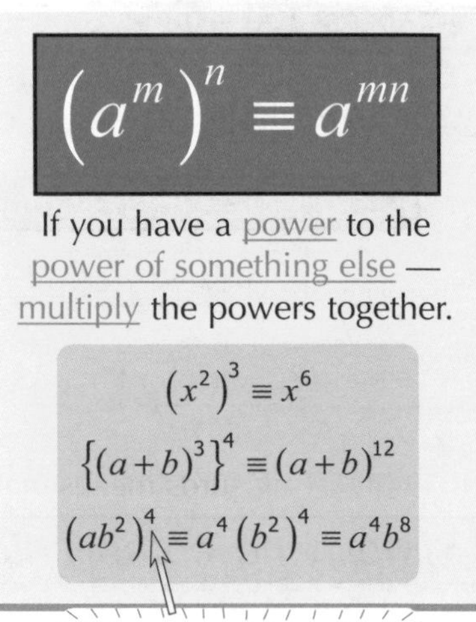

$(a^m)^n \equiv a^{mn}$
If you have a power to the power of something else — multiply the powers together.
$(x^2)^3 \equiv x^6$
$\{(a+b)^3\}^4 \equiv (a+b)^{12}$
$(ab^2)^4 \equiv a^4(b^2)^4 \equiv a^4b^8$

This power applies to both bits inside the brackets.

Other important stuff about *Indices*

You can't get very far without knowing this sort of stuff. Learn it — you'll definitely be able to use it.

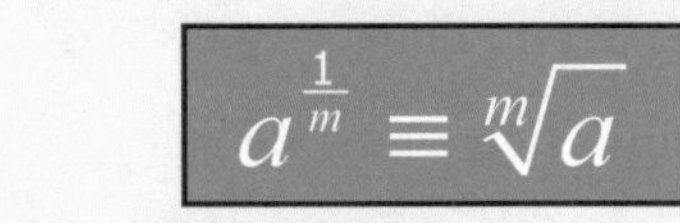

$a^{\frac{1}{m}} \equiv \sqrt[m]{a}$

You can write roots as powers...

Examples:

$x^{\frac{1}{5}} \equiv \sqrt[5]{x}$

$4^{\frac{1}{2}} = \sqrt{4} = 2$

$125^{\frac{1}{3}} = \sqrt[3]{125} = 5$

$$a^{\frac{m}{n}} \equiv \sqrt[n]{a^m} \equiv \left(\sqrt[n]{a}\right)^m$$

A power that's a fraction like this is the root of a power — or the power of a root.

Examples:

$9^{\frac{3}{2}} = \left(9^{\frac{1}{2}}\right)^3 = \left(\sqrt{9}\right)^3 = 3^3 = 27$

$16^{\frac{3}{4}} = \left(16^{\frac{1}{4}}\right)^3 = \left(\sqrt[4]{16}\right)^3 = 2^3 = 8$

It's often easier to work out the root first, then raise it to the power.

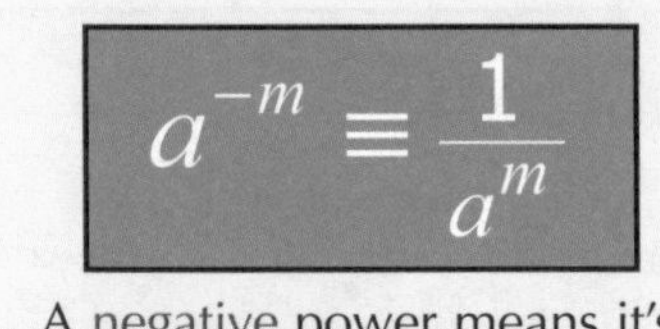

$a^{-m} \equiv \frac{1}{a^m}$

A negative power means it's on the bottom line of a fraction.

Examples:

$x^{-2} \equiv \frac{1}{x^2}$

$(x+1)^{-1} \equiv \frac{1}{x+1}$

$2^{-3} = \frac{1}{2^3} = \frac{1}{8}$

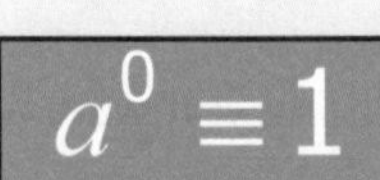

$a^0 \equiv 1$

This works for any number or letter.

Examples:

$x^0 \equiv 1$

$(a+b)^0 \equiv 1$

$2^0 = 1$

Did someone say Powers?

Indices, indices — de fish all live indices...

What can I say that I haven't said already? Blah, blah, important. Blah, blah, learn these. Blah, blah, use them all the time. Mmm, that's about all that needs to be said really. So I'll be quiet and let you get on with what you need to do.

Surds

This page is for AQA Core 1, AQA Core 2, OCR Core 1, OCR Core 2, Edexcel Core 1, Edexcel Core 2

A surd is a number like $\sqrt{2}$, $\sqrt[3]{12}$ or $5\sqrt{3}$ — one that's written with the $\sqrt{\ }$ sign. They're important because you can give exact answers where you'd otherwise have to round to a certain number of decimal places.

Surds are sometimes the only way to give an **Exact Answer**

Put $x = \sqrt{2}$ into a calculator and you'll get something like x = 1.414213562...
But square 1.414213562 and you get 1.999999999.

And no matter how many decimal places you use, you'll never get exactly 2.
The only way to write the exact, spot on answer is to use surds.

There are basically **Three Rules** for using **Surds**

There are three rules you'll need to know to be able to use surds properly. Check out the 'Rules of Surds' box below.

Example: (i) Simplify $\sqrt{12}$ and $\sqrt{\frac{3}{16}}$. (ii) Show that $\frac{9}{\sqrt{3}} = 3\sqrt{3}$. (iii) Find $\left(2\sqrt{5} + 3\sqrt{6}\right)^2$.

(i) Simplifying surds means making the number in the $\sqrt{\ }$ sign smaller, or getting rid of a fraction in the $\sqrt{\ }$ sign.

$$\sqrt{12} = \sqrt{4\times3} = \sqrt{4}\times\sqrt{3} = 2\sqrt{3}$$

Using $\sqrt{ab} \equiv \sqrt{a}\times\sqrt{b}$

$$\sqrt{\frac{3}{16}} = \frac{\sqrt{3}}{\sqrt{16}} = \frac{\sqrt{3}}{4}$$

Using $\sqrt{\frac{a}{b}} \equiv \frac{\sqrt{a}}{\sqrt{b}}$.

(ii) For questions like these, you have to write a number (here, it's 3) as $3 = \left(\sqrt{3}\right)^2 = \sqrt{3}\times\sqrt{3}$.

$$\frac{9}{\sqrt{3}} = \frac{3\times3}{\sqrt{3}} = \frac{3\times\sqrt{3}\times\sqrt{3}}{\sqrt{3}} = 3\sqrt{3}$$

Cancelling $\sqrt{3}$ from the top and bottom lines.

(iii) Multiply surds very carefully — it's easy to make a silly mistake.

$$\begin{aligned}\left(2\sqrt{5}+3\sqrt{6}\right)^2 &= \left(2\sqrt{5}+3\sqrt{6}\right)\left(2\sqrt{5}+3\sqrt{6}\right)\\ &= \left(2\sqrt{5}\right)^2 + 2\times\left(2\sqrt{5}\right)\times\left(3\sqrt{6}\right)+\left(3\sqrt{6}\right)^2\\ &= \left(2^2\times\sqrt{5}^2\right)+\left(2\times2\times3\times\sqrt{5}\times\sqrt{6}\right)+\left(3^2\times\sqrt{6}^2\right)\\ &= 20+12\sqrt{30}+54\\ &= 74+12\sqrt{30}\end{aligned}$$

$= 4\times5 = 20$

$12\sqrt{5}\sqrt{6} = 12\sqrt{30}$

$= 9\times6 = 54$

Rules of Surds

There's not really very much to remember.

$$\sqrt{ab} \equiv \sqrt{a}\sqrt{b}$$

$$\sqrt{\frac{a}{b}} \equiv \frac{\sqrt{a}}{\sqrt{b}}$$

$$a \equiv \left(\sqrt{a}\right)^2 \equiv \sqrt{a}\sqrt{a}$$

Remove surds from the bottom of fractions by **Rationalising the Denominator**

Surds are pretty darn complicated.

So they're the last thing you want at the bottom of any fractions you've got.

But have no fear — Rationalise the Denominator...

Yup, you heard... (it means getting rid of the surds from the bottom of a fraction).

Example: Rationalise the denominator of $\frac{1}{1+\sqrt{2}}$

Multiply the top and bottom by the denominator (but change the sign in front of the surd).

$$\frac{1}{1+\sqrt{2}}\times\frac{1-\sqrt{2}}{1-\sqrt{2}}$$

$$\frac{1-\sqrt{2}}{\left(1+\sqrt{2}\right)\left(1-\sqrt{2}\right)} = \frac{1-\sqrt{2}}{1^2+\sqrt{2}-\sqrt{2}-\sqrt{2}^2}$$

This works because: $(a+b)(a-b) \equiv a^2 - b^2$ So the surds cancel out.

$$\frac{1-\sqrt{2}}{1-2} = \frac{1-\sqrt{2}}{-1} = -1+\sqrt{2}$$

Surely the pun is mightier than the surd...

There's not much to surds really — but they cause a load of hassle. Think of them as just ways to save you the bother of getting your calculator out and pressing buttons — then you might grow to know and love them. The box of rules in the middle is the vital stuff. Learn them till you can write them down without thinking — then get loads of practice with them.

Exam Questions

The only way to really get good at AS maths is to practise exam questions like the ones here — well go on then...

Exam Questions

1 Prove that $\frac{\sqrt{6}+\sqrt{2}}{\sqrt{6}-\sqrt{2}}$ can be written in the form $a+b\sqrt{3}$, where a and b are both integers.

Find the values of a and b.

(6 marks)

2 Given that $3^{p-1}=\frac{1}{27}$ and that $3^q=81$:

(a) Find the exact values of p and q.

(2 marks)

(b) Hence find the exact value of $(3^q)^{\frac{1}{2}}+3^{-p}$.

(2 marks)

3 Express $\frac{x}{(x-3)(x-2)}+\frac{8}{x^2-4}$ as a single fraction in its simplest form.

(4 marks)

4 Given that $27^x=3^{2y+1}$, find x in the form $x = py + q$, where p and q are exact fractions.

(2 marks)

5 **(a)** Write down the exact value of $36^{-½}$.

(2 marks)

(b) Simplify $\frac{a^6\times a^3}{\sqrt{a^4}}\div a^{\frac{1}{2}}$.

(2 marks)

(c) Express $(5\sqrt{5}+2\sqrt{3})^2$ in the form $a+b\sqrt{c}$, where a, b and c are integers to be found.

(4 marks)

Sketching Quadratic Graphs

This page is for AQA Core 1, OCR Core 1, Edexcel Core 1

If a question doesn't seem to make sense, or you can't see how to go about solving a problem, try drawing a graph. It sometimes helps if you can actually see what the problem is, rather than just reading about it.

Sketch the graphs of the following quadratic functions:

① $y = 2x^2 - 4x + 3$

② $y = 8 - 2x - x^2$

Quadratic graphs are ***Always*** *u-shaped or n-shaped*

A The first thing you need to know is whether the graph's going to be u-shaped or n-shaped (upside down). To decide, look at the coefficient of x^2.

$y = 2x^2 - 4x + 3$

The coefficient of x^2 here is positive... ...so the graph's u-shaped. +ve

$y = 8 - 2x - x^2$

The coefficient of x^2 here is negative... ...so the graph's upside down (n-shaped). −ve

B Now find the places where the graph crosses the axes (both the y-axis and the x-axis).

(i) Put x=0 to find where it meets the y-axis.

$y = 2x^2 - 4x + 3$

$y = (2\times0^2) - (4\times0) + 3$ so $y = 3$ ← That's where it crosses the y-axis

(ii) Solve y=0 to find where it meets the x-axis.

$2x^2 - 4x + 3 = 0$

$b^2 - 4ac = -8 < 0$ ← You could use the formula. But first check $b^2 - 4ac$ to see if y = 0 has any roots.

So it has no solutions, and doesn't cross the x-axis.

For more info, see page 16.

(i) Put x=0.

$y = 8 - 2x - x^2$

$y = 8 - (2\times0) - 0^2$ so $y = 8$

(ii) Solve y=0.

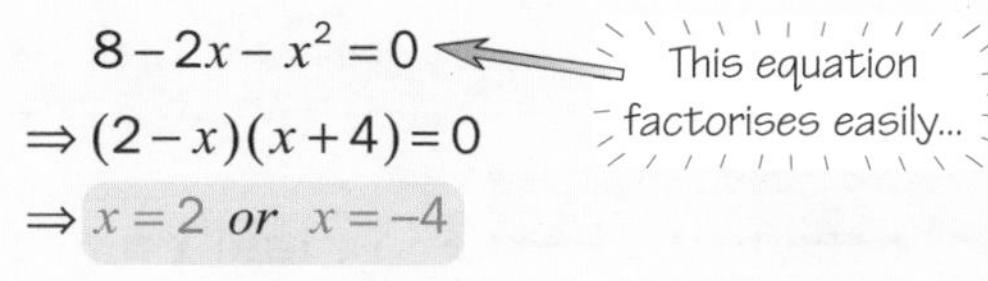

$8 - 2x - x^2 = 0$ ← This equation factorises easily...

$\Rightarrow (2 - x)(x + 4) = 0$

$\Rightarrow x = 2$ or $x = -4$

C Finally, find the minimum or maximum.

Since $y = 2(x-1)^2 + 1$ ← By completing the square (see page 12).

the minimum value is $y = 1$, which occurs at $x = 1$

The minimum or maximum of the graph is always at $x = \frac{-b}{2a}$

The maximum value is halfway between the roots — the graph's symmetrical.

The maximum value is at $x = -1$

So the maximum is $y = 8 - (2\times-1) - (-1)^2$

i.e. the graph has a maximum at the point (–1,9).

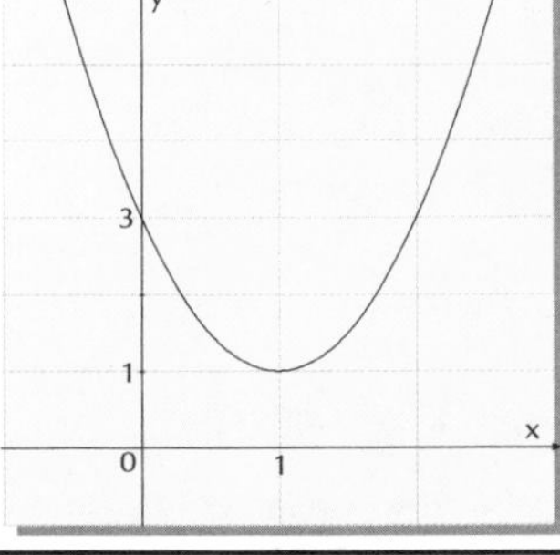

Sketching Quadratic Graphs

A) **up or down** — decide which direction the curve points in.

B) **axes** — find where the curve crosses them.

C) **max / min** — find the turning point.

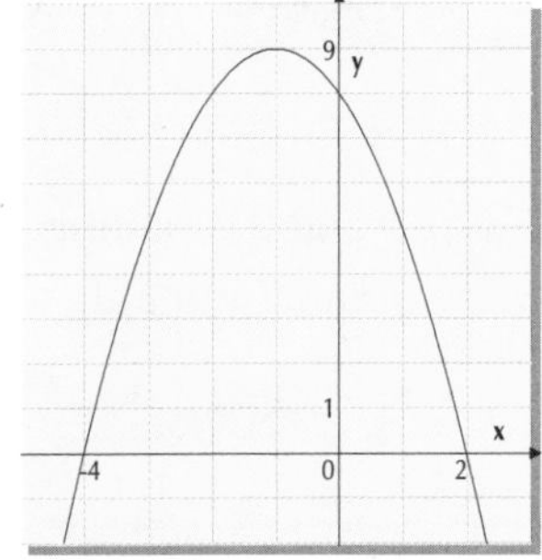

Van Gogh, Monet — all the greats started out sketching graphs...

So there's three steps here to learn. Simple enough. You can do the third step (finding the max/min point) by either a) completing the square, which is covered a bit later, or b) using the fact that the graph's symmetrical — so once you've found the points where it crosses the x-axis, the point halfway between them will be the max/min. It's all laughs here...

Factorising a Quadratic

This page is for AQA Core 1, OCR Core 1, Edexcel Core 1

Factorising a quadratic means putting it into two brackets — and is useful if you're trying to draw a graph of a quadratic or solve a quadratic equation. It's pretty easy if a = 1 (in $ax^2 + bx + c$ form), but can be a real pain otherwise.

$$x^2 - x - 12 = (x-4)(x+3)$$

Factorising's not so bad when **a=1**

Example: Solve $x^2 - 8 = 2x$ by factorising.

A) Put into $ax^2 + bx + c = 0$ Form

$x^2 - 2x - 8 = 0$ ⟸ So a = 1, b = −2, c = −8.

Write down the two brackets with x's in: $x^2 - 2x - 8 = (x \quad)(x \quad)$

B) Find the Two Numbers

Find two numbers that multiply together to make 'c' but which also add or subtract to give 'b' (you can ignore any minus signs for now).

1 and 8 multiply to give 8 — and add / subtract to give 9 and 7.
2 and 4 multiply to give 8 — and add / subtract to give 6 and 2.

This is the value for 'b' you're after — so this is the right combination: 2 and 4.

You could do these two steps in one. You'd be looking for two numbers that multiply together to make −8 and add to make −2.

E.g. −1 × 8 = −8
1 × −8 = −8
−2 × 4 = −8
2 × −4 = −8

C) Find the Signs

Now all you have to do is put in the plus or minus signs.

$x^2 - 2x - 8 = (x \quad 4)(x \quad 2)$

$x^2 - 2x - 8 = (x+2)(x-4)$

If 'b' is negative, then the signs must be different — and the larger number (e.g. 4) must be negative.

It must be +2 and −4 because 2 × (−4) = −8 and 2 + (−4) = 2 − 4 = −2

D) Solve the Equation

All you've done so far is to factorise the equation — you've still got to solve it.

$(x+2)(x-4) = 0$

$\Rightarrow x + 2 = 0$ or $x - 4 = 0$

$\Rightarrow x = -2$ or $x = 4$

Don't forget this last step. The factors aren't the answer.

Factorising Quadratics

A) Rearrange the equation into the standard ax^2+bx+c form.

B) Write down the two brackets: (x)(x)

C) Find two numbers that multiply to give 'c' and add / subtract to give 'b' (ignoring signs).

D) Put the numbers in the brackets and choose their signs.

Another **Example...**

Example: Solve $x^2 + 4x - 21 = 0$ by factorising.

This equation is already in the standard format — you can write down the brackets straight away.

$x^2 + 4x - 21 = (x \quad)(x \quad)$

1 and 21 multiply to give 21 — and add / subtract to give 22 and 20.
3 and 7 multiply to give 21 — and add / subtract to give 10 and 4.

This is the value of 'b' you're after — 3 and 7 are the right numbers.

$x^2 + 4x - 21 = (x+7)(x-3)$

And solving the equation to find x gives... $\Rightarrow x = -7$ or $x = 3$

Scitardauq Gnisirotcaf — you should know it backwards...

Factorising quadratics — this is very basic stuff. You've really got to be comfortable with it. If you're even slightly rusty, you need to practise it until it's second nature. Remember why you're doing it — you don't factorise simply for the pleasure it gives you — it's so you can solve quadratic equations. Well, that's the theory anyway...

Factorising a Quadratic

This page is for AQA Core 1, OCR Core 1, Edexcel Core 1

It's not over yet...

Factorising a quadratic when $a \neq 1$

These can be a real pain. The basic method's the same as on the previous page — but it can be a bit more awkward.

Example: Factorise $3x^2+4x-15$

A Write Down Two Brackets

As before, write down two brackets — but instead of just having x in each, you need two things that will multiply to give $3x^2$.

$3x^2+4x-15=(3x\quad)(x\quad)$

It's got to be 3x and x here.

B The Fiddly Bit

You need to find two numbers that multiply together to make 15 — but which will give you 4x when you multiply them by x and 3x, and then add / subtract them.

$(3x\quad 1)(x\quad 15) \Rightarrow x$ and $45x$ which then add or subtract to give 46x and 44x.

$(3x\quad 15)(x\quad 1) \Rightarrow 15x$ and $3x$ which then add or subtract to give 18x and 12x.

$(3x\quad 3)(x\quad 5) \Rightarrow 3x$ and $15x$ which then add or subtract to give 18x and 12x.

$(3x\quad 5)(x\quad 3) \Rightarrow 5x$ and $9x$ which then add or subtract to give 14x and 4x.

This is the value you're after — so this is the right combination.

C Add the Signs

You know the brackets must be like these... $(3x\quad 5)(x\quad 3)=3x^2+4x-15$

'c' is negative — that means the signs in the brackets are different.

So all you have to do is put in the plus or minus signs.

You've only got two choices — if you're unsure, just multiply them out to see which one's right.

$(3x+5)(x-3)=3x^2-4x-15$

or...

$(3x-5)(x+3)=3x^2+4x-15$ ⇐ So it's this one.

Sometimes it's best just to **Cheat** and use the **Formula**

Here's two final points to bear in mind:

1) It won't always factorise.

2) Sometimes factorising is so messy that it's easier to just use the quadratic formula...

So if the question doesn't tell you to factorise, don't assume it will factorise.
And if it's something like this thing below, don't bother trying to factorise it...

Example: Solve $6x^2+87x-144=0$

This will actually factorise, but there's 2 possible bracket forms to try.

$(6x\quad)(x\quad)$ or $(3x\quad)(2x\quad)$ And for each of these, there's 8 possible ways of making 144 to try.

And you can quote me on that...

"He who can properly do quadratic equations is considered a god."
Plato

"Quadratic equations are the music of reason."
James J Sylvester

Completing the Square

This page is for AQA Core 1, OCR Core 1, Edexcel Core 1

Completing the Square is a handy little trick that you should definitely know how to use. It can be a bit fiddly — but it gives you loads of information about a quadratic really quickly.

Take any old quadratic and put it in a Special Form

Completing the square can be really confusing. For starters, what does "Completing the Square" mean? What is the square? Why does it need completing? Well, there is some logic to it:

1) The square is something like this: $(x + \text{something})^2$ It's basically the factorised equation (with the factors both the same), but there's something missing...

2) ...So you need to 'complete' it by adding a number to the square to make it equal to the original equation. $(x + \text{something})^2 + \text{d}$

You'll start with something like this... $2x^2 + 8x - 5$ → ...sort the x-coefficients... $2(x+2)^2 + ?$ → ...and you'll end up with something like this. $2(x+2)^2 - 13$

Lovely!

Make completing the square a bit Easier

There are only a few stages to completing the square — if you can't be bothered trying to understand it, just learn how to do it. But I reckon it's worth spending a bit more time to get your head round it properly.

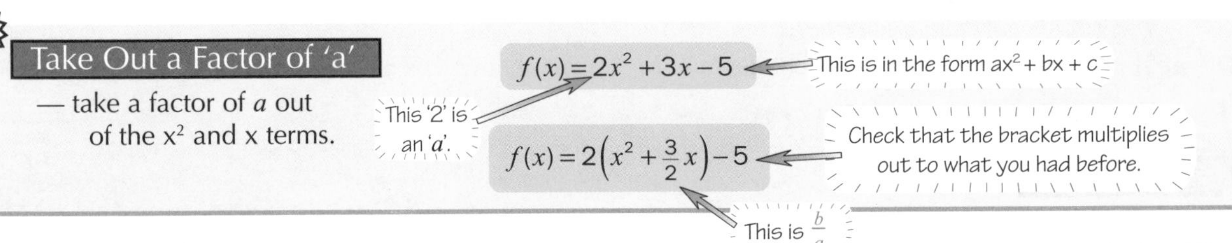

A **Take Out a Factor of 'a'** — take a factor of a out of the x^2 and x terms.

$f(x) = 2x^2 + 3x - 5$ ← This is in the form $ax^2 + bx + c$

This '2' is an 'a'.

$f(x) = 2\left(x^2 + \frac{3}{2}x\right) - 5$ ← Check that the bracket multiplies out to what you had before.

This is $\frac{b}{a}$

B **Rewrite the Bracket** — rewrite the bracket as one bracket squared.

The number in the brackets is always half the old number in front of the x. $\frac{b}{2a}$

$f(x) = 2\left(x + \frac{3}{4}\right)^2 + \text{d}$ ← d is a number you have to find to make the new form equal to the old one.

Don't forget the 'squared' sign.

C **Complete the Square** — find d.

To do this, make the old and new equations equal each other...

$$2\left(x + \frac{3}{4}\right)^2 + \text{d} = 2x^2 + 3x - 5$$

...and you can find d.

$$2x^2 + 3x + \frac{9}{8} + \text{d} = 2x^2 + 3x - 5$$

The x^2 and x bits are the same on both sides so they can disappear.

$$\frac{9}{8} + \text{d} = -5$$

$$\Rightarrow \text{d} = -\frac{49}{8}$$

Completing the Square

A) THE BIT IN THE BRACKETS IS ALWAYS — $a\left(x + \frac{b}{2a}\right)^2$

B) CALL THE NUMBER AT THE END d — $a\left(x + \frac{b}{2a}\right)^2 + \text{d}$

C) MAKE THE TWO FORMS EQUAL — $ax^2 + bx + c = a\left(x + \frac{b}{2a}\right)^2 + \text{d}$

D **So the Answer is:** $f(x) = 2x^2 + 3x - 5 = 2\left(x + \frac{3}{4}\right)^2 - \frac{49}{8}$

Complete your square — it'd be root not to...

Remember — you're basically trying to write the expression as one bracket squared, but it doesn't quite work. So you have to add a number (d) to make it work. It's a bit confusing at first, but once you've learnt it you won't forget it in a hurry.

Completing the Square

This page is for AQA Core 1, OCR Core 1, Edexcel Core 1

Once you've completed the square, you can very quickly say loads about a quadratic function. And it all relies on the fact that a squared number can never be less than zero... ever.

Completing the square can sometimes be ***Useful***

This is a quadratic written as a completed square. As it's a quadratic function and the coefficient of x^2 is positive, it's a u-shaped graph.

This is a square — it can never be negative. The smallest it can be is 0.

$$f(x) = 3x^2 - 6x - 7 = 3(x-1)^2 - 10$$

A **Find the Minimum** — make the bit in the brackets equal to zero.

When the squared bit is zero, f(x) reaches its minimum value. This means the graph reaches its lowest point.

$$f(x) = 3(x-1)^2 - 10$$

This number here is the minimum.

$$f(1) = 3(1-1)^2 - 10$$

f(1) means using x=1 in the function

$$f(1) = 3(0)^2 - 10 = -10$$

So the minimum is -10, when x=1

B **Where Does f(x) Cross the x-axis?** — i.e. find d.

Make the completed square function equal zero.

$$3(x-1)^2 - 10 = 0$$

Solve it to find where f(x) crosses the x-axis.

$$\Rightarrow (x-1)^2 = \frac{10}{3}$$

da-de-dah ... rearranging again.

$$\Rightarrow x - 1 = \pm\sqrt{\frac{10}{3}}$$

$$\Rightarrow x = 1 \pm \sqrt{\frac{10}{3}}$$

So f(x) crosses the x-axis when... $x = 1 + \sqrt{\frac{10}{3}}$ or $1 - \sqrt{\frac{10}{3}}$

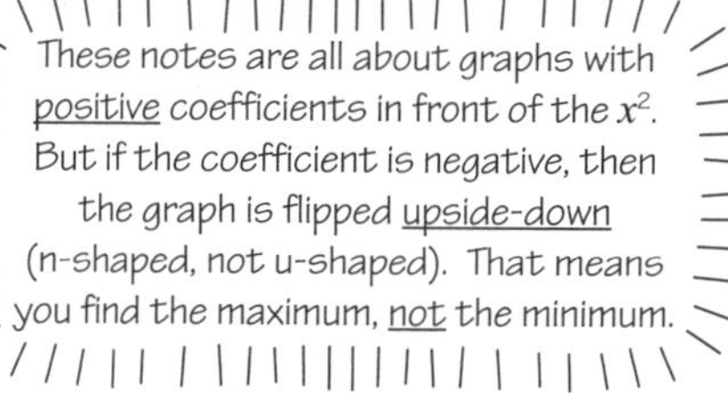
These notes are all about graphs with positive coefficients in front of the x^2.
But if the coefficient is negative, then the graph is flipped upside-down (n-shaped, not u-shaped). That means you find the maximum, not the minimum.

With this information, you can easily sketch the graph...

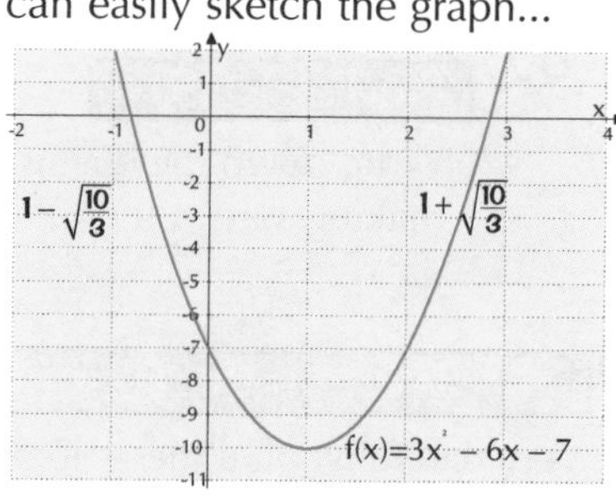
y
x
$1-\sqrt{\frac{10}{3}}$
$1+\sqrt{\frac{10}{3}}$
$f(x)=3x^2-6x-7$

Some functions don't have ***Real Roots***

By completing the square, you can also quickly tell if the graph of a quadratic function ever crosses the x-axis. It'll only cross the x-axis if the function changes sign (i.e. goes from positive to negative or vice versa). Take this function...

Find the Roots

$$f(x) = x^2 + 4x + 7$$

$$f(x) = (x+2)^2 + 3$$

This number's positive.

The smallest this bit can be is zero (at $x = -2$).

$(x + 2)^2$ is never less than zero so f(x) is never less than three.

This means that:

a) f(x) can never be negative.
b) The graph of f(x) never crosses the x-axis.

If the coefficient of x^2 is negative, you can do the same sort of thing to check whether f(x) ever becomes positive.

Don't forget — two wrongs don't make a root...

You'll be pleased to know that that's the end of me trying to tell you how to do something you probably really don't want to do. Now you can push it to one side and run off to roll around in a bed of nettles... much more fun.

The Quadratic Formula

This page is for AQA Core 1, OCR Core 1, Edexcel Core 1

Unlike factorising, the quadratic formula always works... no ifs, no buts, no butts, no nothing...

The **Quadratic Formula** — a reason to be cheerful, but careful...

If you want to solve a quadratic equation $ax^2+bx+c=0$, then the answers are given by this formula:

$$x=\frac{-b\pm\sqrt{b^2-4ac}}{2a}$$

The formula's a godsend — but use the power wisely...

If any of the coefficients (i.e. if a, b or c) in your quadratic equation are negative — be especially careful.

Always take things nice and slowly — don't try to rush it.

It's a good idea to write down what a, b and c are before you start plugging them into the formula.

There are a couple of minus signs in the formula — which can catch you out if you're not paying attention.

I shall teach you the ways of the **Formula**

Example: Solve the quadratic equation $3x^2-4x=8$, leaving your answers in surd form.

The mention of surds is a big clue that you should use the formula.

A Rearrange the Equation

Get the equation in the standard $ax^2+bx+c=0$ form.

$3x^2-4x=8$

$3x^2-4x-8=0$

B Find *a*, *b* and *c*

Write down the coefficients a, b and c — making sure you don't forget minus signs.

$3x^2-4x-8=0$

$a=3 \quad b=-4 \quad c=-8$

C Stick Them in the Formula

Very carefully, plug these numbers into the formula. It's best to write down each stage as you do it.

$$x=\frac{-b\pm\sqrt{b^2-4ac}}{2a}$$

$$=\frac{-(-4)\pm\sqrt{(-4)^2-4\times3\times(-8)}}{2\times3}$$

$$=\frac{4\pm\sqrt{16+96}}{6}$$

$$=\frac{4\pm\sqrt{112}}{6}$$

$$=\frac{4\pm\sqrt{16\times7}}{6}=\frac{4\pm4\sqrt{7}}{6}=\frac{2}{3}\pm\frac{2}{3}\sqrt{7}$$

$$x=\frac{2}{3}+\frac{2}{3}\sqrt{7} \text{ or } \frac{2}{3}-\frac{2}{3}\sqrt{7}$$

The $\pm$ sign means that we have two different expressions for x — which you get by replacing the $\pm$ with + and –.

Using this magic formula, I shall take over the world... ha ha ha...

Okay, maybe it's not quite that good... but it's really important. So learn it properly — which means spending enough time until you can just say it out loud the whole way through, with no hesitations. Or perhaps you could try singing it as loud as you can to the tune of your favourite cheesy song. Sha-la-la-la-la-la-la-ha... La-di-da... Sha-la-la-la-la-la-la-ha... La-di-da... Sha-la-la-la-la-la-la-ha...

The Quadratic Formula

This page is for AQA Core 1, OCR Core 1, Edexcel Core 1

By using part of the quadratic formula, you can quickly tell if a quadratic equation has two solutions, one solution, or no solutions at all. Tell me more, I hear you cry...

How Many Roots? Check the $b^2 - 4ac$ bit...

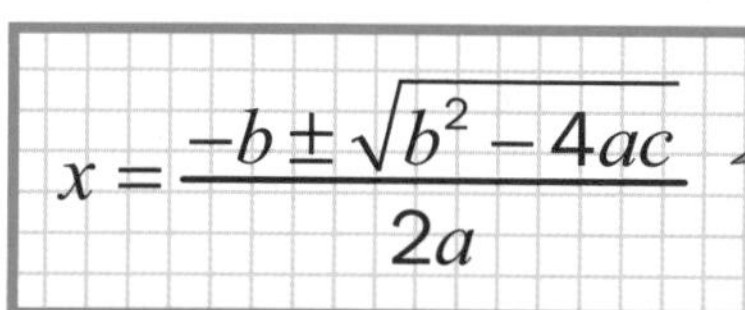

When you try to find the roots of a quadratic function, this bit in the square-root sign ($b^2 - 4ac$) can be positive, zero, or negative. It's this that tells you if a quadratic function has two roots, one root, or no roots.

The $b^2 - 4ac$ bit is called the discriminant.

Because — if the discriminant is positive, the formula will give you two different values — when you add or subtract the $\sqrt{b^2 - 4ac}$ bit.

But if it's zero, you'll only get one value, since adding or subtracting zero doesn't make any difference.

And if it's negative, you don't get any (real) values because you can't take the square root of a negative number.

Well, not here. In later modules, you can actually take the square root of negative numbers and get 'imaginary' numbers. That's why we say no 'real' roots — because there are 'imaginary' roots!

It's good to be able to picture what this means:

A root is just when y = 0, so it's where the graph touches or crosses the x-axis.

$b^2 - 4ac > 0$ Two roots	$b^2 - 4ac = 0$ One root	$b^2 - 4ac < 0$ No roots
So the graph crosses the x-axis twice and these are the roots:	The graph just touches the x-axis from above (or from below if the x^2 coefficient is negative).	The graph doesn't touch the x-axis at all.

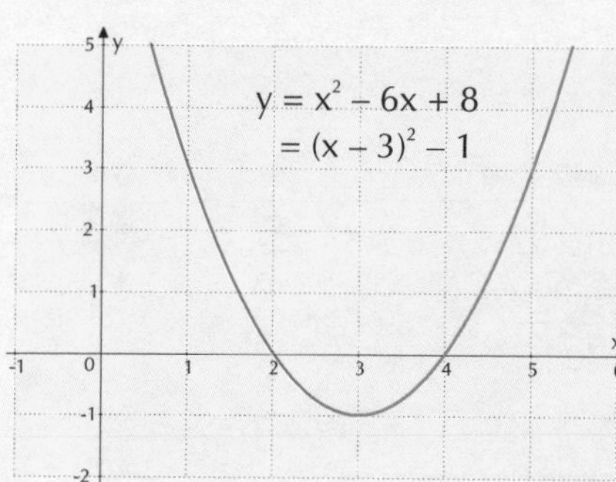

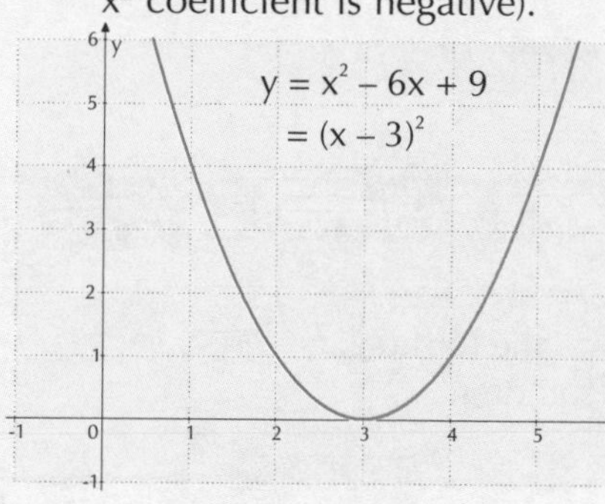

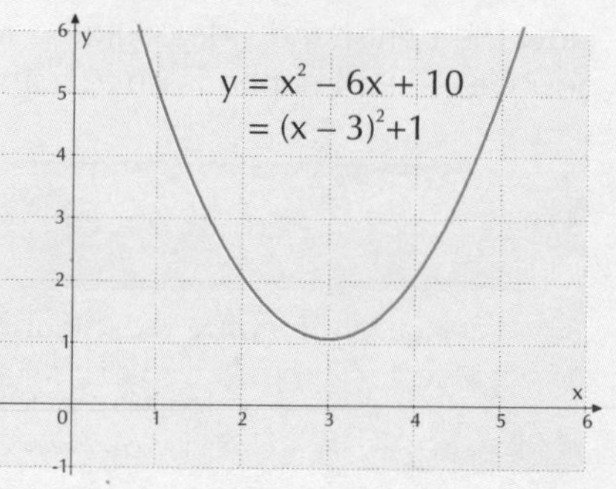

Example: Find the range of values of k for which: a) f(x)=0 has 2 distinct roots, b) f(x)=0 has 1 root, c) f(x) has no real roots, where $f(x) = 3x^2 + 2x + k$.

First of all, work out what the discriminant is: $b^2 - 4ac = 2^2 - 4 \times 3 \times k$
$= 4 - 12k$

These calculations are exactly the same. You don't need to do them if you've done a) because the only difference is the equality symbol.

a) Two distinct roots means:

$b^2 - 4ac > 0 \Rightarrow 4 - 12k > 0$
$\Rightarrow 4 > 12k$
$\Rightarrow k < \frac{1}{3}$

b) One root means:

$b^2 - 4ac = 0 \Rightarrow 4 - 12k = 0$
$\Rightarrow 4 = 12k$
$\Rightarrow k = \frac{1}{3}$

c) No roots means:

$b^2 - 4ac < 0 \Rightarrow 4 - 12k < 0$
$\Rightarrow 4 < 12k$
$\Rightarrow k > \frac{1}{3}$

ha ha ha ha haaaaaa... ha ha ha... ha ha ha ... ha ha ha........

So for questions about "how many roots", think discriminant — i.e. $b^2 - 4ac$. And don't get the inequality signs (> and <) the wrong way round. It's obvious, if you think about it.

The Quadratic Formula

r AQA Core 1, OCR Core 1, Edexcel Core 1

All the stuff you know about the quadratic formula can be proved. And you've got to be able to prove it. Remember you've got to get the equations in the form $ax^2 + bx + c = 0$ before you do anything.

Prove the quadratic formula by **Completing the Square**

The quadratic formula is a wonderful thing. And don't just accept my word that the formula's true — prove it.

Say you've got a quadratic equation in standard form... $ax^2+bx+c=0$ ...and you need to find x.

The first thing to do is complete the square.

$$ax^2+bx+c=0$$

$$a\left(x^2+\frac{b}{a}x\right)+c=0$$

$$a\left(x+\frac{b}{2a}\right)^2-\frac{b^2}{4a}+c=0$$

Take a common factor of 'a' out of the first two terms — then it's easier to see how to complete the square.

What you're trying to do is find x — get it on its own on one side.

This is much better, because there's only one x now — so you can rearrange the formula and find what x actually is.

$$a\left(x+\frac{b}{2a}\right)^2=\frac{b^2}{4a}-c=\frac{b^2-4ac}{4a}$$

$$\left(x+\frac{b}{2a}\right)^2=\frac{b^2-4ac}{4a^2}$$

Divide both sides by a.

Take the last two terms over to the right-hand side, and then add them together as fractions.

Now the right-hand side could be negative, zero or positive — it all depends on a, b and c.

$$x+\frac{b}{2a}=\pm\sqrt{\frac{b^2-4ac}{4a^2}}=\pm\frac{\sqrt{b^2-4ac}}{2a}$$

$$x=-\frac{b}{2a}\pm\frac{\sqrt{b^2-4ac}}{2a}$$

$$x=\frac{-b\pm\sqrt{b^2-4ac}}{2a}$$

And this is the quadratic formula that you've come to know and love.

If you're good enough at algebra, you might be able to get away with not learning all this proof, and doing all the rearranging yourself.
Just remember:

You start with $ax^2+bx+c=0$

And end with $x=\frac{-b\pm\sqrt{b^2-4ac}}{2a}$

Forget maths, do French instead...

The stuff on this page can really only be used one way — and that's by learning it. Not what you want to hear, but it's the truth. And the truth hurts sometimes, don't it? Well listen buster, you've just got to get your head down, and get on with it — no slacking. Unless you've already passed AS Maths, and you're just reading this for fun...

Linear Inequalities

This page is for AQA Core 1, OCR Core 1, Edexcel Core 1

Solving inequalities is very similar to solving equations. You've just got to be really careful that you keep the inequality sign pointing the right way.

Find the ranges of x that satisfy these inequalities:
(i) $x-3<-1+2x$ (ii) $8x+2\geq 2x+17$ (iii) $4-3x\leq 16$ (iv) $36x<6x^2$

Sometimes the inequality sign **Changes Direction**

Like I said, these are pretty similar to solving equations — because whatever you do to one side, you have to do to the other. But multiplying or dividing by negative numbers changes the direction of the inequality sign.

Adding or **Subtracting** doesn't change the direction of the inequality sign

Example: If you add or subtract something from both sides of an inequality, the inequality sign doesn't change direction.

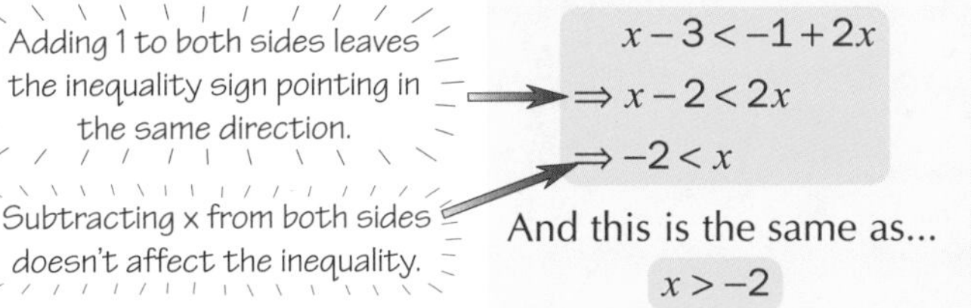

$x-3<-1+2x$
$\Rightarrow x-2<2x$
$\Rightarrow -2<x$

And this is the same as...
$x>-2$

Multiplying or **Dividing** by something **Positive** doesn't affect the inequality sign

Example: Multiplying or dividing both sides of an inequality by a positive number doesn't affect the direction of the inequality sign.

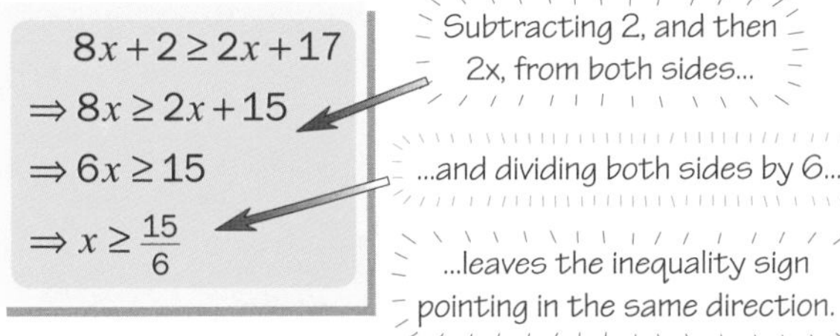

Subtracting 2, and then 2x, from both sides...
...and dividing both sides by 6...
...leaves the inequality sign pointing in the same direction.

$8x+2\geq 2x+17$
$\Rightarrow 8x\geq 2x+15$
$\Rightarrow 6x\geq 15$
$\Rightarrow x\geq \frac{15}{6}$

But **Change** the inequality if you **Multiply** or **Divide** by something **Negative**

But multiplying or dividing both sides of an inequality by a negative number changes the direction of the inequality.

Example:

$4-3x\leq 16$
$\Rightarrow -3x\leq 12$
$\Rightarrow x\geq -4$

Subtract 4 from both sides.

Then divide both sides by -3 — but change the direction of the inequality.

The reason for the sign changing direction is because it's just the same as swapping everything from one side to the other:
$-3x\leq -12 \quad \Rightarrow 12\leq 3x \quad \Rightarrow x\geq 4$

Don't divide both sides by **Variables** — like x and y

You've got to be really careful when you divide by things that might be negative — well basically, don't do it.

Example: $36x<6x^2$

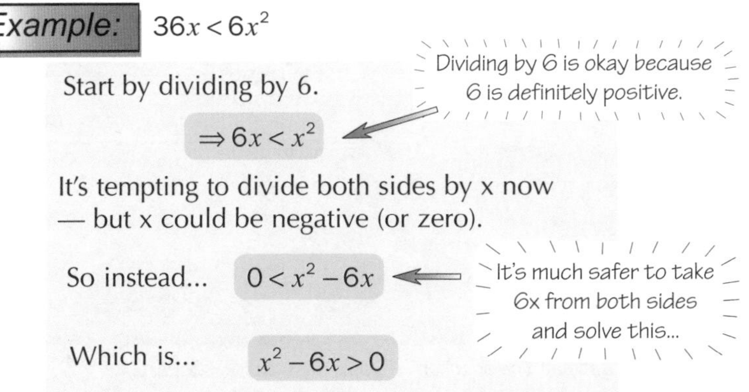

Dividing by 6 is okay because 6 is definitely positive.
It's much safer to take 6x from both sides and solve this...

Start by dividing by 6.

$\Rightarrow 6x<x^2$

It's tempting to divide both sides by x now — but x could be negative (or zero).

So instead... $0<x^2-6x$

Which is... $x^2-6x>0$

See the next page for more on solving quadratic inequalities.

Two types of inequality sign

There are two kinds of inequality sign:

Type 1: $<$ **— less than**
$>$ **— greater than**

Type 2: $\leq$ **— less than or equal to**
$\geq$ **— greater than or equal to**

Whatever type the question uses — use the same kind all the way through your answer.

So no one knows we've arrived safely — splendid...

So just remember — inequalities are just like normal equations except that you have to reverse the sign when multiplying or dividing by a negative number. And don't divide both sides by variables. (You should know not to do this with normal equations anyway because the variable could be zero.) OK — lecture's over.

Quadratic Inequalities

This page is for AQA Core 1, OCR Core 1, Edexcel Core 1

With quadratic inequalities, you're best off drawing the graph and taking it from there.

Draw a Graph to solve a Quadratic inequality

Example: Find the ranges of x which satisfy these inequalities:

① $-x^2+2x+4\geq 1$

First rewrite the inequality with zero on one side.

$-x^2+2x+3\geq 0$

Then draw the graph of $y=-x^2+2x+3$:

So find where it crosses the x-axis (i.e. where y=0):

$-x^2+2x+3=0 \Rightarrow x^2-2x-3=0$

$\Rightarrow (x+1)(x-3)=0$

$\Rightarrow x=-1$ or $x=3$

And the coefficient of x^2 is negative, so the graph is n-shaped. So it looks like this:

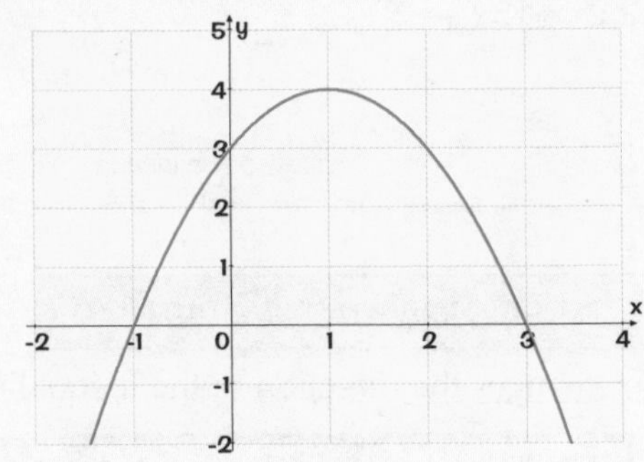

You're interested in when this is positive or zero, i.e. when it's above the x-axis.

From the graph, this is when x is between –1 and 3 (including those points). So your answer is...

$-x^2+2x+4\geq 1$ when $-1\leq x\leq 3$.

② $2x^2-x-3>0$

This one already has zero on one side, so draw the graph of $y=2x^2-x-3$.

Find where it crosses the x-axis:

$2x^2-x-3=0$

$\Rightarrow (2x-3)(x+1)$

$\Rightarrow x=\frac{3}{2}$ or $x=-1$

Factorise it to find the roots.

And the coefficient of x^2 is positive, so the graph is u-shaped. And looks like this:

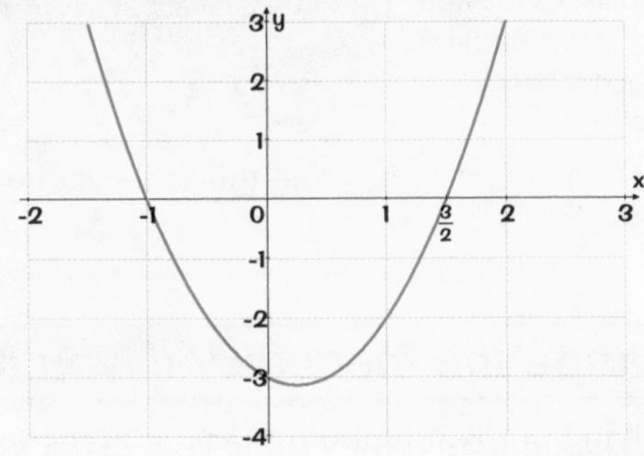

You need to say when this is positive. Looking at the graph, there are two parts of the x-axis where this is true — when x is less than –1 and when x is greater than 3/2. So your answer is:

$2x^2-x-3>0$ when $x<-1$ or $x>\frac{3}{2}$.

Example (revisited)**:**

On the last page you had to solve $36x<6x^2$.

$36x<6x^2$

equation 1 ⟹ $\Rightarrow 6x<x^2$

$\Rightarrow 0<x^2-6x$

So draw the graph of

$y=x^2-6x=x(x-6)$

And this is positive when $x<0$ or $x>6$.

If you divide by x in equation 1, you'd only get half the solution — you'd miss the $x<0$ part.

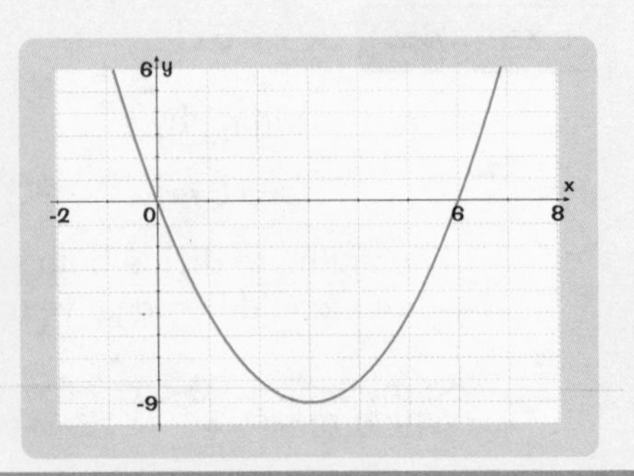

That's nonsense — I can see perfectly...

Call me sad, but I reckon these questions are pretty cool. They look a lot more difficult than they actually are and you get to draw a picture. Wow! When you do the graph, the important thing is to find where it crosses the x-axis (you don't need to know where it crosses the y-axis) and make sure you draw it the right way up. Then you just need to decide which bit of the graph you want. It'll either be the range(s) of x where the graph is below the x-axis or the range(s) where it's above. And this depends on the inequality sign.

'Almost' Quadratic Equations

This page is for OCR Core 1
Sometimes you'll be asked to solve equations that look really difficult, like the ones on this page. But with a bit of rearrangement and fiddling you can get them to look just like an ordinary quadratic you can solve.

Some **Nasty-looking** equations are just **Quadratics**

$$x^4 + 6x^2 + 9 = 0$$

Arrrgh. How on earth are you supposed to solve something like that? Well the answer is... with great difficulty — that's if you don't spot that you can turn it into quadratic form like this:

$$(x^2)^2 + 6(x^2) + 9 = 0$$

It still looks weird. But, if those x^2's were y's:

$$y^2 + 6y + 9 = 0$$

Now it's a just a simple quadratic that you could solve in your sleep — or the exam, which would probably be more useful.

$$4\sin^2 t - 3\sin t - 1 = 0$$

...it looks hard

$$4(\sin t)^2 - 3(\sin t) - 1 = 0$$

...still looks hard

$$4y^2 - 3y - 1 = 0$$

...looks easy.

Just make a **Substitution** to **Simplify**

Example: $2x^6 - 11x^3 + 5 = 0$

1. **Spot That It's a Quadratic**

Put it in the form: a(something)² + b(same thing) + (number) = 0.

$$2(x^3)^2 - 11(x^3) + 5 = 0$$

Now substitute x^3 for y to make it like a normal quadratic.

2. **Substitute**

let $x^3 = y$ $\quad 2y^2 - 11y + 5 = 0$

And solve this quadratic to find the values of y.

3. **Solve it**

$$2y^2 - 11y + 5 = 0$$
$$(2y - 1)(y - 5) = 0$$
$$y = \tfrac{1}{2}, \text{ or } 5$$

Now you've got the values of y, you can get the values of x.

4. **Find the Original Unknown x**

$y = \tfrac{1}{2}, \text{ or } 5$ but... $y = x^3$ (This comes from stage 2.)

Which means... $x^3 = \tfrac{1}{2}, \text{ or } 5$

So the answer is... $x = \sqrt[3]{\tfrac{1}{2}}$ or $\sqrt[3]{5}$

Disguised Quadratics

1) **Put the equation in the FORM:**
a(something)²+b(same thing)+(number)=0
2) **SUBSTITUTE y for the something in the brackets to get a normal-looking quadratic.**
3) **SOLVE the quadratic in the usual way — i.e. by factorising or using the quadratic formula.**
4) **Stick your answers in the substitution equation to get the values for the ORIGINAL unknown.**

Almost quadratics — almost worthwhile, almost interesting, almost...

Quadratics with delusions of grandeur. Whatever next. Anyway, I haven't really got anything to add to what's already on this page. What you need to do is spot what substitution you can use in order to make your life easier — then things will be, well, easier.

The Remainder and Factor Theorems

This page is for AQA Core 1, OCR Core 2, Edexcel Core 2
Algebraic division — Mmmm, fun it's not. It's something you have to know, though — so if you can't do it yet, get on and learn it. Then you can move on to brighter things.

Do *Polynomial Division* by means of *Subtraction*

$$(2x^3 - 3x^2 - 3x + 7) \div (x - 2) = ?$$

The trick with this is to see how many times you can subtract $(x - 2)$ from $2x^3 - 3x^2 - 3x + 7$.
The idea is to keep subtracting lumps of $(x - 2)$ until you've got rid of all the powers of x.

Do the subtracting in *Stages*

At each stage, always try to get rid of the highest power of x.
Then start again with whatever you've got left.

① Start with $2x^3 - 3x^2 - 3x + 7$, and subtract $2x^2$ lots of $(x - 2)$ to get rid of the x^3 term.

$$(2x^3 - 3x^2 - 3x + 7) - 2x^2(x - 2)$$
$$(2x^3 - 3x^2 - 3x + 7) - 2x^3 + 4x^2$$
$$= x^2 - 3x + 7$$

$2x^3 \div x = 2x^2$

This is what's left — so now you have to get rid of the x^2 term.

② Now start again with $x^2 - 3x + 7$.
The highest power of x is the x^2 term.
So subtract x lots of $(x - 2)$ to get rid of that.

$$(x^2 - 3x + 7) - x(x - 2)$$
$$(x^2 - 3x + 7) - x^2 + 2x$$
$$= -x + 7$$

Now start again with this — and get rid of the x term.

③ All that's left now is $-x + 7$.
Get rid of the $-x$ by subtracting -1 times $(x - 2)$.

$$(-x + 7) - (-1(x - 2))$$
$$(-x + 7) + x - 2$$
$$= 5$$

There are no more powers of x to get rid of — so stop here.

The remainder's 5.

Interpreting the results...

Time to work out exactly what all that meant.

Started with: $2x^3 - 3x^2 - 3x + 7$

Subtracted: $2x^2(x - 2) + x(x - 2) - 1(x - 2)$

$= (x - 2)(2x^2 + x - 1)$

Remainder: $= 5$

So... $2x^3 - 3x^2 - 3x + 7 = (x - 2)(2x^2 + x - 1) + 5$

or... $\dfrac{2x^3 - 3x^2 - 3x + 7}{(x - 2)} = 2x^2 + x - 1$ with remainder 5.

Algebraic Division

$$(ax^3 + bx^2 + cx + d) \div (x - k) = ?$$

1) SUBTRACT a multiple of $(x - k)$ to get rid of the highest power of x.
2) REPEAT step 1 until you've got rid of all the powers of x.
3) WORK OUT how many lumps of $(x - k)$, you've subtracted, and the REMAINDER.

The *Remainder Theorem* is an easy way to work out *Remainders*

When you divide $f(x)$ by $(x - a)$, the remainder is $f(a)$.

So in the example above, you could have worked out the remainder dead easily.

1) $f(x) = 2x^3 - 3x^2 - 3x + 7$.
2) You're dividing by $(x - 2)$, so $a = 2$.
3) So the remainder must be $f(2) = (2 \times 8) - (3 \times 4) - (3 \times 2) + 7 = 5$.

Careful now... when you're dividing by something like $(x + 7)$, a is negative — so here, $a = -7$.

The Fantastic Factor Theorem

This page is for AQA Core 1, OCR Core 2, Edexcel Core 2

The factor theorem is fantastic — no doubt about it. It can really help you factorise cubics. But not only that, at this week's special CGP price of only £27.49, thanks to its twin rotating blades, it allows a closer factorisation than ever before. So for smoother, closer, and faster factorising, choose the CGP Fantastic Factor Theorem every time.*

The **Factor Theorem** is just the Remainder Theorem with a **Zero Remainder**

If you get a remainder of zero when you divide f(x) by (x – a), then (x – a) must be a factor. That's the Factor Theorem.

> The Factor Theorem:
>
> If f(x) is a polynomial, and f(a) = 0, then (x – a) is a factor of f(x).
>
> In other words: If you know the roots, you also know the factors — and vice versa.

Example: Show that $(x+1)$ is a factor of $f(x)=x^3+4x^2-7x-10$

The question's giving you a big hint here. If you show that f(–1) = 0, then the factor theorem says that (x + 1) is a factor.

$$f(x)=x^3+4x^2-7x-10$$

$$f(-1)=-1+4+7-10=0$$

So, by the factor theorem, (x+1) is a factor of f(x).

See pages 10-11 for more on factorising.

You can use the Factor Theorem when **a ≠ 1**

You might be asked to show that (ax – b) (i.e. something with a number in front of the x) is a factor of a polynomial. But don't start panicking just yet — you can still use the factor theorem.

Example: Show that $(2x-1)$ is a factor of $f(x)=2x^2-9x+4$

Notice that 2x – 1 = 0 when x = ½. Plug this value of x into f(x).

$$f(x)=2x^2-9x+4$$

$$f\left(\tfrac{1}{2}\right)=2\left(\tfrac{1}{2}\right)^2-\left(9\times\tfrac{1}{2}\right)+4$$

$$f\left(\tfrac{1}{2}\right)=\tfrac{1}{2}-\tfrac{9}{2}+4=0$$

So by the factor theorem, (x – ½) is a factor.

And if that's a factor, then 2 (x – ½) = (2x – 1) is also a factor.

Explanation of a tricky bit...

If you multiply one factor by a number, you've got to divide the other factor by the same number.

e.g. $f(x)=\left(x-\tfrac{1}{2}\right)(p)$

$f(x)=(2x-1)\left(\tfrac{1}{2}p\right)$

e.g. $8\times6=48$

$(8\times3)(6\div3)=48$

$(24)(2)=48$

(x–1) is a **Factor** if the coefficients add up to **0**

This is a useful thing to remember.
It works for all polynomials — no exceptions.
It could save a fair whack of time in the exam.

Example: Factorise the polynomial $f(x)=6x^2-7x+1$

The coefficients (6, –7 and 1) add up to 0. That means f(1) = 0.
And that applies to any polynomial at all... always.

So by the factor theorem, if f (1) = 0, (x – 1) is a factor. Easy.

Just factorise it like an easy quadratic to get this:

$$f(x)=6x^2-7x+1=(6x-1)(x-1)$$

* Plus Postage and Packaging, at £28.49. Allow at least 28 months for delivery.

You've lost that Factor Theorem, whoah, yeah that Factor Theorem...

You know, I had NO idea how great a factor I could get. Not until I tried CGP's Fantastic Factor Theorem. I used to have problems with factorisation when dating, and in my general life. But now I've put all that behind me. Now I've got the confidence to talk about factors wherever I go. I can really experience the joy of factorising. And it's all thanks to CGP's Fantastic Factor Theorem: Try it today — it changed my life, and it could change yours too.

Factorising Cubics

This page is for AQA Core 1, Edexcel Core 1, Edexcel Core 2

Factorising a quadratic function is okay — but you might also be asked to factorise a cubic (something with x^3 in it). And that takes a bit more time — there are more steps, so there are more chances to make mistakes.

Factorising a cubic given One Factor

$$f(x) = 2x^3 + x^2 - 8x - 4$$

Factorising a cubic means exactly what it meant with a quadratic — putting brackets in.
When they ask you to factorise a cubic equation, they'll usually tell you one of the factors.

Example: Given that $(x + 2)$ is a factor of $f(x) = 2x^3 + x^2 - 8x - 4$, express $f(x)$ as the product of three linear factors.

1) The first step is to find a quadratic factor. So write down the factor you know, along with another set of brackets.

$$(x+2)(\qquad\qquad) = 2x^3 + x^2 - 8x - 4$$

Put the x^2 bit in this new set of brackets.
These have to multiply together to give you this.

$$(x+2)(2x^2\qquad\quad) = 2x^3 + x^2 - 8x - 4$$

2) Find the number for the second set of brackets.
These have to multiply together to give you this.

$$(x+2)(2x^2\qquad -2) = 2x^3 + x^2 - 8x - 4$$

3) These multiplied give you $-2x$, but there's $-8x$ in $f(x)$ — so you need an 'extra' $-6x$.
And that's what this $-3x$ is for.

$$(x+2)(2x^2 - 3x - 2) = 2x^3 - x^2 - 8x - 4$$

You only need $-3x$ because it's going to be multiplied by 2, which makes $-6x$.

If you wanted to solve a cubic, you'd do it exactly the same way — put it in the form $ax^3 + bx^2 + cx + d = 0$ and factorise.

4) Before you go any further, check that there are the same number of x^2's on both sides.

$4x^2$ from here...

$$(x+2)(2x^2 - 3x - 2) = 2x^3 - x^2 - 8x - 4$$

...and $-3x^2$ from here... ...add together to give this $-x^2$.

If this is okay, factorise the quadratic into two linear factors.

$$(2x^2 - 3x - 2) = (2x+1)(x-2)$$

And so... $2x^3 + x^2 - 8x - 4 = (x+2)(2x+1)(x-2)$

Factorising Cubics

1) Find a factor, (if you need to) by finding $f(0)$, $f(\pm1)$, $f(\pm2)$,... until you find $f(k) = 0$. Then $(x - k)$ is a factor.
2) Put in the x^2 term.
3) Put in the constant.
4) Put in the x term by comparing the number of x's on both sides.
5) Check there are the same number of x^2's on both sides.
6) Factorise the quadratic you've found — if that's possible.

Factorising a cubic given No Factors

If they don't give you the first factor, you have to find it yourself. But it's okay — they'll give you an easy one.
The best way to find a factor is to guess — use trial and error.

Find f(1) If the answer is zero, you know $(x - 1)$ is a factor.
If the answer isn't zero, find $f(-1)$. If that's zero, then $(x + 1)$ is a factor.

If that doesn't work, keep trying small numbers (find $f(2)$, $f(-2)$, $f(3)$, $f(-3)$ and so on) until you find a number that gives you zero when you put it in the cubic. Call that number k.

$(x - k)$ is a factor of the cubic (from the Factor Theorem).

I love the smell of fresh factorised cubics in the morning...

If you're dead keen on algebraic division, you could use it here as well. Just divide the cubic by the factor you've found or been given, and you'll end up with a quadratic and remainder 0. If you've done it right, that is. Just depends which way you prefer looking at it.

Exam Questions

After all those pages of things to learn, guess these questions will come as quite a relief...

Exam Questions

1 Given that the equation $x^2 - 3px + 2p = 0$, where p is a positive integer, has real roots:

(a) Prove that $p(9p - 8) \geq 0$

(3 marks)

(b) Given that $p = 2$, find the roots of the equation, giving your answer in surd form.

(4 marks)

2 Solve the inequality $2x < 35 - x^2$.

(3 marks)

3 Solve the inequality $x^2 + 6x + 9 > 0$.

(3 marks)

4 $f(x) = x^5 + x^4 - 19x^3 - 25x^2 + 66x + 72$

Use the Factor Theorem to determine whether or not $(x + 1)$ is a factor of $f(x)$.

(2 marks)

5 **(a)** Either algebraically, or by sketching the graphs, solve the inequality

$$4x + 7 > 7x + 4$$

(2 marks)

(b) Find the values of k, such that

$(x - 5)(x - 3) > k$ for all possible values of x.

(3 marks)

(c) Find the range of x that satisfies the inequality

$$(x + 3)(x - 2) < 2$$

(3 marks)

6 **(a)** Express $x^2 - 7x + 17$ in the form $(x - a)^2 + b$, where a and b are constants.

Hence state the maximum value of $f(x) = \dfrac{1}{x^2 - 7x + 17}$.

(3 marks)

(b) Find the possible values of b if the equation $g(x) = 0$ is to have only one root, where $g(x)$ is given by $g(x) = 3x^2 + bx + 12$.

(3 marks)

Simultaneous Equations

This page is for AQA Core 1, OCR Core 1, Edexcel Core 1

Solving simultaneous equations means finding the answers to two equations at the same time — i.e. finding values for x and y for which both equations are true. And it's one of those things that you'll have to do again and again — so it's definitely worth practising them until you feel really confident.

① $3x + 5y = -4$
② $-2x + 3y = 9$

This is how simultaneous equations are usually shown. It's a good idea to label them as equation ① and equation ② — so you know which one you're working with.

But they'll look different sometimes, maybe like this. Make sure you rearrange them as 'ax + by = c'.

$4 + 5y = -3x$
$-2x = 9 - 3y$

rearrange as ax + by = c

$3x + 5y = -4$
$-2x + 3y = 9$

Solving them by Elimination

Elimination is a lovely method. It's really quick when you get the hang of it — you'll be doing virtually all of it in your head.

Example:

① $3x + 5y = -4$
② $-2x + 3y = 9$

To get the x's to match, you need to multiply the first equation by 2 and the second by 3:

① × 2 $6x + 10y = -8$
② × 3 $-6x + 9y = 27$

Add the equations together to eliminate the x's.

① + ② $19y = 19$
$y = 1$

So y is 1. Now stick that value for y into one of the equations to find x:

$y = 1$ in ① $\Rightarrow 3x + 5 = -4$
$3x = -9$
$x = -3$

So the solution is x = –3, y = 1.

A Match the Coefficients

Multiply the equations by numbers that will make either the x's or the y's match in the two equations. (Ignoring minus signs.)

Go for the lowest common multiple (LCM). e.g. LCM of 2 and 3 is 6.

B Eliminate to Find One Variable

If the coefficients are the same sign, you'll need to subtract one equation from the other.

If the coefficients are different signs, you need to add the equations.

C Find the Variable You Eliminated

When you've found one variable, put its value into one of the original equations so you can find the other variable.

But you should always...

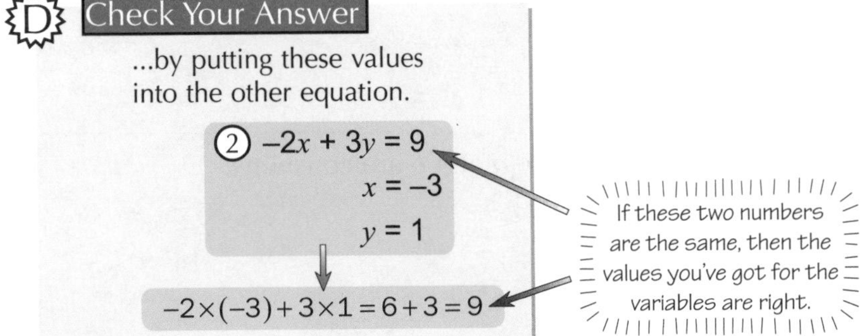

D Check Your Answer

...by putting these values into the other equation.

② $-2x + 3y = 9$
$x = -3$
$y = 1$

$-2 \times (-3) + 3 \times 1 = 6 + 3 = 9$

If these two numbers are the same, then the values you've got for the variables are right.

Elimination Method

1) Match the coefficients
2) Eliminate and then solve for one variable
3) Find the other variable (that you eliminated)
4) Check your answer

Eliminate your social life — do AS-level maths...

This is a fairly basic method that won't be new to you. So make sure you know it. The only possibly tricky bit is matching the coefficients — work out the lowest common multiple of the coefficients of x, say, then multiply the equations to get this number in front of each x.

Simultaneous Equations with Quadratics

This page is for AQA Core 1, OCR Core 1, Edexcel Core 1

Elimination is great for simple equations. But it won't always work. Sometimes one of the equations has not just x's and y's in it — but bits with x^2 and y^2 as well. When this happens, you can only use the substitution method.

Use Substitution if one equation is Quadratic

Example:

$-x+2y=5$ — Ⓛ ← The linear equation — with only x's and y's in.

$x^2+y^2=25$ — Ⓠ ← The quadratic equation — with some x^2 and y^2 bits in.

Rearrange the linear equation so that either x or y is on its own on one side of the equals sign.

Ⓛ $-x+2y=5$

$\Rightarrow x=2y-5$

Substitute this expression into the quadratic equation...

Sub into Ⓠ: $x^2+y^2=25$

$\Rightarrow (2y-5)^2+y^2=25$

...and then rearrange this into the form $ax^2+bx+c=0$, so you can solve it — either by factorising or using the quadratic formula.

$\Rightarrow (4y^2-20y+25)+y^2=25$

$\Rightarrow 5y^2-20y=0$

$\Rightarrow 5y(y-4)=0$

$\Rightarrow y=0$ or $y=4$

One Quadratic and One Linear Eqn

1) **Isolate variable in linear equation**
 Rearrange the linear equation to get either x or y on its own.
2) **Substitute into quadratic equation**
 — to get a quadratic equation in just one variable.
3) **Solve to get values for one variable**
 — either by factorising or using the quadratic formula.
4) **Stick these values in the linear equation**
 — to find corresponding values for the other variable.

Finally put both these values back into the linear equation to find corresponding values for x:

When y = 0: $-x+2y=5$ Ⓛ

$\Rightarrow x=-5$

When y = 4: $-x+2y=5$ Ⓛ

$\Rightarrow -x+8=5$

$\Rightarrow x=3$

So the solutions to the simultaneous equations are: $x=-5$, $y=0$ and $x=3$, $y=4$.

As usual, check your answers by putting these values back into the original equations.

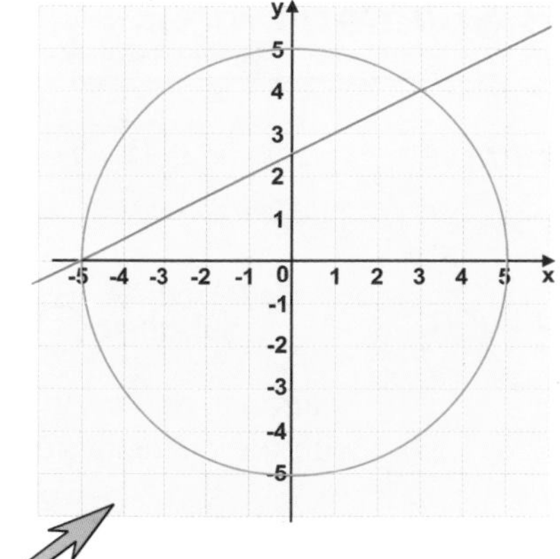

Check Your Answers

x = -5, y = 0: $-(-5)+2\times0=5$ ✓

$(-5)^2+0^2=25$ ✓

x = 3, y = 4: $-(3)+2\times4=5$ ✓

$3^2+4^2=25$ ✓

y = x² — a match-winning substitution...

The quadratic equation above is actually a circle about the origin with radius 5. (Don't worry — you don't need to know about equations of circles in this section). The linear equation is just a standard straight line. So what you're actually finding here are the two points where the line passes through the circle. And these turn out to be (–5,0) and (3,4). See the graph. (I thought you might appreciate seeing a graph that wasn't a line or a parabola for a change.)

Geometric Interpretation

This page is for AQA Core 1, OCR Core 1, Edexcel Core 1

When you have to interpret something geometrically — you have to draw a picture and 'say what you see'.

Two Solutions — Two points of Intersection

Example:

$y = x^2 - 4x + 5$ — ①

$y = 2x - 3$ — ②

Solution:

Substitute expression for y from ② into ①: $2x - 3 = x^2 - 4x + 5$

Rearrange and solve:

$x^2 - 6x + 8 = 0$

$(x-2)(x-4) = 0$

$x = 2$ or $x = 4$

In ② gives:

$x = 2 \Rightarrow y = 2 \times 2 - 3 = 1$

$x = 4 \Rightarrow y = 2 \times 4 - 3 = 5$

There's 2 pairs of solutions: x=2, y=1 and x=4, y=5

Geometric Interpretation:

So from solving the simultaneous equations, you know that the graphs meet in two places — the points (2,1) and (4,5).

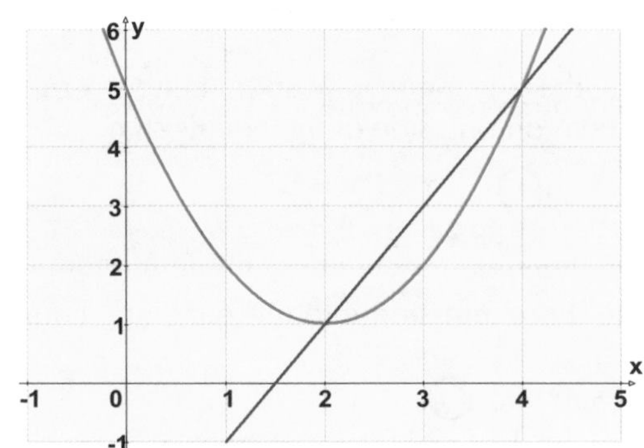

One Solution — One point of Intersection

Example:

$y = x^2 - 4x + 5$ — ①

$y = 2x - 4$ — ②

Solution:

Substitute ② in ①: $2x - 4 = x^2 - 4x + 5$

Rearrange and solve:

$x^2 - 6x + 9 = 0$

$(x-3)^2 = 0$

$x = 3$

Double root i.e. you only get 1 solution from the quadratic.

In Equation ② gives:

$y = 2 \times 3 - 4$

$y = 2$

There's 1 solution: x=3, y=2

Geometric Interpretation:

Since the equations have only one solution, the two graphs only meet at one point — (3,2). The straight line is a tangent to the curve.

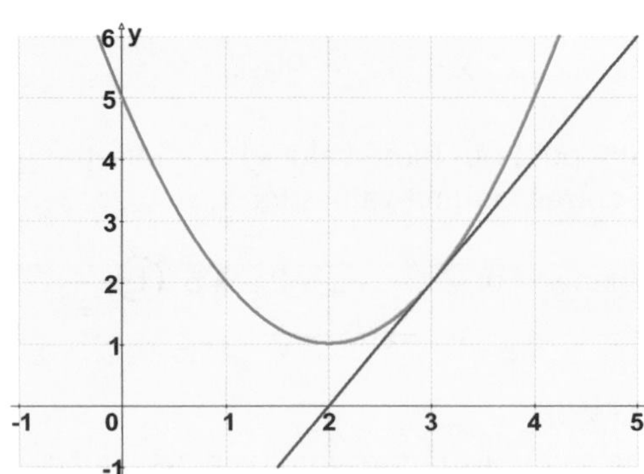

No Solutions means the Graphs Never Meet

Example:

$y = x^2 - 4x + 5$ — ①

$y = 2x - 5$ — ②

Solution:

Substitute ② in ①: $2x - 5 = x^2 - 4x + 5$

Rearrange and try to solve with the quadratic formula:

$x^2 - 6x + 10 = 0$

$b^2 - 4ac = (-6)^2 - 4.10$

$= 36 - 40 = -4$

$b^2 - 4ac < 0$, so the quadratic has no roots.

So the simultaneous equations have no solutions.

Geometric Interpretation:

The equations have no solutions — the graphs never meet.

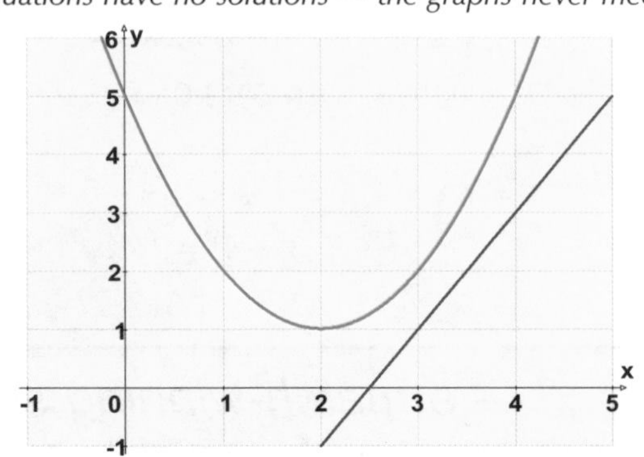

Geometric Interpretation? Frankly my dear, I don't give a damn...

Wow, one solution, two solutions... or no solutions — I mean... WOW.

Coordinate Geometry

This page is for AQA Core 1, OCR Core 1, Edexcel Core 1

Welcome to geometry club... nice — today I shall be mostly talking about straight lines...

Finding the equation of a line **Through Two Points**

If you get through your exam without having to find the equation of a line through two points, I'm a Dutchman.

Example: Find the equation of the line that passes through the points (–3, 10) and (1, 4), and write it in the forms:

$$y - y_1 = m(x - x_1) \qquad y = mx + c \qquad ax + by + c = 0$$

— where a, b and c are integers.

You might be asked to write the equation of a line in any of these forms — but they're all similar.
Basically, if you find an equation in one form — you can easily convert it into either of the others.

The **Easiest** to find is $y - y_1 = m(x - x_1)$...

Point 1 is (–3, 10) and Point 2 is (1, 4)

Label the Points Label Point 1 as (x_1, y_1) and Point 2 as (x_2, y_2).

Point 1 — $(x_1, y_1) = (-3, 10)$

Point 2 — $(x_2, y_2) = (1, 4)$

It doesn't matter which way round you label them.

Find the Gradient Find the gradient of the line m — this is $m = \frac{y_2 - y_1}{x_2 - x_1}$.

$$m = \frac{4-10}{1-(-3)} = \frac{-6}{4} = -\frac{3}{2}$$

Write Down the Equation Write down the equation of the line, using the coordinates x_1 and y_1 — this is just $y - y_1 = m(x - x_1)$

$x_1 = -3$ and $y_1 = 10$ $\Longrightarrow$ $y - 10 = -\frac{3}{2}(x - (-3))$

$$y - 10 = -\frac{3}{2}(x + 3)$$

...and **Rearrange** this to get the other two forms:

For the form $y = mx + c$, take everything except the y over to the right.

$$y - 10 = -\frac{3}{2}(x + 3)$$
$$\Rightarrow y = -\frac{3}{2}x - \frac{9}{2} + 10$$
$$\Rightarrow y = -\frac{3}{2}x + \frac{11}{2}$$

To find the form $ax + by + c = 0$, take everything over to one side — and then get rid of any fractions.

$$y = -\frac{3}{2}x + \frac{11}{2}$$
$$\Rightarrow \frac{3}{2}x + y - \frac{11}{2} = 0$$
$$\Rightarrow 3x + 2y - 11 = 0$$

Multiply the whole equation by 2 to get rid of the 2's on the bottom line.

Equations of Lines

1) **LABEL the points (x_1, y_1) and (x_2, y_2).**
2) **GRADIENT — find it and call it m.**
3) **WRITE DOWN THE EQUATION using $y - y_1 = m(x - x_1)$**
4) **CONVERT to one of the other forms, if necessary.**

If you end up with an equation like $\frac{3}{2}x - \frac{4}{3}y + 6 = 0$, where you've got a 2 and a 3 on the bottom of the fractions — multiply everything by the lowest common multiple of 2 and 3, i.e. 6.

There ain't nuffink to this geometry lark, Mister...

This is the sort of stuff that looks hard but is actually pretty easy. Finding the equation of a line in that first form really is a piece of cake — the only thing you have to be careful of is when a point has a negative coordinate (or two). In that case, you've just got to make sure you do the subtractions properly when you work out the gradient. See, this stuff ain't so bad...

Coordinate Geometry

This page is for AQA Core 1, OCR Core 1

Find the midpoint by ***Averaging*** *each of the coordinates*

Don't complain. It doesn't get any easier than this.

Example: Find the midpoint of AB, where A and B are (–3, 10) and (1, 4) respectively.

Find the midpoint by taking the average of the x- and y-coordinates:

Label the points (x_1, y_1) and (x_2, y_2).

Average x-coordinate = $\frac{x_1+x_2}{2} = \frac{-3+1}{2} = -1$

Average y-coordinate = $\frac{y_1+y_2}{2} = \frac{10+4}{2} = 7$

These are the midpoint coordinates.

So the midpoint has coordinates: (–1, 7)

Use ***Pythagoras*** *to find the* ***Length*** *of a line segment*

Example: Find the length of AB, where A and B are (2, 12) and (6, 7) respectively.

Find the length by treating the line segment as the hypotenuse of a right-angled triangle.

Label the points (x_1, y_1) and (x_2, y_2).

Length of side "x" of the triangle = $x_2 - x_1 = 6 - 2 = 4$

Length of side "y" of the triangle = $y_2 - y_1 = 7 - 12 = -5$

So, length of line segment = $\sqrt{(-5)^2+4^2} = \sqrt{25+16} = 6.4$

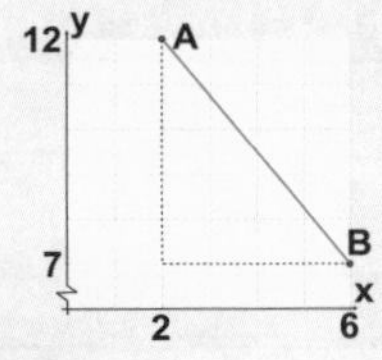

Finding where lines meet means solving ***Simultaneous Equations***

Okay, you can complain now. This is no fun at all.

Two lines... Line l_1: $5x+2y-9=0$ or $y=-\frac{5}{2}x+\frac{9}{2}$ Line l_2: $3x+4y-4=0$ or $y=-\frac{3}{4}x+1$

Example: Find where the line l_1 meets the line l_2.

$5x+2y-9=0$ — ①

$3x+4y-4=0$ — ②

Finding where the lines meet means solving these simultaneous equations.

$10x+4y-18=0$ — ③ = 2 × ①

$7x-14=0$ — ③–②

$\Rightarrow x=2$

Putting this back into equation ② then gives...

$(3\times2)+4y-4=0$

$\Rightarrow 6+4y-4=0$

$\Rightarrow 4y=-2$

$\Rightarrow y=-\frac{1}{2}$

Can't remember how to do simultaneous equations? Have a look at pages 24 & 25.

So the lines meet at the point $(2, -\frac{1}{2})$.

If you've got the equations in the form y = mx + c, make the right-hand sides of both equations equal.

Line l_1: $y=-\frac{5}{2}x+\frac{9}{2}$ Line l_2: $y=-\frac{3}{4}x+1$

$-\frac{5}{2}x+\frac{9}{2}=-\frac{3}{4}x+1$

Solve this equation to find a value for x.

$\Rightarrow -\frac{7}{4}x=-\frac{7}{2}$

$\Rightarrow x=2$

Then put this value of x into one of the equations to find the y-coordinate...

$y=-\frac{5}{2}\times2+\frac{9}{2}$

It doesn't matter which of the equations you use.

$y=-\frac{1}{2}$

So the lines meet at the point $(2, -\frac{1}{2})$.

And I think to myself — what a wonderful page...

What an absolutely superb page. There it is, above all these words that you never read. It's fuller than a student at an all-you-can-eat curry house — absolutely jam- (or madras-) packed with useful things about simultaneous equations, lengths and midpoints. Learn this lot, get a few more marks, get the grades you need, and get yourself into some more all-you-can-eat curry houses.

Coordinate Geometry

This page is for AQA Core 1, OCR Core 1, Edexcel Core 1

This page is based around two really important facts that you've got to know — one about parallel lines, one about perpendicular lines. It's really a page of unparalleled excitement...

Two more lines...

Line l_1: $3x-4y-7=0$, $y=\frac{3}{4}x-\frac{7}{4}$

Line l_2: $x-3y-3=0$, $y=\frac{1}{3}x-1$

...and two points...

Point A $(3,-1)$

Point B $(-2,4)$

Parallel** lines have equal **Gradient

That's what makes them parallel — the fact that the gradients are the same.

Example: Find the line parallel to l_1 that passes through the point A (3, –1).

Parallel lines have the same gradient.

The original equation is this: $y=\frac{3}{4}x-\frac{7}{4}$

So the new equation will be this: $y=\frac{3}{4}x+c$

We know that the line passes through A, so at this point x will be 3, and y will be –1.

We just need to find c.

Stick these values into the equation to find c.

$$-1=\frac{3}{4}\times 3+c$$

$$\Rightarrow c=-1-\frac{9}{4}=-\frac{13}{4}$$

So the equation of the line is... $y=\frac{3}{4}x-\frac{13}{4}$

And if you're only given the ax + by + c = 0 form it's even easier:

The original line is: $3x-4y-7=0$

So the new line is: $3x-4y-k=0$

Then just use the values of x and y at the point A to find k...

$$3\times 3-4\times(-1)-k=0$$
$$\Rightarrow 13-k=0$$
$$\Rightarrow k=13$$

So the equation is: $3x-4y-13=0$

*The gradient of a **Perpendicular** line is: **–1 ÷ the Other Gradient***

Finding perpendicular lines (or 'normals') is just as easy as finding parallel lines — as long as you remember the gradient of the perpendicular line is –1 ÷ the gradient of the other one.

Example: Find the line perpendicular to l_2 that passes through the point B (–2, 4).

l_2 has equation: $y=\frac{1}{3}x-1$

So if the equation of the new line is y=mx+c, then

$$m=-1\div\frac{1}{3}$$
$$\Rightarrow m=-3$$

Since the gradient of a perpendicular line is: –1 ÷ the other one.

Also...

$$4=(-3)\times(-2)+c$$
$$\Rightarrow c=4-6=-2$$

Putting the coordinates of B(–2, 4) into y = mx + c.

So the equation of the line is... $y=-3x-2$

Or if you start with: l_2 $x-3y-3=0$

To find a perpendicular line, swap these two numbers around, and change the sign of one of them. (So here, 1 and –3 become 3 and 1.)

So the new line has equation... $3x+y+d=0$

Or you could have used – 3x – y + d = 0.

But...

$$3\times(-2)+4+d=0$$
$$\Rightarrow d=2$$

Using the coordinates of point B.

And so the equation of the perpendicular line is... $3x+y+2=0$

Wowzers — parallel lines on the same graph dimension...

This looks more complicated than it actually is, all this tangent and normal business. All you're doing is finding the equation of a straight line through a certain point — the only added complication is that you have to find the gradient first. And there's another way to remember how to find the gradient of a normal — just remember that the gradients of perpendicular lines multiply together to make –1.

Circles

This page is for AQA Core 1, OCR Core 1, Edexcel Core 2

I always say a beautiful shape deserves a beautiful formula, and here you've got one of my favourite double acts...

Equation of a circle: $(x - a)^2 + (y - b)^2 = r^2$

The equation of a circle looks complicated, but it's all based on Pythagoras' theorem.
Take a look at the circle below, with centre (6, 4) and radius 3.

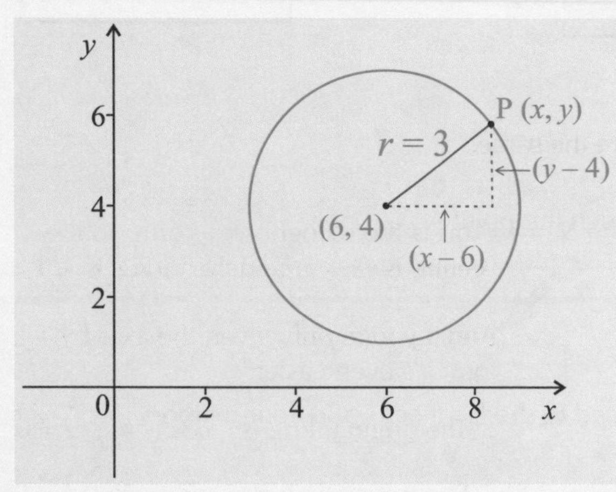

Joining a point P (x, y) on the circumference of the circle to its centre (6, 4), we can create a right-angled triangle.

Now let's see what happens if we use Pythagoras' theorem:

$$(x - 6)^2 + (y - 4)^2 = 3^2$$

or: $$(x - 6)^2 + (y - 4)^2 = 9$$

This is the equation for the circle. It's as easy as that.

In general, a circle with radius r and centre (a, b) has the equation: $(x - a)^2 + (y - b)^2 = r^2$

This formula is not in the formula book.

Example:

i) What is the centre and radius of the circle with equation $(x - 2)^2 + (y + 3)^2 = 16$?

ii) Write down the equation of the circle with centre (–4, 2) and radius 6.

Solution:

i) Comparing $(x - 2)^2 + (y + 3)^2 = 16$ with the general form:

$$(x - a)^2 + (y - b)^2 = r^2$$

then $a = 2$, $b = -3$ and $r = 4$.

So the centre (a, b) is: (2, –3)
and the radius (r) is: 4.

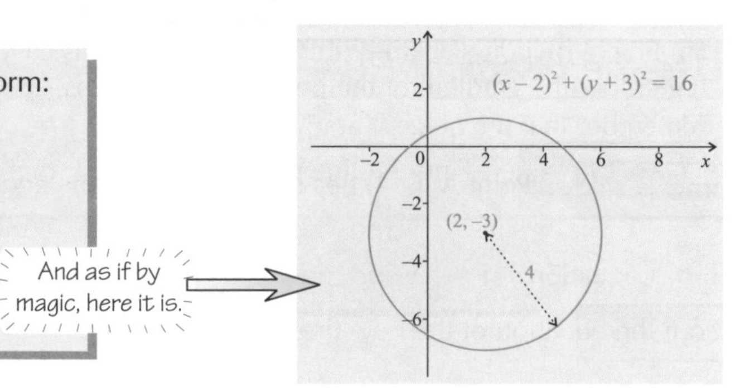

ii) The question says, 'Write down...', so you know you don't need to do any working.
The centre of the circle is (–4, 2), so $a = -4$ and $b = 2$.
The radius is 6, so $r = 6$.
Using the general equation for a circle $(x - a)^2 + (y - b)^2 = r^2$ you can write: $(x + 4)^2 + (y - 2)^2 = 36$

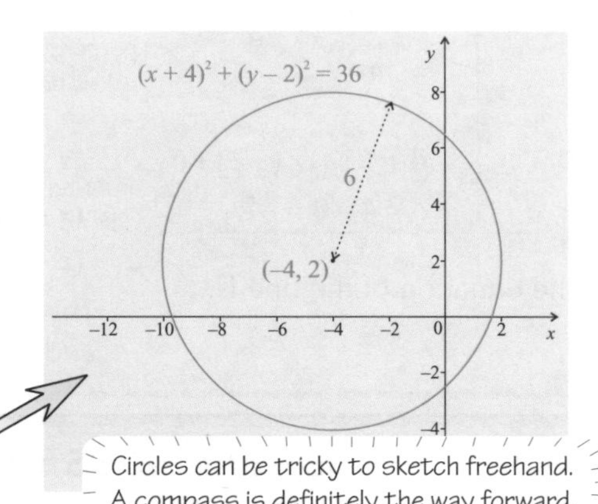

This is pretty much all you need to learn. Everything on the next page uses stuff you should know already.

Circles

This page is for AQA Core 1, OCR Core 1, Edexcel Core 2

Rearrange the equation into the familiar form

Sometimes you'll be given an equation for a circle that doesn't look much like $(x-a)^2 + (y-b)^2 = r^2$.
This is a bit of a pain, because it means you can't immediately tell what the **radius** is or where the **centre** is.
But all it takes is a bit of **rearranging**.

Let's take the equation: $x^2 + y^2 - 6x + 4y + 4 = 0$

You need to get it into the form $(x-a)^2 + (y-b)^2 = r^2$

This is just like completing the square.

$x^2 + y^2 - 6x + 4y + 4 = 0$

$x^2 - 6x + y^2 + 4y + 4 = 0$

$(x-3)^2 - 9 + (y+2)^2 - 4 + 4 = 0$

$(x-3)^2 + (y+2)^2 = 9$ ⟹ This is the recognisable form, so the centre is (3, –2) and the radius is $\sqrt{9} = 3$.

Have a look at pages 12-13 for more on completing the square.

Don't forget the Properties of Circles

You will have seen the circle rules at GCSE. You'll sometimes need to dredge them up in your memory for these circle questions. Here's a reminder of a few useful ones.

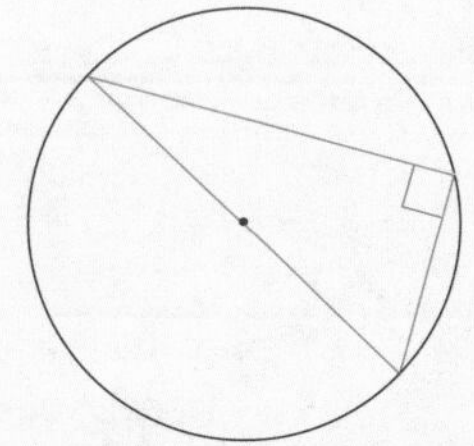

The angle in a semicircle is a right angle.

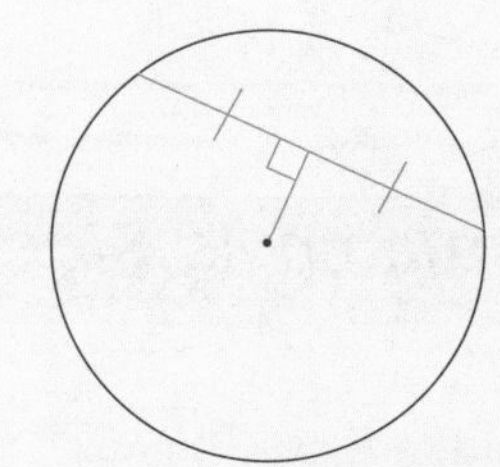

The perpendicular from the centre to a chord bisects the chord.

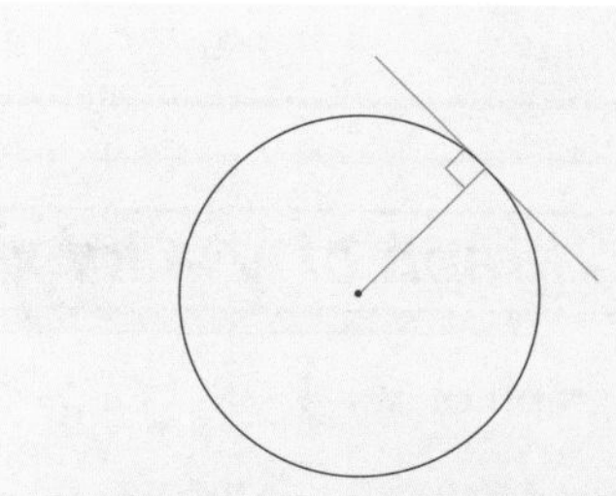

A radius and tangent to the same point will meet at right angles.

Use the Gradient Rule for Perpendicular Lines

Remember that the tangent at a given point will be perpendicular to the radius and the normal at that same point.

Example: Point A (6, 4) lies on a circle with the equation $x^2 + y^2 - 4x - 2y - 20 = 0$.

i) Find the centre and radius of the circle.

ii) Find the equation of the tangent to the circle at A.

Solution:

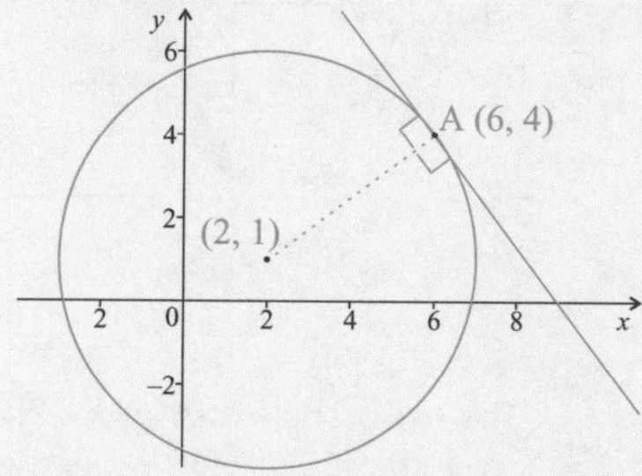

i) Rearrange the equation to show it as the sum of 2 squares:

$x^2 + y^2 - 4x - 2y - 20 = 0$

$x^2 - 4x + y^2 - 2y - 20 = 0$

$(x-2)^2 - 4 + (y-1)^2 - 1 - 20 = 0$

$(x-2)^2 + (y-1)^2 = 25$

This shows the centre is (2, 1) and the radius is 5.

ii) The tangent is at right angles to the radius at (6, 4).

Gradient of radius at (6, 4) = $\frac{4-1}{6-2} = \frac{3}{4}$

Gradient of tangent = $\frac{-1}{\frac{3}{4}} = -\frac{4}{3}$

Using $y - y_1 = m(x - x_1)$

$y - 4 = -\frac{4}{3}(x-6)$

$3y - 12 = -4x + 24$

$3y + 4x - 36 = 0$

So the chicken comes from the egg, and the egg comes from the chicken...

Well folks, at least it makes a change from all those straight lines and quadratics.
I reckon if you know the formula and what it means, you should be absolutely fine with questions on circles.

Curve Sketching

This page is for AQA Core 1, OCR Core 1, Edexcel Core 1

A picture speaks a thousand words... and graphs are what pass for pictures in maths. They're dead useful in getting your head round tricky questions, and time spent learning how to sketch graphs is time well spent.

The graph of $y = kx^n$ is a different shape for different k and n

Usually, you only need a rough sketch of a graph — so just knowing the basic shapes of these graphs will do.

n positive and even

You get a u-shape or an n-shape.

n positive and odd

You get a 'corner-to-corner' shape.

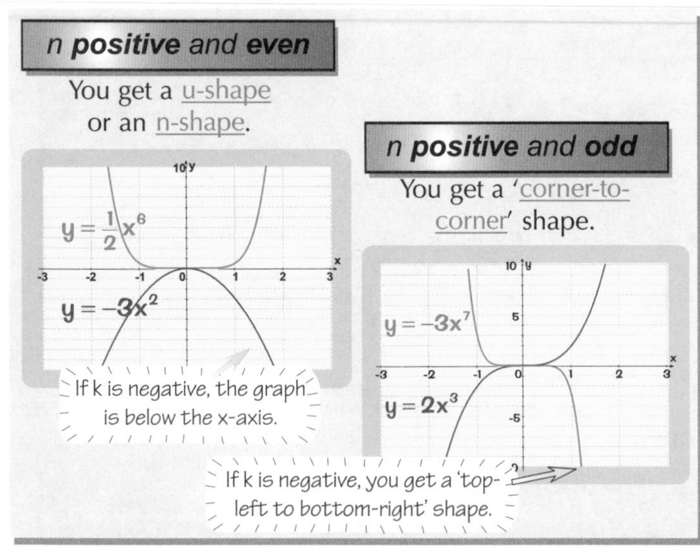
$y = \frac{1}{2}x^6$
$y = -3x^2$
If k is negative, the graph is below the x-axis.
$y = -3x^7$
$y = 2x^3$
If k is negative, you get a 'top-left to bottom-right' shape.

n negative and even

You get a graph with two bits next to each other.

n negative and odd

You get a graph with two bits opposite each other.

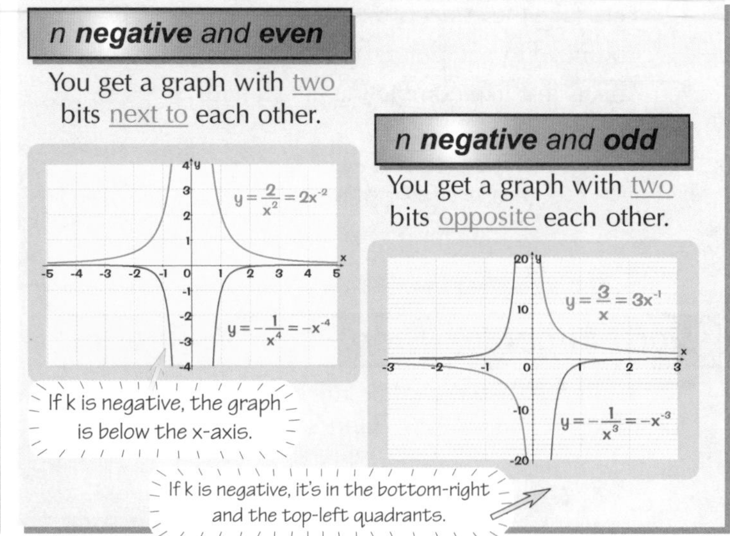
$y = \frac{2}{x^2} = 2x^{-2}$
$y = -\frac{1}{x^4} = -x^{-4}$
If k is negative, the graph is below the x-axis.
$y = \frac{3}{x} = 3x^{-1}$
$y = -\frac{1}{x^3} = -x^{-3}$
If k is negative, it's in the bottom-right and the top-left quadrants.

The graph of $y = k\sqrt{x}$ is a Parabola on its Side

The graph of $y = k\sqrt{x}$ is a parabola on its side.

This makes sense really, because if $y = k\sqrt{x}$, then $x = \frac{1}{k^2}y^2$ — and this is just a normal quadratic with the x and y switched round.

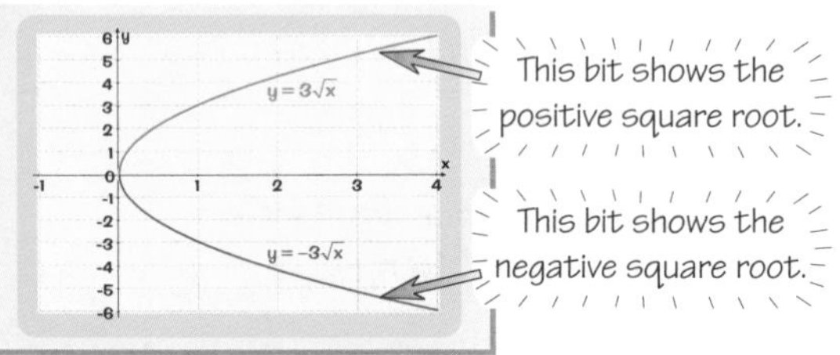
$y = 3\sqrt{x}$
$y = -3\sqrt{x}$
This bit shows the positive square root.
This bit shows the negative square root.

If you know the Factors of a cubic — the graph's easy to Sketch

A cubic function has an x^3 term in it, and all cubics have 'bottom-left to top-right' shape — or a 'top-left to bottom-right' shape if the coefficient of x^3 is negative.

If you know the factors of a cubic, the graph is easy to sketch — just find where the function is zero.

Example: Sketch the graphs of the following cubic functions.

(i) $f(x) = x(x-1)(2x+1)$ (ii) $g(x) = (1-x)(x^2-2x+2)$ (iii) $h(x) = (x-3)^2(x+1)$ (iv) $m(x) = (2-x)^3$

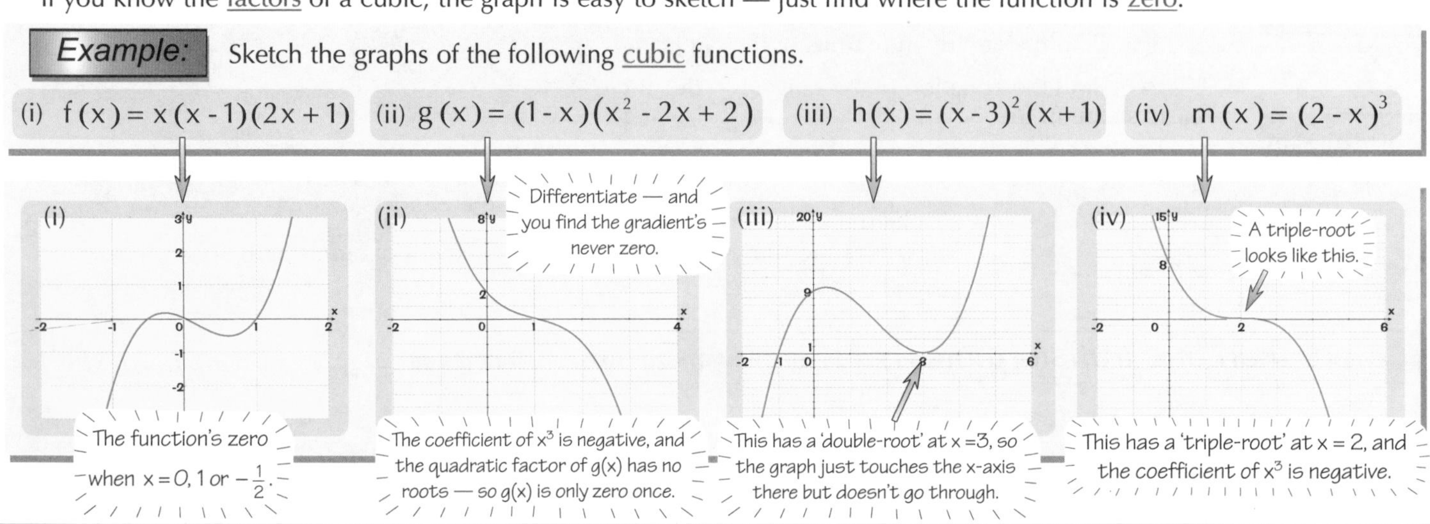
(i)
(ii)
Differentiate — and you find the gradient's never zero.
(iii)
(iv)
A triple-root looks like this.
The function's zero when $x = 0, 1$ or $-\frac{1}{2}$.
The coefficient of x^3 is negative, and the quadratic factor of g(x) has no roots — so g(x) is only zero once.
This has a 'double-root' at x = 3, so the graph just touches the x-axis there but doesn't go through.
This has a 'triple-root' at x = 2, and the coefficient of x^3 is negative.

Graphs, graphs, graphs — you can never have too many graphs...

It may seem like a lot to remember, but graphs can really help you get your head round a question — a quick sketch can throw a helluva lot of light on a problem that's got you completely stumped. So being able to draw these graphs won't just help with an actual graph-sketching question — it could help with loads of others too. Got to be worth learning.

Graph Transformations

This page is for AQA Core 1, AQA Core 2, OCR Core 1, Edexcel Core 1

Suppose you start with any old function f(x). Then you can transform (change) it in three ways — by translating it, stretching or reflecting it.

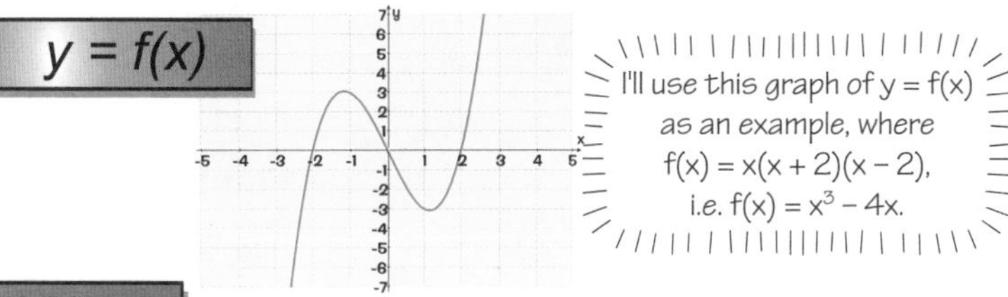
y = f(x)
I'll use this graph of y = f(x) as an example, where f(x) = x(x + 2)(x − 2), i.e. f(x) = x³ − 4x.

Translations are caused by Adding things

y = f(x)+a

Adding a number to the whole function shifts the graph up or down.

1) If $a > 0$, the graph goes upwards.
2) If $a < 0$, the graph goes downwards.

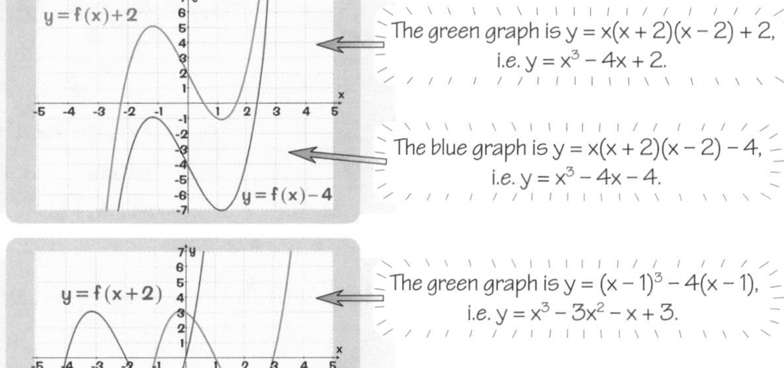
y = f(x)+2
y = f(x)−4
The green graph is y = x(x + 2)(x − 2) + 2, i.e. y = x³ − 4x + 2.
The blue graph is y = x(x + 2)(x − 2) − 4, i.e. y = x³ − 4x − 4.

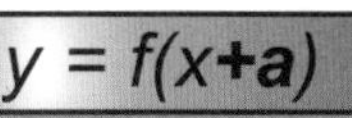

y = f(x+a)

Writing 'x + a' instead of 'x' means the graph moves sideways.

1) If $a > 0$, the graph goes to the left.
2) If $a < 0$, the graph goes to the right.

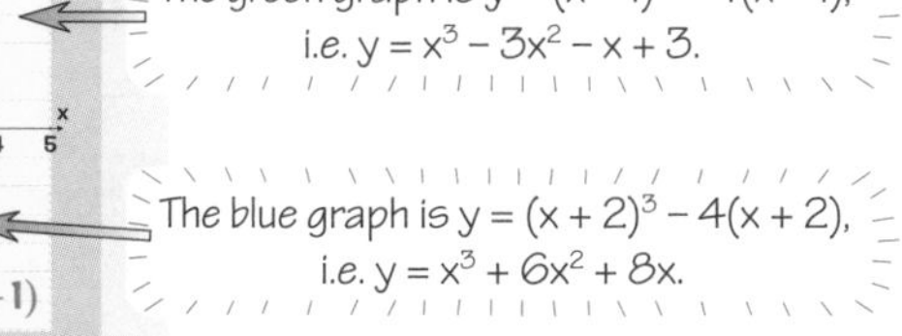
y = f(x+2)
y = f(x−1)
The green graph is y = (x − 1)³ − 4(x − 1), i.e. y = x³ − 3x² − x + 3.
The blue graph is y = (x + 2)³ − 4(x + 2), i.e. y = x³ + 6x² + 8x.

Stretches and Reflections are caused by Multiplying things

y = af(x)

Multiplying the whole function stretches, squeezes or reflects the graph vertically.

1) Negative values of 'a' reflect the basic shape in the x-axis.
2) If $a > 1$ or $a < -1$ (i.e. $|a| > 1$) the graph is stretched vertically.
3) If $-1 < a < 1$ (i.e. $|a| < 1$) the graph is squashed vertically.

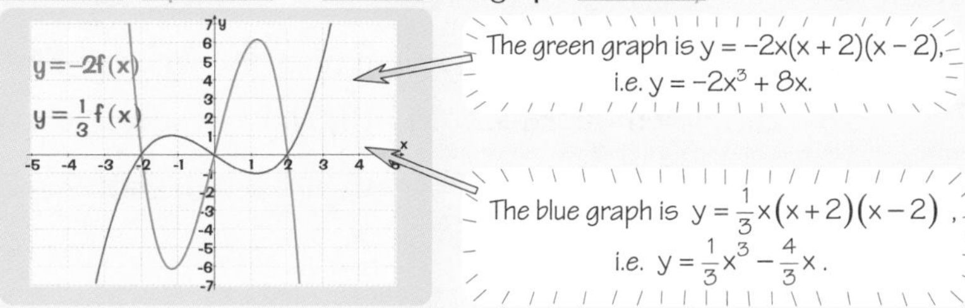
y = −2f(x)
y = ⅓f(x)
The green graph is y = −2x(x + 2)(x − 2), i.e. y = −2x³ + 8x.
The blue graph is y = ⅓x(x + 2)(x − 2), i.e. y = ⅓x³ − 4/3 x.

y = f(ax)

Writing 'ax' instead of 'x' stretches, squeezes or reflects the graph horizontally.

1) Negative values of 'a' reflect the basic shape in the y-axis.
2) If $a > 1$ or $a < -1$ (i.e. if $|a| > 1$) the graph is squashed horizontally.
3) If $-1 < a < 1$ (i.e. if $|a| < 1$) the graph is stretched horizontally.

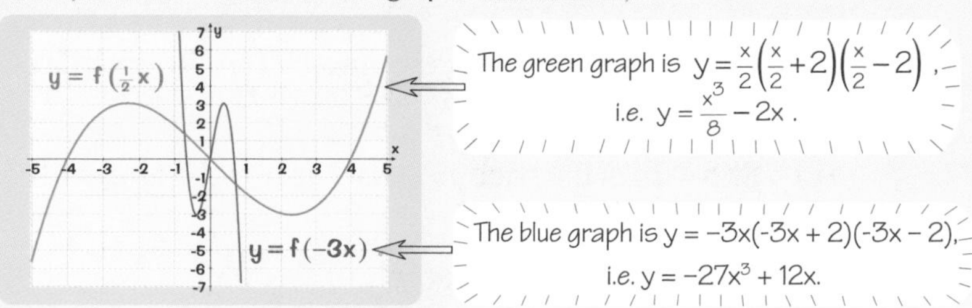
y = f(½x)
y = f(−3x)
The green graph is y = x/2 (x/2 + 2)(x/2 − 2), i.e. y = x³/8 − 2x.
The blue graph is y = −3x(−3x + 2)(−3x − 2), i.e. y = −27x³ + 12x.

More than one transformation at a time: y = af(bx + c)+d

Example: $y = af(bx+c)+d$

$a=\frac{1}{3}$ ⇒ The graph is squashed vertically.

$b=\frac{1}{2}$ ⇒ The graph is stretched horizontally.

$c=\frac{1}{2}$ ⇒ The graph is moved horizontally (to the left).

$d=1$ ⇒ The graph is shifted vertically (upwards).

This is a combination of all these transformations together.

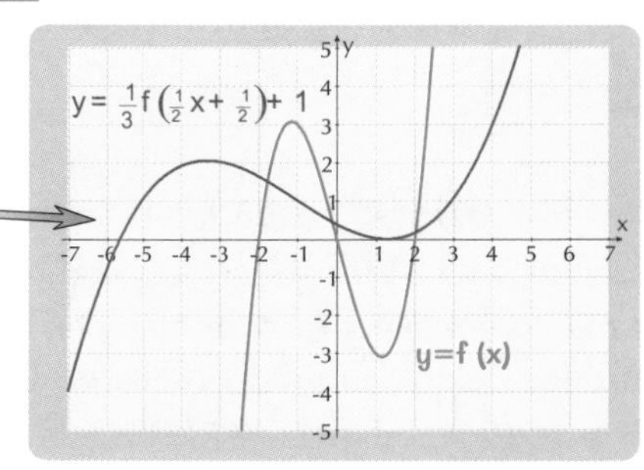
y = ⅓f(½x + ½) + 1
y=f (x)

Exam Questions

Exam Questions

1

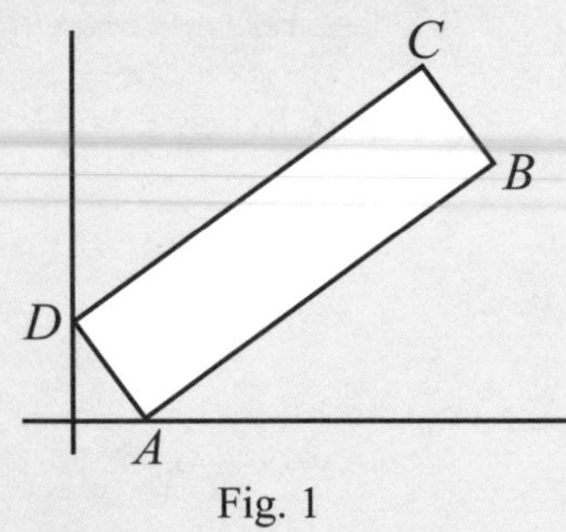

Fig. 1

Diagram not drawn to scale.

Rectangle ABCD is shown in figure 1.
Point A has coordinates (2, 0) and point D has coordinates (0, 5).

(a) Find the equation of the line through AD. *(2 marks)*

(b) Hence, find the equation of the line through DC. *(2 marks)*

(c) Given that the x-coordinate of point C is 10, find the area of rectangle ABCD. *(4 marks)*

2 Find the equation of the line which passes through point A(4, 7) and which is parallel to the line $14 - 2y + 5x = 0$.
Give your answer in the form $ay + bx + c = 0$. *(3 marks)*

3 **(a)** Solve the simultaneous equations

$$x + y = -4 \qquad y = -x^2 + 8x - 12$$

(4 marks)

(b) Interpret your solution to part (a) geometrically. *(1 mark)*

4 **(a)** Find the coordinates of the point A, when A lies at the intersection of the lines l_1 and l_2, and when the equations of l_1 and l_2 respectively are

$$x - y + 1 = 0 \text{ and } 2x + y - 8 = 0.$$

(3 marks)

(b) The points B and C have coordinates (6, -4) and $\left(-\frac{4}{3}, -\frac{1}{3}\right)$ respectively, and D is the midpoint of AC.
Find the equation of the line BD in the form $ax + by + c = 0$, where a, b and c are integers. *(5 marks)*

(c) Show that the triangle ABD is a right-angled triangle. *(3 marks)*

5 This diagram shows the graph of $y = f(x)$. Using a different set of axes for each one, sketch the following graphs.
Show as much of the graph as you can, labelling the axes wherever appropriate.

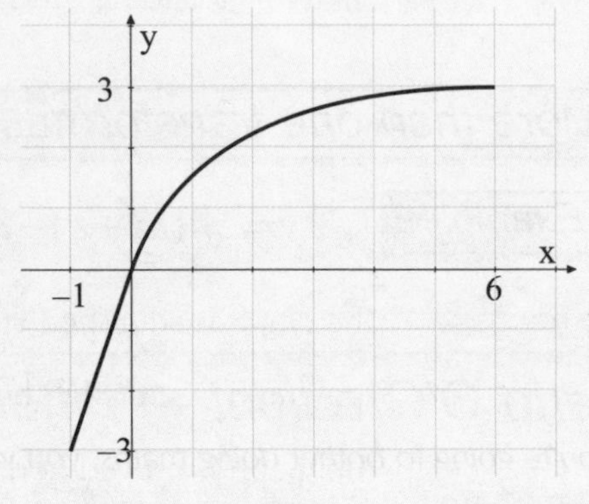

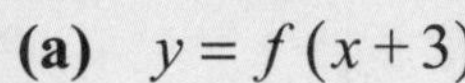

(a) $y = f(x+3)$ *(2 marks)*

(b) $y = f(x)+3$ *(2 marks)*

(c) $y = f(-3x)$ *(2 marks)*

(d) $y = 3f(x)$ *(2 marks)*

Differentiation

This page is for AQA Core 1, AQA Core 2, OCR Core 1, Edexcel Core 1

Brrrrrr... differentiation is a bad one — it really is. Not because it's that hard, but because in the exams it comes up all over the place, every time. So if you don't know it perfectly, you're asking for trouble.

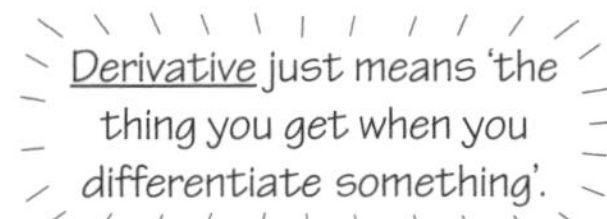
Derivative just means 'the thing you get when you differentiate something'.

$$\frac{d}{dx}\left(x^n\right) = nx^{n-1}$$

$\frac{d}{dx}$ just means 'the derivative of the thing in the brackets'.

Use this formula to differentiate Powers of x

Example: Differentiate y when:

i) $y = x^5$ ii) $y = 6x^3$ iii) $y = 24x$ iv) $y = 5$

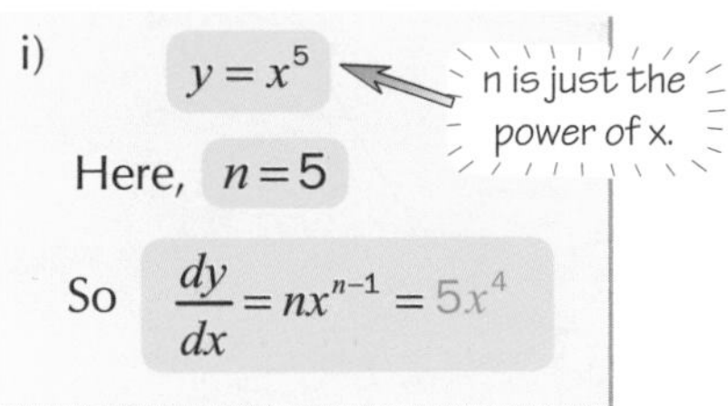
i) $y = x^5$
n is just the power of x.
Here, $n = 5$
So $\frac{dy}{dx} = nx^{n-1} = 5x^4$

ii) $y = 6x^3$

$$\frac{dy}{dx} = 6(3x^2)$$

i.e. $\frac{dy}{dx} = 18x^2$

$\frac{d(af(x))}{dx} = a\frac{d(f(x))}{dx}$ — i.e. you ignore the number and only worry about the x bit.

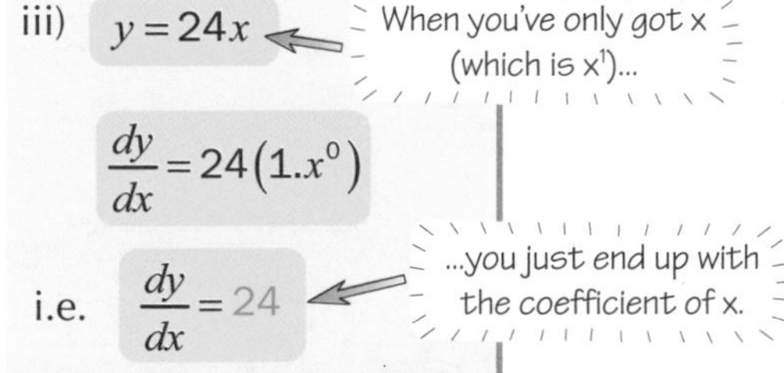
iii) $y = 24x$
When you've only got x (which is x^1)...
$\frac{dy}{dx} = 24(1.x^0)$
i.e. $\frac{dy}{dx} = 24$
...you just end up with the coefficient of x.

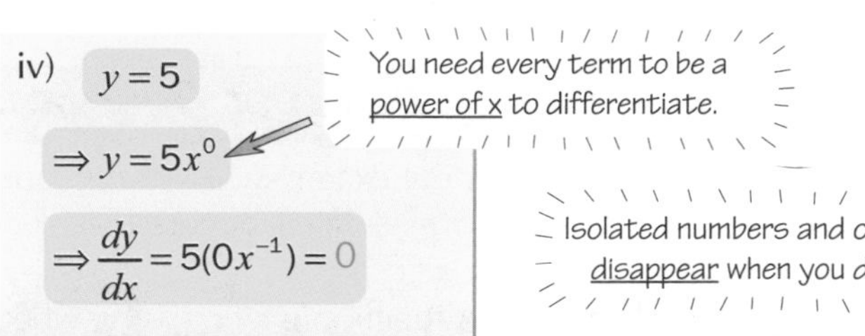
iv) $y = 5$
You need every term to be a power of x to differentiate.
$\Rightarrow y = 5x^0$
$\Rightarrow \frac{dy}{dx} = 5(0x^{-1}) = 0$

Isolated numbers and constants just disappear when you differentiate.

Differentiate each term in an equation Separately

This formula is better than cake — even better than that really nice sticky black chocolate one from that place in town. Even if there are loads of terms in the equation, it doesn't matter. Differentiate each bit separately and you'll be fine.

You can do this because: $\frac{d(f(x)+g(x))}{dx} = \frac{d(f(x))}{dx} + \frac{d(g(x))}{dx}$

Example: Differentiate y when: i) $y = 6x^4 + 4x^3 - 2x + 1$ ii) $y = (x+2)(x+3)$

i) For equations like this...

$$y = 6x^4 + 4x^3 - 2x + 1$$

...just differentiate each term separately.

$$\frac{dy}{dx} = 6(4x^3) + 4(3x^2) - 2 + 0$$

$$\frac{dy}{dx} = 24x^3 + 12x^2 - 2$$

ii) You can't differentiate it until it's written as separate terms which are all powers of x.

$$y = (x+2)(x+3)$$

So multiply out the brackets...

$$y = x^2 + 5x + 6$$

Then it's easy: $\frac{dy}{dx} = 2x + 5$

Dario O'Gradient — differentiating Crewe from the rest...

If you're going to bother doing maths, you've got to be able to differentiate things. Simple as that. But luckily, once you can do the simple stuff, you should be all right. Big long equations are just made up of loads of simple little terms, so they're not really that much harder. Learn the formula, and make sure you can use it by practising all day and all night forever.

Differentiation

This page is for AQA Core 1, OCR Core 1, Edexcel Core 1, Edexcel Core 2

But what's the point of differentiation? I dunno. Search me...

Differentiating a function gives you its Gradient

Differentiating a function (a curve, say) gives you an expression for the gradient of the curve. Then you can find the gradient of the curve at any point by substituting for x.

Example: Find the gradient of the graph $y = x^2$ at $x = 1$ and $x = -2$...

Differentiate to get the gradient:

$$\frac{dy}{dx} = 2x$$

Now when $x = 1$, $\frac{dy}{dx} = 2$

And so the gradient of the graph at x = 1 is 2.

And when $x = -2$, $\frac{dy}{dx} = -4$

So the gradient of the graph at x = –2 is –4.

The gradient of a curve is the same as the gradient of the tangent at that point.

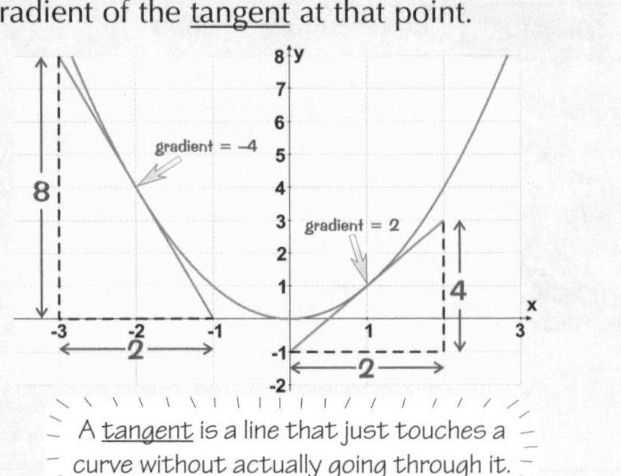

gradient = –4
gradient = 2
8
4
2
2
A tangent is a line that just touches a curve without actually going through it.

Find out if a function is Increasing or Decreasing

You can also use differentiation to work out exactly where a function is increasing or decreasing. And you can use it to find out its rate of change — that's how fast it's decreasing or increasing.

A function is increasing when...
...the gradient is positive.

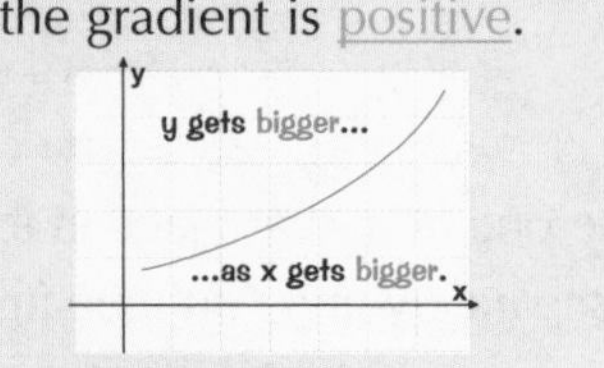

y gets bigger...
...as x gets bigger.

A function is decreasing when...
...the gradient is negative.

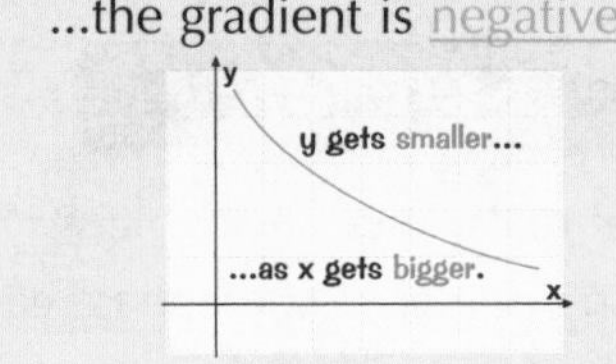

y gets smaller...
...as x gets bigger.

The bigger the gradient...
...the faster y changes with x.

A small change in x means a big change in y.

A big change in x means a small change in y.

Example: Find where the following function is increasing and decreasing: $f(x) = 3x^2 - 6x$.

This is a question about gradients — so differentiate.

$$f(x) = 3x^2 - 6x$$
$$\Rightarrow f'(x) = 6x - 6$$

f'(x) (pronounced, 'f-dash of x' or 'f-prime of x') is another way to write the derivative.

This is an increasing function when

$$6x - 6 > 0$$
$$\Rightarrow x > 1$$

This is a decreasing function when

$$6x - 6 < 0$$
$$\Rightarrow x < 1$$

Differentiation and Gradients

To find the gradient of a curve at a certain point:

1) **Differentiate the equation** of the curve.
2) **Work out the derivative** at the point.

An increasing function has a **positive** gradient.
A decreasing function has a **negative** gradient.

Help me Differentiation — You're my only hope...

There's not much hard maths on this page — but there are a couple of very important ideas that you need to get your head round pretty darn soon. Understanding that differentiating gives the gradient of the graph is more important than washing regularly — AND THAT'S IMPORTANT. The other thing on the page — that you can tell whether a function is getting bigger or smaller by looking at the derivative — is also vital. Sometimes the examiners ask you to find where a function is increasing or decreasing — so you'd just have to find where the derivative was positive or negative.

Differentiation

To find a stationary point, you need to find where the graph 'levels off' — that means where the gradient becomes zero.

Stationary Points are when the gradient is Zero

This section is for AQA Core 1, OCR Core 1, Edexcel Core 2

Example: Find the stationary points on the curve $y = 2x^3 - 3x^2 - 12x + 5$, and work out the nature of each one.

A stationary point can be...

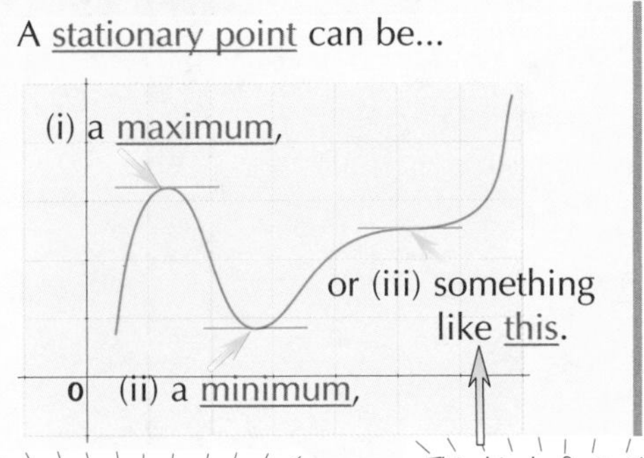

(i) a maximum,
or (iii) something like this.
0
(ii) a minimum,

At stationary points, the gradient = 0, which means $\frac{dy}{dx} = 0$.

This kind of stationary point is called a 'point of inflection'.

A turning point is a another name for a maximum or a minimum.

You need to find where $\frac{dy}{dx} = 0$. So first, differentiate the function.

$$y = 2x^3 - 3x^2 - 12x + 5$$
$$\Rightarrow \frac{dy}{dx} = 6x^2 - 6x - 12$$

This is the expression for the gradient.

And then set this derivative equal to zero.

$$6x^2 - 6x - 12 = 0$$
$$\Rightarrow x^2 - x - 2 = 0$$
$$\Rightarrow (x-2)(x+1) = 0$$
$$\Rightarrow x = 2 \text{ or } x = -1$$

See pages 10 to 16 for more about solving quadratics.

So the graph has two stationary points, at $x = 2$ and $x = -1$.

Decide if it's a Maximum or a Minimum by differentiating Again

This section is for AQA Core 1, OCR Core 1, Edexcel Core 2

Once you've found where the stationary points are, you have to decide whether each of them is a maximum or minimum — this is all a question means when it says, '...determine the nature of the turning points'.

To decide whether a stationary point is a maximum or a minimum — just differentiate again to find $\frac{d^2y}{dx^2}$, or f''(x).

If $\frac{d^2y}{dx^2} < 0$, it's a maximum.

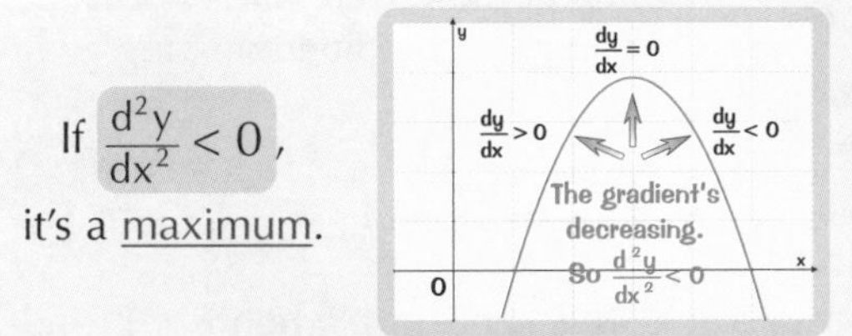

If $\frac{d^2y}{dx^2} > 0$, it's a minimum.

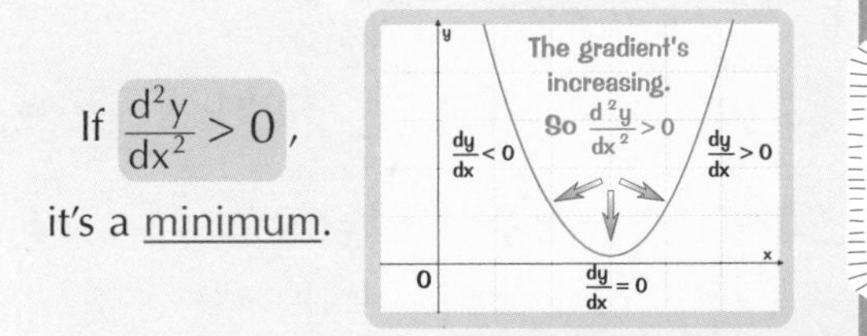

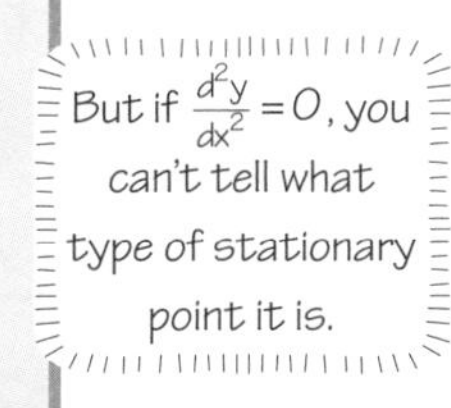

You've just found that $\frac{dy}{dx} = 6x^2 - 6x - 12$.

So differentiating again gives $\frac{d^2y}{dx^2} = 12x - 6$.

Stick in the x-coordinates of the stationary points.

At $x = -1$, $\frac{d^2y}{dx^2} = -18$, which is negative — so $x = -1$ is a maximum.

And at $x = 2$, $\frac{d^2y}{dx^2} = 18$, which is positive — so $x = 2$ is a minimum.

And since a cubic graph (where the coefficient of x^3 is positive) goes from bottom-left to top-right...

...you can draw a rough sketch of the graph, even though the roots would be hard to find.

Stationary Points

1) Find stationary points by solving $\frac{dy}{dx} = 0$. (That's $f'(x) = 0$)

2) Differentiate again to decide whether this is a maximum or a minimum.

3) If $\frac{d^2y}{dx^2} < 0$ — it's a maximum.

$\frac{d^2y}{dx^2}$ is exactly the same as $f''(x)$

If $\frac{d^2y}{dx^2} > 0$ — it's a minimum.

An anagram of differentiation is "Perfect Insomnia Cure"...

No joke this, is it — this differentiation business — but it's a dead important topic in maths. It's so important to know how to find whether a stationary point is a max or a min — but it can get a bit confusing. Try remembering MINMAX — which is short for 'MINUS means a MAXIMUM'. Or make up some other clever way to remember what means what.

Curve Sketching

This page is for AQA Core 1, OCR Core 1, Edexcel Core 2

You'll even be asked to do some drawing in the exam... but don't get too excited — it's just drawing graphs... great.

Find where the curve crosses the **Axes**

Sketch the graph of $f(x) = \frac{x^2}{2} - 2\sqrt{x}$, for $x \geq 0$.

The curve crosses the y-axis when x = 0 — so put x = 0 in the expression for y.

When x = 0, f(x) = 0 — and so the curve goes through the origin.

The curve crosses the x-axis when f(x) = 0. So solve

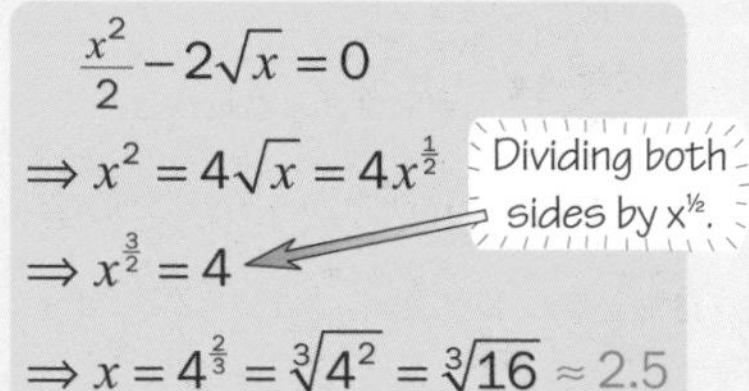

$$\frac{x^2}{2} - 2\sqrt{x} = 0$$
$$\Rightarrow x^2 = 4\sqrt{x} = 4x^{\frac{1}{2}}$$
$$\Rightarrow x^{\frac{3}{2}} = 4$$
$$\Rightarrow x = 4^{\frac{2}{3}} = \sqrt[3]{4^2} = \sqrt[3]{16} \approx 2.5$$

Dividing both sides by $x^{½}$.

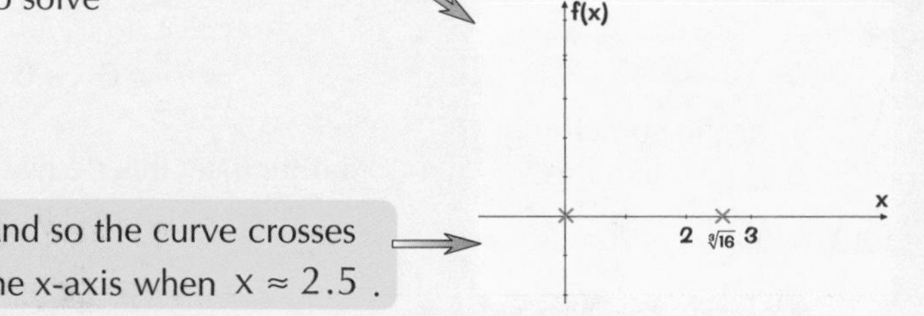

And so the curve crosses the x-axis when $x \approx 2.5$.

...**Differentiate** to find **Gradient** info...

Differentiating the function gives...

$$f(x) = \frac{1}{2}x^2 - 2x^{\frac{1}{2}}$$
$$\Rightarrow f'(x) = \frac{1}{2}(2x) - 2\left(\frac{1}{2}x^{-\frac{1}{2}}\right) = x - x^{-\frac{1}{2}} = x - \frac{1}{\sqrt{x}}$$

Using the derivative — you can find stationary points and tell when the graph goes 'uphill' and 'downhill'.

1) So there's a stationary point when...

$$x - \frac{1}{\sqrt{x}} = 0$$
$$\Rightarrow x = \frac{1}{\sqrt{x}}$$
$$\Rightarrow x^{\frac{3}{2}} = 1 \Rightarrow x = 1$$

And at x = 1, $f(x) = \frac{1}{2} - 2 = -\frac{3}{2}$.

2) The gradient's negative when...

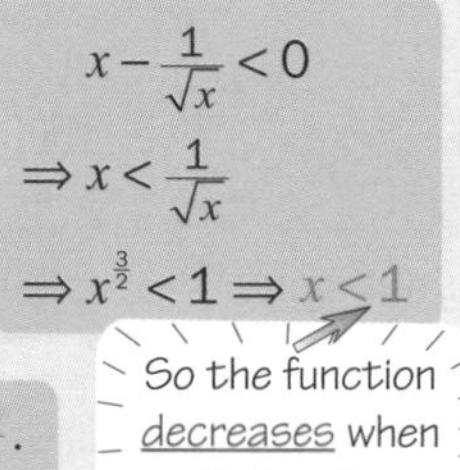

$$x - \frac{1}{\sqrt{x}} < 0$$
$$\Rightarrow x < \frac{1}{\sqrt{x}}$$
$$\Rightarrow x^{\frac{3}{2}} < 1 \Rightarrow x < 1$$

So the function decreases when $0 \leq x < 1$...

3) The gradient's positive when...

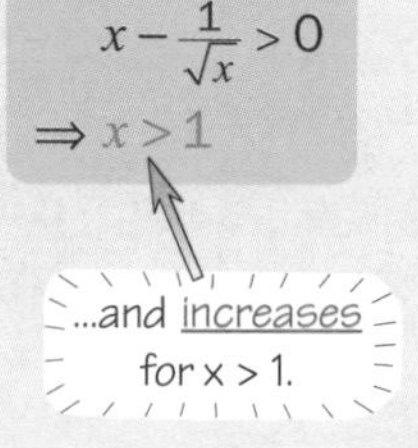

$$x - \frac{1}{\sqrt{x}} > 0$$
$$\Rightarrow x > 1$$

...and increases for x > 1.

This is the quickest way to check if something's a max or a min.

You could check that x = 1 is a minimum by differentiating again.

$$f''(x) = 1 - \left(-\frac{1}{2}x^{-\frac{3}{2}}\right) = 1 + \frac{1}{2\sqrt{x^3}}$$

This is positive when x = 1, and so this is definitely a minimum.

...and find out what happens when x gets **Big**

You can also try and decide what happens as x gets very big — in both the positive and negative directions.

Factorise f(x) by taking the biggest power outside the brackets...

$$\frac{x^2}{2} - 2\sqrt{x} = x^2\left(\frac{1}{2} - 2x^{-\frac{3}{2}}\right) = x^2\left(\frac{1}{2} - \frac{2}{x^{\frac{3}{2}}}\right)$$

As x gets large, this bit disappears — and the bit in brackets gets closer to $\frac{1}{2}$.

And so as x gets larger, f(x) gets closer and closer to $\frac{1}{2}x^2$ — and this just keeps growing and growing.

And the graph looks like this...

Curve-sketching's important — but don't take my word for it...

Curve-sketching — an underrated skill, in my opinion. As Shakespeare once wrote, 'Those who can do fab sketches of graphs and stuff are likely to get pretty good grades in maths exams, no word of a lie'. Well, he probably would've written something like that if he was into maths. And he would've written it because graphs are helpful when you're trying to work out what a question's all about — and once you know that, you can decide the best way forward. And if you don't believe me, remember the saying of the ancient Roman Emperor Julius Caesar, 'If in doubt, draw a graph'.

Finding Tangents and Normals

This page is for AQA Core 1, OCR Core 1, Edexcel Core 1

What's a tangent? Beats me. Oh no, I remember, it's one of those thingies on a curve. Ah, yes... I remember now...

Tangents *Just* touch a curve

To find the equation of a tangent or a normal to a curve, you first need to know its gradient — so differentiate. Then complete the line's equation using the coordinates of one point on the line.

Example: Find the tangent to the curve y = (4 – x)(x + 2) at the point (2, 8).

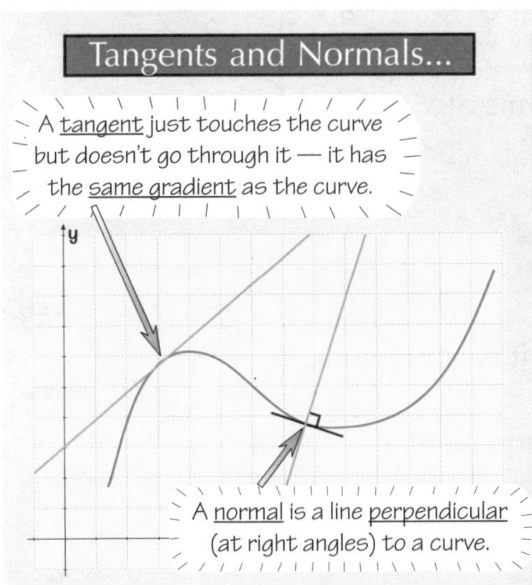

Tangents and Normals...
A tangent just touches the curve but doesn't go through it — it has the same gradient as the curve.
y
A normal is a line perpendicular (at right angles) to a curve.

To find the curve's (and the tangent's) gradient, first write the equation in a form you can differentiate...

$$y = 8 + 2x - x^2$$

...and then differentiate it.

$$\frac{dy}{dx} = 2 - 2x$$

The gradient of the tangent will be the gradient of the curve at x = 2.

At x = 2, $\frac{dy}{dx} = -2$,

So the tangent has equation,

$$y - y_1 = -2(x - x_1)$$

in $y - y_1 = m(x - x_1)$ form. See page 27.

And since it passes through the point (2,8), this becomes

$y - 8 = -2(x - 2)$, or $y = -2x + 12$.

You can also write it in $y = mx + c$ form.

Normals are at *Right Angles* to a curve

Example: Find the normal to the curve $y = \frac{(x+2)(x+4)}{6\sqrt{x}}$ at the point (4, 4).

Write the equation of the curve in a form you can differentiate.

$$y = \frac{x^2 + 6x + 8}{6x^{\frac{1}{2}}} = \frac{1}{6}x^{\frac{3}{2}} + x^{\frac{1}{2}} + \frac{4}{3}x^{-\frac{1}{2}}$$

Dividing everything on the top line by everything on the bottom line.

Differentiate it...

$$\frac{dy}{dx} = \frac{1}{6}\left(\frac{3}{2}x^{\frac{1}{2}}\right) + \frac{1}{2}x^{-\frac{1}{2}} + \frac{4}{3}\left(-\frac{1}{2}x^{-\frac{3}{2}}\right)$$
$$= \frac{1}{4}\sqrt{x} + \frac{1}{2\sqrt{x}} - \frac{2}{3\sqrt{x^3}}$$

Find the gradient at the point you're interested in. At x = 4,

$$\frac{dy}{dx} = \frac{1}{4} \times 2 + \frac{1}{2 \times 2} - \frac{2}{3 \times 8} = \frac{2}{3}$$

So the gradient of the normal is $-\frac{3}{2}$.

Because the gradient of the normal multiplied by the gradient of the curve must be –1.

And the equation of the normal is $y - y_1 = -\frac{3}{2}(x - x_1)$.

Finally, since the normal goes through the point (4, 4), the equation of the normal must be $y - 4 = -\frac{3}{2}(x - 4)$, or after rearranging, $y = -\frac{3}{2}x + 10$.

Finding tangents and normals

1) **Differentiate the function.**
2) **Find the gradient, m, of the tangent or normal. This is,**
   for a tangent: the gradient of the curve
   for a normal: $\frac{-1}{\text{gradient of the curve}}$
3) **Write the equation of the tangent or normal in the form $y - y_1 = m(x - x_1)$, or $y = mx + c$.**
4) **Complete the equation of the line using the coordinates of a point on the line.**

Repeat after me... "I adore tangents and normals..."

Examiners can't stop themselves saying the words 'Find the tangent...' and 'Find the normal...'. They love the words. These phrases are music to their ears. They can't get enough of them. I just thought it was my duty to tell you that. And so now you know, you'll definitely be wanting to learn how to do the stuff on this page. Of course you will.

Exam Questions

Exam Questions

1 The volume of a closed circular tin (figure 1) is 450 cm^3.
The radius of the tin is x cm, and its height is y cm.
The surface area of the tin is S.

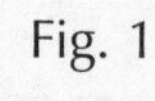
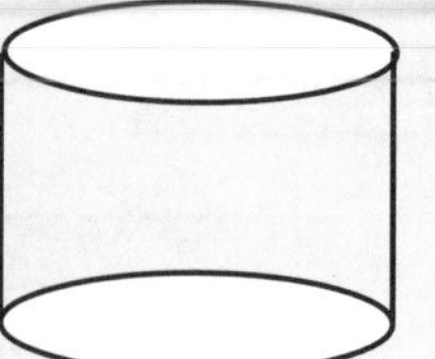

Fig. 1

(a) Show that the surface area of the tin can be expressed in terms of x as:

$$S = 2\pi x^2 + \frac{900}{x}$$

(4 marks)

(b) Find the minimum value of the surface area and prove that it is a minimum.

(8 marks)

(c) Find the height of the tin when the surface area is at a minimum.

(2 marks)

2 **(a)** Find $\frac{dy}{dx}$ for the curve $y = x^3 - x^2 - 8x + 24$.

(1 mark)

(b) Use your result to find the x-coordinates of the turning points of $y = x^3 - x^2 - 8x + 24$.

(3 marks)

(c) Use your result from part **(a)** to find an equation for the tangent to the curve $y = x^3 - x^2 - 8x + 24$ at the point (1, 16).

(2 marks)

3 **(a)** Show that the graph of $f(x) = -x^2(x-2)$ has turning points at the origin and the point $\left(\frac{4}{3}, \frac{32}{27}\right)$.

(3 marks)

(b) Sketch the graph of $f(x)$, marking in any turning points and points where it crosses the axes.

(2 marks)

Sequences

A sequence is a list of numbers that follow a certain pattern. Sequences can be finite or infinite (infinity — oooh), and they're usually generated in one of two ways. And guess what? You have to know everything about them.

A Sequence can be defined by its nth Term

You may well have covered this stuff at GCSE — if so, you've got no excuses for mucking it up.

With some sequences, you can work out any value (the nth term) from its position in the sequence (n). And often, you can even work out a formula for the nth term.

Example: Find the nth term of the sequence 5, 8, 11, 14, 17, ...

1st	2nd	3rd	4th	5th
5	8	11	14	17

+3 +3 +3 +3

Each term is 3 more than the one before it. That means that you need to start by multiplying n by 3.

Take the first term (where n = 1). If you multiply n by 3, you still have to add 2 to get 5.

The same goes for n = 2. To get 8 you need to multiply n by 3, then add 2.
Every term in the sequence is worked out exactly the same way.

So the nth term is $3n + 2$

You can define a sequence by a Recurrence Relation too

Don't be put off by the fancy name — recurrence relations are pretty easy really.

The main thing to remember is:
x_n just means the nth term of the sequence

The next term in the sequence is x_{n+1}. You need to describe how to work out x_{n+1} if you're given x_n.
What you're actually doing is working out x_{n+1} as a function of x_n, so you can write $x_{n+1} = f(x_n)$.

Example: Find the recurrence relation of the sequence 5, 8, 11, 14, 17, ...

From the example above, you know that each term equals the one before it, plus 3.

This is written like this: $\mathbf{x_{n+1} = x_n + 3}$

In everyday language, $x_{n+1} = x_n + 3$ means that the second term equals the first term plus 3.

BUT $x_{n+1} = x_n + 3$ on its own isn't enough to describe 5, 8, 11, 14, 17,...
For example, the sequence 87, 90, 93, 96, 99, ... also has each term being 3 more than the one before.

The recurrence relation needs to be more specific, so you've got to give one term in the sequence.
You almost always give the first value, x_1.

Putting all of this together gives 5, 8, 11, 14, 17,... as $x_{n+1} = x_n + 3,\ x_1 = 5$

Sequences

This page is for AQA Core 2, OCR Core 2, Edexcel Core 1

Some sequences involve **Multiplying**

You've done the easy 'adding' business. Now it gets really tough — multiplying. Are you sure you're ready for this...

Example: A sequence is defined by $a_{k+1} = 2a_k - 1$, $a_2 = 5$. List the first five terms.

OK, you're told the second term, $a_2 = 5$. Just plug that value into the equation, and carry on from there.

$a_3 = 2 \times 5 - 1 = 9$ ⟸ From the equation $a_k = a_2$ so $a_{k+1} = a_3$

$a_4 = 2 \times 9 - 1 = 17$ ⟸ Now use a_3 to find $a_{k+1} = a_4$ and so on...

$a_5 = 2 \times 17 - 1 = 33$

Now to find the first term, a_1:

$a_2 = 2a_1 - 1$ ⟸ Just make $a_k = a_1$

$5 = 2a_1 - 1$

$2a_1 = 6$

$a_1 = 3$

So the first five terms of the sequence are 3, 5, 9, 17, 33 .

Some Sequences have a **Certain Number** of terms — others go on **Forever**

Some sequences are only defined for a certain number of terms.

It's the $1 \le k \le 20$ bit that tells you it's finite.

For example, $a_{k+1} = a_k + 3$, $a_1 = 1$, $1 \le k \le 20$ will be 1, 4, 7, 10, ..., 58 and will contain 20 terms.
This is a finite sequence.

Other sequences don't have a specified number of terms and could go on forever.

For example, $u_{k+1} = u_k + 2$, $u_1 = 5$, will be 5, 7, 9, 11, 13, ... and won't have a final term.
This is an infinite sequence.

While others are periodic, and just revisit the same values over and over again.

For example, $u_k = u_{k-3}$, $u_1 = 1$, $u_2 = 4$, $u_3 = 2$, will be 1, 4, 2, 1, 4, 2, 1, 4, 2,...
This is a periodic sequence with period 3.

Practice Questions

1) A sequence has an nth term of $n^2 + 3$. Find a) the first four terms, and b) the 20th term.

2) A sequence is defined by $a_{k+1} = 3a_k - 2$. If $a_1 = 4$, find a_2, a_3 and a_4.

*3) A sequence of integers, $u_1, u_2, u_3, \ldots$ is given by $u_{n+1} = u_n + 5$, $u_1 = 3$.
Write an expression for the nth term of this sequence.*

Like maths teachers, sequences can go on and on and on and on...

If you know the formula for the nth term, you can work out any term using a single formula, so it's kind of easy. If you only know a recurrence relation, then you can only work out the next term. So if you want the 20th term, and you only know the first one, then you have to use the recurrence relation 19 times. (So it'd be quicker to work out a formula really.)

Arithmetic Progressions

This page is for AQA Core 2, OCR Core 2, Edexcel Core 1

Right, you've got basic sequences tucked under your belt now — time to step it up a notch (sounds painful). When the terms of a sequence progress by adding a fixed amount each time, this is called an arithmetic progression.

It's all about Finding the nth Term

The first term of a sequence is given the symbol **a**. The amount you add each time is called the common difference, called **d**. The position of any term in the sequence is called **n**.

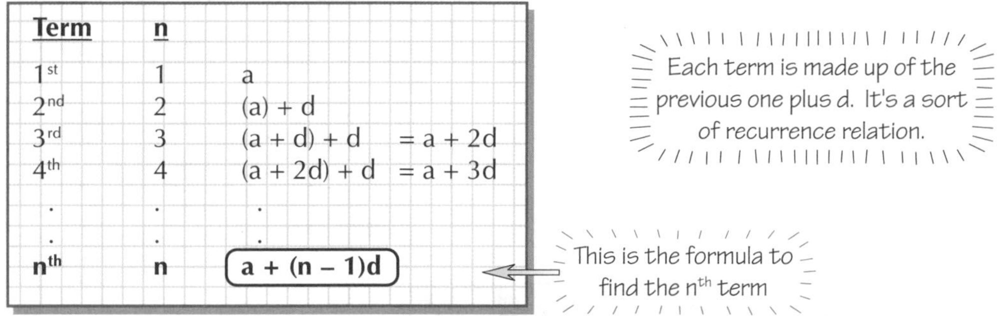

Term	n		
1st	1	a	
2nd	2	(a) + d	
3rd	3	(a + d) + d	= a + 2d
4th	4	(a + 2d) + d	= a + 3d
.	.	.	
.	.	.	
n^th	**n**	**a + (n – 1)d**	

Each term is made up of the previous one plus d. It's a sort of recurrence relation.

This is the formula to find the nth term

Example: Find the 20th term of the arithmetic progression 2, 5, 8, 11,... and find the formula for the nth term.

Here a = 2 and d = 3

To get d, just find the difference between two terms next to each other — e.g. 11 – 8 = 3

So 20th term $= a + (20 - 1)d$
$= 2 + 19 \times 3$
$= 59$

The general term is the nth term, i.e. $a + (n - 1)d$
$= 2 + (n - 1)3$
$= 3n - 1$

A Sequence becomes a Series when you Add the Terms to Find the Total

S_n is the total of the first n terms of the arithmetic progression:

$$S_n = a + (a + d) + (a + 2d) + (a + 3d) + \ldots + (a + (n - 1)d)$$

There's a really neat version of the same formula too:

$$S_n = n \times \frac{(a + l)}{2}$$

The l stands for the last value in the progression. You work it out as $l = a + (n - 1)d$

Nobody likes formulas, so think of it as the average of the first and last terms multiplied by the number of terms.

Example: Find the sum of the series with first term 3, last term 87 and common difference 4.

Here you know a, d and l, but you don't know n yet.

Use the information about the last value, l: $a + (n - 1)d = 87$

Then plug in the other values: $3 + 4(n - 1) = 87$

$4n - 4 = 84$

$4n = 88$

$n = 22$

So $S_{22} = 22 \times \frac{(3+87)}{2}$ $\qquad S_{22} = 990$

Arithmetic Progressions

This page is for AQA Core 2, OCR Core 2, Edexcel Core 1

They **Won't** always give you the **Last Term**...

...but don't panic — there's a formula to use when the last term is unknown. But you knew I'd say that, didn't you?

You know $l = a + (n-1)d$ and $S_n = n\frac{(a+l)}{2}$.

Plug l into S_n and rearrange to get the formula in the box:

$$S_n = \frac{n}{2}[2a + (n-1)d]$$

Example: For the sequence -5, -2, 1, 4, 7, ... find the sum of the first 20 terms.

So a = -5 and d = 3. The question says n = 20 too.

$$S_{20} = \frac{20}{2}[2\times -5 + (20-1)\times 3]$$

$$= 10[-10 + 19\times 3]$$

$$S_{20} = 470$$

There's **Another** way of **Writing Series**, too

So far, the letter S has been used for the sum. The Greeks did a lot of work on this — their capital letter for S is sigma, or Σ. This is used today, together with the general term, to mean the sum of the series.

Example: Find $\sum_{n=1}^{15}(2n+3)$

Starting with n=1... ...and ending with n=15

This means you have to find the sum of the first 15 terms of the series with n^{th} term $2n + 3$.

The first term ($n = 1$) is 5, the second term ($n = 2$) is 7, the third is 9, ... and the last term ($n = 15$) is 33. In other words, you need to find 5 + 7 + 9 + ... + 33. This gives $a = 5$, $d = 2$, $n = 15$ and $l = 33$.

You know all of a, d, n and l, so you can use either formula:

$$S_n = n\frac{(a+l)}{2}$$

$$S_{15} = 15\frac{(5+33)}{2}$$

$$S_{15} = 15 \times 19$$

$$S_{15} = 285$$

It makes no difference which method you use.

$$S_n = \frac{n}{2}[2a + (n-1)d]$$

$$S_{15} = \frac{15}{2}[2\times 5 + 14\times 2]$$

$$S_{15} = \frac{15}{2}[10 + 28]$$

$$S_{15} = 285$$

Practice Questions

1) *Find the sum of the series that begins with 5, 8, ... and ends with 65.*
2) *A series has first term 7 and 5th term 23.*
   Find a) the common difference, b) the 15th term, and c) the sum of the first ten terms.
3) *A series has seventh term 36 and tenth term 30. Find the sum of the first five terms and the nth term.*
4) *Find* $\sum_{n=1}^{20}(3n-1)$ 5) *Find* $\sum_{n=1}^{10}(48-5n)$

This sigma notation is all Greek to me... *(Ho ho ho)*

A sequence is just a list of numbers (with commas between them) — a series on the other hand is when you add all the terms together. It doesn't sound like a big difference, but mathematicians get all hot under the collar when you get the two mixed up. Remember that BlackADDer was a great TV series — not a TV sequence. (Sounds daft, but I bet you remember it now.)

Arithmetic Progressions

This page is for AQA Core 2, OCR Core 2, Edexcel Core 1

Use **Arithmetic Progressions** to add up the **First n Whole Numbers**

The sum of the first n natural numbers looks like this: $S_n = 1 + 2 + 3 + \ldots + (n-2) + (n-1) + n$

So a = 1, l = n and also n = n. Now just plug those values into the formula:

$$S_n = n \times \frac{(a+l)}{2} \implies S_n = \frac{1}{2}n(n+1)$$

Natural numbers are just positive whole numbers.

Example: Add up all the whole numbers from 1 to 100.

Sounds pretty hard, but all you have to do is stick it into the formula:

$S_{100} = ½ \times 100 \times 101$. So $S_{100} = 5050$

Series **Don't** have to start with **n = 1**

Instead of adding up numbers from 1 to 100, what do you do if you want to add the natural numbers from 50 to 100? This just means the sum from 1 to 100, but without the first 49 whole numbers.

You can write this as
$$\sum_{n=50}^{100} n = \sum_{n=1}^{100} n - \sum_{n=1}^{49} n$$
$$= 5050 - \frac{49 \times 50}{2}$$
$$= 5050 - 1225$$
$$= 3825$$

From the example above.

Using $S_n = \frac{n(n+1)}{2}$

Subtract any **Series** if it **Doesn't** Start at **n = 1**

Example: Find $\sum_{n=7}^{20}(4n-1)$

$$\sum_{n=7}^{20}(4n-1) = \sum_{n=1}^{20}(4n-1) - \sum_{n=1}^{6}(4n-1)$$
$$= \frac{20(3+79)}{2} - \frac{6(3+23)}{2}$$
$$= 820 - 78$$
$$= 742$$

Using $S_n = n\frac{(a+l)}{2}$

NB: $\sum_{n=1}^{20}(4n-1)$ and $\sum_{n=1}^{6}(4n-1)$ could both have been worked out a different way:

$$\sum_{n=1}^{6}(4n-1) = \sum_{n=1}^{6} 4n - \sum_{n=1}^{6} 1$$
$$= 4\sum_{n=1}^{6} n - \sum_{n=1}^{6} 1$$
$$= 4\left(\frac{6 \times 7}{2}\right) - 6$$
$$= 78$$

Because $\sum_{n=1}^{6} 4n = 4\sum_{n=1}^{6} n$

$\sum_{n=1}^{6} 1 = 1+1+1+1+1+1 = 6$

1) *Sample exam question:*

a) An arithmetic series has first term a and common difference d.

Prove that the sum of the first n terms, S_n, is given by the formula $S_n = \frac{n}{2}[2a + (n-1)d]$. [4 marks]

b) Evaluate $\sum_{n=9}^{32}(2n-5)$. [4 marks]

Geometric Progressions

This page is for AQA Core 2, OCR Core 2, Edexcel Core 2

So arithmetic progressions mean you add a number to get the next term.
Geometric progressions are a bit different — you multiply by a number to get the next term.

Geometric Progressions Multiply by a Constant each time

Geometric progressions work like this: the next term in the sequence is obtained by multiplying the previous one by a constant value. Couldn't be easier.

$u_1 = a \quad = a$

$u_2 = a \times r \quad = ar$

$u_3 = a \times r \times r \quad = ar^2$

$u_4 = a \times r \times r \times r \quad = ar^3$

The first term (u_1) is called 'a'.

The number you multiply by each time is called 'the common ratio', symbolised by 'r'.

Here's the formula describing any term in the geometric progression:

$$u_n = ar^{n-1}$$

Example: There is a chessboard with a 1p piece on the first square, 2p on the second square, 4p on the third, 8p on the forth and so on until the board is full. Calculate how much money is on the board.

This is a geometric progression, where you get the next term in the sequence by multiplying the previous one by 2.

So a = 1 (because you start with 1p on the first square) and r = 2.

So $u_1 = 1$, $u_2 = 2$, $u_3 = 4$, $u_4 = 8$, ...

You often have to work out the Sum of the Terms

Just like before, S_n stands for the sum of the first n terms.
In the example above, you're told to work out S_{64} (because there are 64 squares on a chessboard).

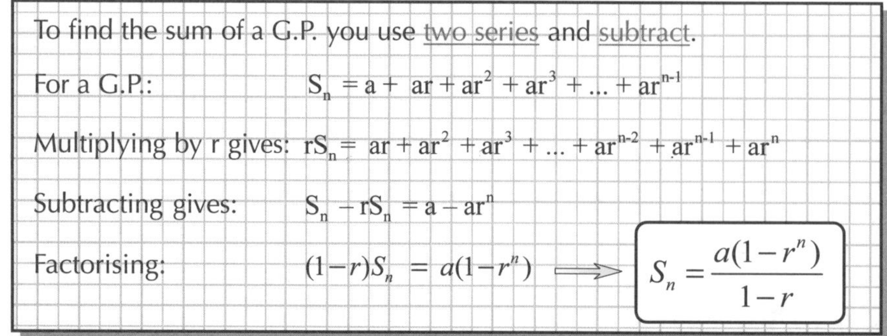

To find the sum of a G.P. you use two series and subtract.

For a G.P.: $S_n = a + ar + ar^2 + ar^3 + ... + ar^{n-1}$

Multiplying by r gives: $rS_n = ar + ar^2 + ar^3 + ... + ar^{n-2} + ar^{n-1} + ar^n$

Subtracting gives: $S_n - rS_n = a - ar^n$

Factorising: $(1-r)S_n = a(1-r^n) \implies S_n = \frac{a(1-r^n)}{1-r}$

If the series were subtracted the other way around you'd get $S_n = \frac{a(r^n-1)}{r-1}$. Both versions are correct.

So, back to the chessboard example: a = 1, r = 2, n = 64 $\quad S_{64} = \frac{1(1-2^{64})}{1-2}$

$S_{64} = 1.84 \times 10^{19}$ pence or £1.84×10^{17}

The whole is more than the sum of the parts — hmm, not in maths, it ain't...

You really need to understand the difference between arithmetic and geometric progressions — it's not hard, but it needs to be fixed firmly in your head. There are only a few formulas for sequences and series (the nth term of a sequence, the sum of the first n terms of a series), but you need to learn them, since they won't be in the formula book they give you.

Geometric Progressions

This page is for AQA Core 2, OCR Core 2, Edexcel Core 2

Geometric progressions can either **Grow** or **Shrink**

In the chessboard example, each term was bigger than the previous one, 1, 2, 4, 8, 16, ...
You can create a series where each term is less than the previous one by using a small value of r.

Example: If $a = 20$ and $r = \frac{1}{5}$, write down the first five terms of the sequence and the 20th term.

Each term is the previous one multiplied by r.

$u_1 = 20$

$u_2 = 20 \times \frac{1}{5} = 4$

$u_3 = 4 \times \frac{1}{5} = 0.8$

$u_4 = 0.8 \times \frac{1}{5} = 0.16$

$u_5 = 0.16 \times \frac{1}{5} = 0.032$

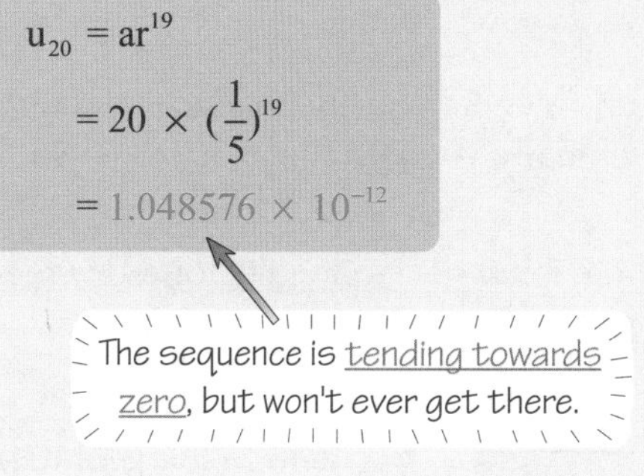
$u_{20} = ar^{19}$
$= 20 \times (\frac{1}{5})^{19}$
$= 1.048576 \times 10^{-12}$
The sequence is tending towards zero, but won't ever get there.

In general, for each term to be smaller than the one before, you need $|r| < 1$.
A sequence with $|r| < 1$ is called convergent.
Any other sequence (like the chessboard example on page 71) is called divergent.

i.e. you need r to be a fraction between −1 and 1

A **Convergent** series has a **Sum** to **Infinity**

In other words, if you just kept on with a convergent series, you'd get closer and closer to a certain number, but you'd never actually reach it.

If $|r| < 1$ and n is very, very big, then r^n will be very, very small — or to put it technically, $r^n \to 0$. (Try working out $(½)^{100}$ on your calculator if you don't believe me.)

This means $(1 - r^n)$ is really, really close to 1.

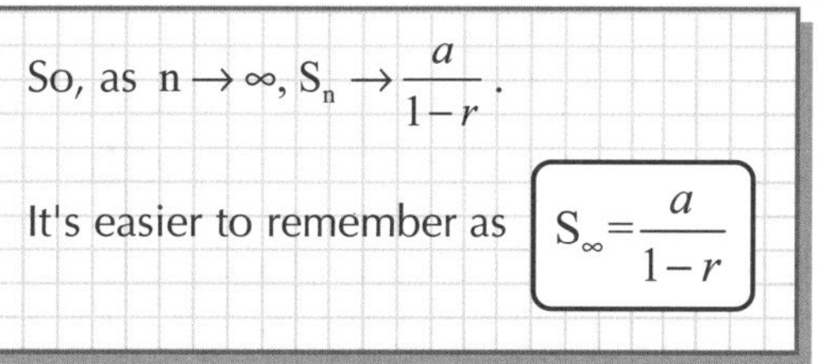
So, as $n \to \infty$, $S_n \to \frac{a}{1-r}$.
It's easier to remember as $S_\infty = \frac{a}{1-r}$

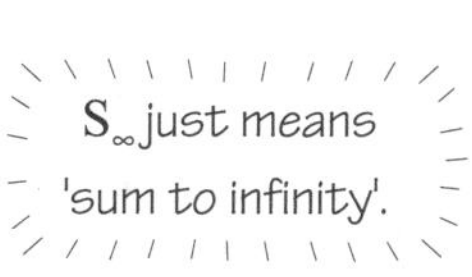
S_∞ just means 'sum to infinity'.

Example: If $a = 2$ and $r = ½$, find the sum to infinity of the geometric series.

These values are getting smaller each time.

$u_1 = 2 \longrightarrow S_1 = 2$

$u_2 = 2 \times \frac{1}{2} = 1 \longrightarrow S_2 = 2 + 1 = 3$

$u_3 = 1 \times \frac{1}{2} = \frac{1}{2} \longrightarrow S_3 = 2 + 1 + \frac{1}{2} = 3\frac{1}{2}$

$u_4 = \frac{1}{2} \times \frac{1}{2} = \frac{1}{4} \longrightarrow S_4 = 2 + 1 + \frac{1}{2} + \frac{1}{4} = 3\frac{3}{4}$

$u_5 = \frac{1}{4} \times \frac{1}{2} = \frac{1}{8} \longrightarrow S_5 = 2 + 1 + \frac{1}{2} + \frac{1}{4} + \frac{1}{8} = 3\frac{7}{8}$

$u_6 = \frac{1}{8} \times \frac{1}{2} = \frac{1}{16} \longrightarrow S_6 = 2 + 1 + \frac{1}{2} + \frac{1}{4} + \frac{1}{8} + \frac{1}{16} = 3\frac{15}{16}$

These values are getting closer (converging) to 4.
So, the sum to infinity is 4.

You can show this graphically:
The line on the graph is getting closer and closer to 4, but it'll never actually get there.

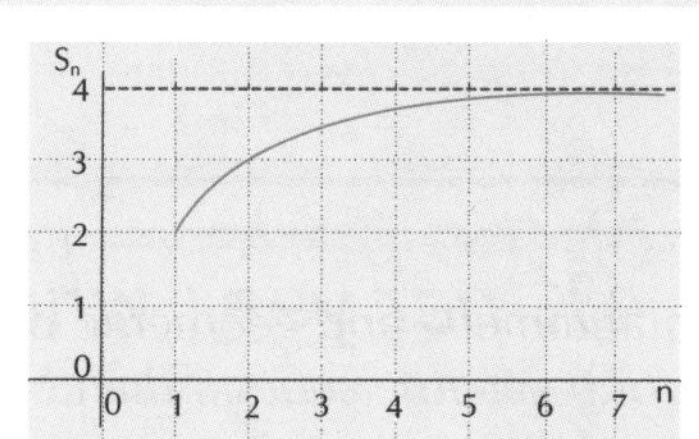
S_n
4
3
2
1
0
0 1 2 3 4 5 6 7 n

Of course, you could have saved yourself a lot of bother by using the sum to infinity formula:

$$S_\infty = \frac{a}{1-r} = \frac{2}{1-\frac{1}{2}} = 4$$

Geometric Progressions

This page is for AQA Core 2, OCR Core 2, Edexcel Core 2

A Divergent series Doesn't have a sum to infinity

Example:

If a = 2 and r = 2, find the sum to infinity of the series.

$u_1 = 2 \Rightarrow S_1 = 2$

$u_2 = 2 \times 2 = 4 \Rightarrow S_2 = 2 + 4 = 6$

$u_3 = 4 \times 2 = 8 \Rightarrow S_3 = 2 + 4 + 8 = 14$

$u_4 = 8 \times 2 = 16 \Rightarrow S_4 = 2 + 4 + 8 + 16 = 30$

$u_5 = 16 \times 2 = 32 \Rightarrow S_5 = 2 + 4 + 8 + 16 + 32 = 62$

As $n \to \infty$, $S_n \to \infty$ in a big way. So big, in fact, that eventually you can't work it out — so don't bother.

There is no sum to infinity for a divergent series.

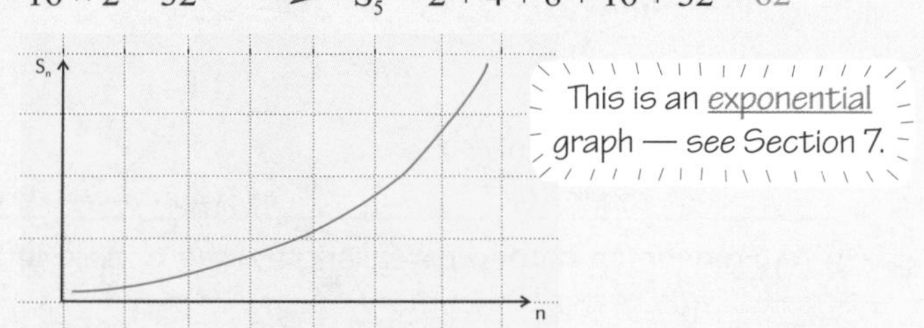
S_n
n
This is an exponential graph — see Section 7.

Example:

When a baby is born, £3000 is invested in an account with a fixed interest rate of 4% per year.
a) What will the account be worth at the start of the seventh year?
b) Will the account have doubled in value by the time the child reaches its 21st birthday?

a) $u_1 = a = 3000$

$u_2 = 3000 + (4\% \text{ of } 3000)$ ← This is the interest.

$= 3000 + (0.04 \times 3000)$

$= 3000(1 + 0.04)$

$= 3000 \times 1.04$ ← So, r = 1.04

$u_3 = u_2 \times 1.04$

$= (3000 \times 1.04) \times 1.04$

$= 3000 \times (1.04)^2$

$u_4 = 3000 \times (1.04)^3$

. . .

. . . ← I've missed out some steps here — check that you understand what's happened.

$u_7 = 3000 \times (1.04)^6$

$= £3795.96$ (to the nearest penny)

b) You need to know when $u_n > 6000$ ← double the original value.

From part a) you can tell that $u_n = 3000 \times (1.04)^{n-1}$

So $3000 \times (1.04)^{n-1} > 6000$

$(1.04)^{n-1} > 2$

To complete this you need to use logs (see section 8):

$\log(1.04)^{n-1} > \log 2$

$(n - 1)\log(1.04) > \log 2$

$$n - 1 > \frac{\log 2}{\log 1.04}$$

$n - 1 > 17.67$

$n > 18.67$ (to 2 d.p.)

So u_{19} (the amount at the start of the 19th year) will be more than double the original amount — plenty of time to buy a Porsche for the 21st birthday.

Practice Questions

1) *For the sequence 2, –6, 18, …, find the 10th term.*
2) *For the sequence 24, 12, 6, …, find a) the common ratio, b) the seventh term, and c) the sum to infinity.*
3) *A G.P. and an A.P. both begin with 2, 6, … Which term of the A.P. will be equal to the fifth term of the G.P.?*
4) *Sample exam question:*

For the series with second term –2 and common ratio –½, find

a) the first term	[3 marks]
b) the first seven terms	[3 marks]
c) the sum of the first seven terms	[3 marks]
d) the sum to infinity	[3 marks]

So tell me — if my savings earn 4% per year, when will I be rich...

Now here's a funny thing — you can have a convergent geometric series if the common ratio is small enough. I find this odd — that I can keep adding things to a sum forever, but the sum never gets really really big.

Binomial Expansions

This page is for AQA Core 2, OCR Core 2, Edexcel Core 2

If you're feeling a bit stressed, just take a couple of minutes to relax before trying to get your head round this page — it's a bit of a stinker in places. Have a cup of tea and think about something else for a couple of minutes. Ready...

*Writing **Binomial Expansions** is all about **Spotting Patterns***

Doing binomial expansions just involves multiplying out the brackets. It would get nasty when you raise the brackets to higher powers — but once again I've got a cunning plan...

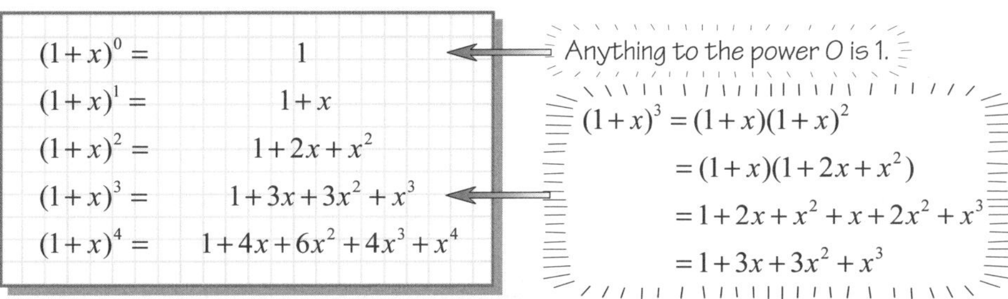

A Frenchman named Pascal spotted the pattern in the coefficients and wrote them down in a triangle.
So it was called 'Pascal's Triangle' (imaginative, eh?).
The pattern's easy — each number is the sum of the two above it.

So, the next line will be: **1 5 10 10 5 1**
giving **(1 + x)^5 = 1 + 5x + 10x^2 + 10x^3 + 5x^4 + x^5.**

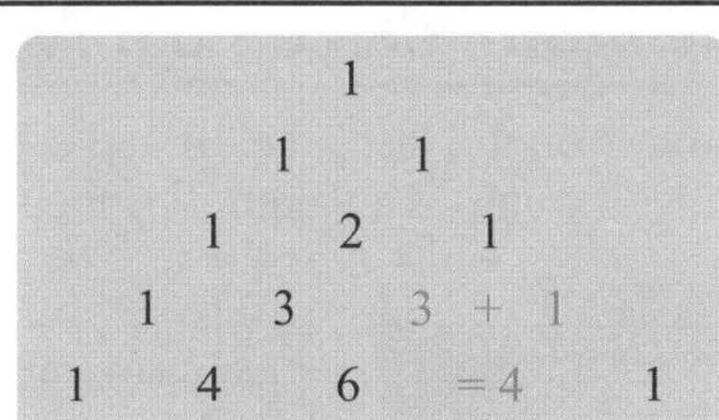

*You **Don't** need to write out Pascal's Triangle for **Higher Powers***

There's a formula for the numbers in the triangle. The formula looks horrible (one of the worst in AS maths) so don't try to learn it letter by letter — look for the patterns in it instead. Here's an example:

Example: Expand $(1 + x)^{20}$, giving the first four terms only.

So you can use this formula for any power, the power is called n. In this example n = 20.

$$(1+x)^n = 1+\frac{n}{1}x+\frac{n(n-1)}{1\times2}x^2+\boxed{\frac{n(n-1)(n-2)}{1\times2\times3}x^3}+\ldots\ldots\ldots\ldots+x^n$$

Here's a closer look at the term in the black box:

$$\frac{n(n-1)(n-2)}{1\times2\times3}x^3$$

There are three things multiplied together on the top row. If n=20, this would be 20×19×18.

Start here. The power of x is 3 and everything else here is based on 3.

There are three integers here multiplied together. 1x2x3 is written as 3! and called 3 factorial.

This means, if n = 20 and you were asked for 'the term in x^7' you should write $\frac{20\times19\times18\times17\times16\times15\times14}{1\times2\times3\times4\times5\times6\times7}x^7$.

This can be simplified to $\frac{20!}{7!13!}x^7$ ⟵ $20\times19\times18\times17\times16\times15\times14=\frac{20!}{13!}$ because it's the numbers from 20 to 1 multiplied together, divided by the numbers from 13 to 1 multiplied together.

Believe it or not, there's an even shorter form: $\frac{20!}{7!13!}$ is written as ${}^{20}C_7$ or $\binom{20}{7}$

$${}^nC_r = \binom{n}{r} = \frac{n!}{r!(n-r)!}$$

So, to finish the example, $(1+x)^{20} = 1+\frac{20}{1}x+\frac{20\times19}{1\times2}x^2+\frac{20\times19\times18}{1\times2\times3}x^3+\ldots = 1+20x+190x^2+1140x^3+\ldots$

Binomial Expansions

This page is for AQA Core 2, OCR Core 2, Edexcel Core 2

Example: What is the term in x^5 in the expansion of $(1 - 3x)^{12}$?

The term in x^5 will be as follows: $\frac{12\times11\times10\times9\times8}{1\times2\times3\times4\times5}(-3x)^5$ ← Watch out — the -3 is included here with the x.

$$= \frac{12!}{5!7!}(-3)^5x^5 = -\frac{12!}{5!7!}\times3^5x^5 = -192456x^5$$

Note that $(-3)^{\text{even}}$ will always be positive and $(-3)^{\text{odd}}$ will always be negative.
Here's another tip — the digits on the bottom of the fraction should always add up to the number on the top.

Some **Binomials** contain **More Complicated Expressions**

The binomials on the last page all had a 1 in the brackets — things get tricky when there's a number other than 1. Don't panic, though. The method is the same as before once you've done a bit of factorising.

Example: What is the coefficient of x^4 in the expansion of $(2 + 5x)^7$?

Factorising $(2+5x)$ gives $2(1+\frac{5}{2}x)$

So, $(2+5x)^7$ gives $2^7(1+\frac{5}{2}x)^7$ ← It's really easy to forget the first bit (here it's 2^7) — you've been warned...

$$(2+5x)^7 = 2^7(1+\frac{5}{2}x)^7$$

$$=2^7[1+7\left(\frac{5}{2}x\right)+\frac{7\times6}{1\times2}\left(\frac{5}{2}x\right)^2+\frac{7\times6\times5}{1\times2\times3}\left(\frac{5}{2}x\right)^3+\frac{7\times6\times5\times4}{1\times2\times3\times4}\left(\frac{5}{2}x\right)^4+...]$$

Here's the one you want. (the x^4 term)

The coefficient of x^4 will be $2^7\times\frac{7!}{4!3!}(\frac{5}{2})^4 = 175000$

Don't forget the 2^7.

So, there's no need to work out all of the terms.
In fact, you could have gone directly to the term in x^4 by using the method on page 49.

> Note: The question asked for the coefficient of x^4 in the expansion, so don't include any x's in your answer. If you'd been asked for the term in x^4 in the expansion, then you should have included the x^4 in your answer. Always read the question very carefully.

1) *Sample exam question:*

a) Write down the first four terms in the expansion of $(1 + ax)^{10}$, $a > 0$. [2 marks]

b) Find the coefficient of x^2 in the expansion of $(2 + 3x)^5$. [2 marks]

c) If the coefficients of x^2 in both expansions are equal, find the value of a. [2 marks]

Pascal was fine at maths but rubbish at music — he only played the triangle...

You can use your calculator to work out these tricky fractions — you use the ${}^nC_r$ button (though it could be called something else on your calculator). So to work out ${}^{20}C_7$, press '20', then press the ${}^nC_r$ button, then press '7', and then finish with '='. Now work out ${}^{15}C_7$ and ${}^{15}C_8$ — you should get the same answers, since they're both $\frac{15!}{7!8!}$.

The Trig Formulas You Need to Know

This page is for AQA Core 2, OCR Core 2, Edexcel Core 2

I thought I'd just stick all the trig formulas you need to learn for the exam on one page. So here they are: Learn them or you're seriously stuffed! Worse than an aubergine.

Radians or Degrees? Choose your weapon...

The main thing is that you know how radians relate to degrees.

$\pi \text{ rad} = 180°$ $2\pi \text{ rad} = 360°$

$1 \text{ rad} = \frac{180°}{\pi} \approx 57.3°$ $\frac{\pi}{2} \text{ rad} = 90°$

You've got to know these ones really well — so you don't need to think about them at all.

To do a really rough check of an answer, take π as 3, so $90° \approx \frac{3}{2}$ radians, and a whole circle's about 6 radians. Always make sure the answer looks about right.

If you need to convert from one to the other:

RADIANS TO DEGREES
Divide by π, multiply by 180.

DEGREES TO RADIANS
Divide by 180, multiply by π.

You can easily work this out, since you know that π rads is 180°.

The Sine Rule and Cosine Rule work for Any triangle

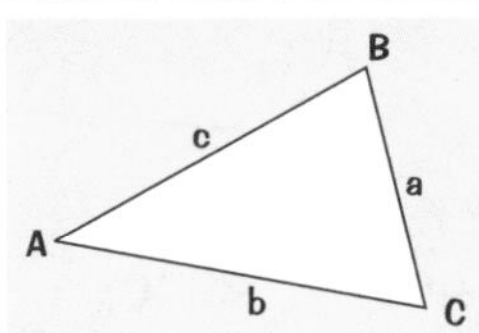

THE SINE RULE

$$\frac{a}{\sin A} = \frac{b}{\sin B} = \frac{c}{\sin C}$$

THE COSINE RULE

$$a^2 = b^2 + c^2 - 2bc\cos A$$

AREA OF ANY TRIANGLE

$$Area = \tfrac{1}{2}ab\sin C$$

Remember these three formulas work for ANY triangle, not just right-angled ones.

Draw triangles to remember sin, cos and tan of π/6, π/3, and π/4 radians

You should know the values of sin, cos and tan at $\pi/6$, $\pi/3$ and $\pi/4$ (that's 30°, 60° and 45° — make sure you can work those out). But to help you remember, you can draw these two groovy triangles. It does make it easier. Honest.

Half an equilateral triangle with sides of length 2.

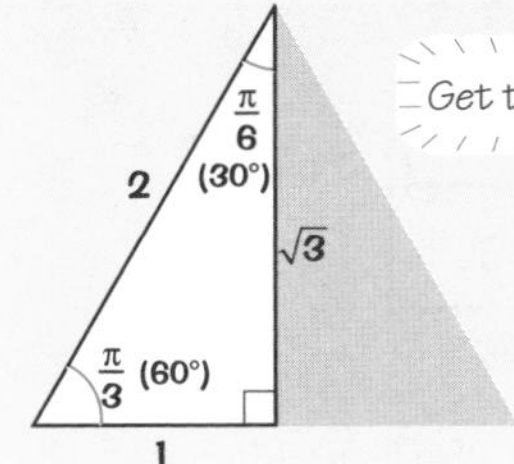

Get the height $\sqrt{3}$ by Pythagoras' theorem: $1^2 + \sqrt{3}^2 = 2^2$.)

Then you can use the triangle to work out sin, cos and tan of $\pi/6$ and $\pi/3$ rad.

Remember: SOH CAH TOA...

$\sin = \frac{opp}{hyp}$ $\cos = \frac{adj}{hyp}$ $\tan = \frac{opp}{adj}$

Right-angled triangle with two sides of length 1.

The $\sqrt{2}$ just comes from Pythagoras.

This triangle gives you sin, cos and tan of $\pi/4$.

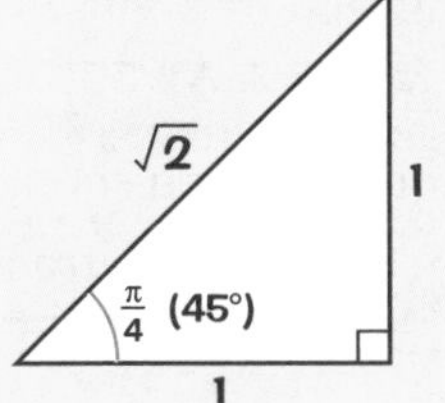

Trig Values from Triangles

$30° = \frac{\pi}{6}$	$60° = \frac{\pi}{3}$	$45° = \frac{\pi}{4}$
$\sin 30° = \frac{1}{2}$	$\sin 60° = \frac{\sqrt{3}}{2}$	$\sin 45° = \frac{1}{\sqrt{2}}$
$\cos 30° = \frac{\sqrt{3}}{2}$	$\cos 60° = \frac{1}{2}$	$\cos 45° = \frac{1}{\sqrt{2}}$
$\tan 30° = \frac{1}{\sqrt{3}}$	$\tan 60° = \sqrt{3}$	$\tan 45° = 1$

The Best has been saved till last...

$$\tan x \equiv \frac{\sin x}{\cos x}$$

$$\sin^2 x + \cos^2 x \equiv 1$$

$$\sin^2 x \equiv 1 - \cos^2 x$$

$$\cos^2 x \equiv 1 - \sin^2 x$$

Work out these two using $\sin^2 + \cos^2 = 1$.

Tri angles — go on... you might like them.

Formulas and trigonometry go together even better than Richard and Judy. I can count 19 on this page. That's not many, so please, just make sure you know them! Practise those triangles until you can scribble them down without thinking. Then use them to get the values in the table — they're what you actually need. You get them by applying the basic trig formulas (e.g. sin = opp/hyp). If you haven't learned them I will cry for you. I will sob. 💧☹💧

Using the Sine and Cosine Rules

This page is for AQA Core 2, OCR Core 2, Edexcel Core 2

This page is about "solving" triangles, which just means finding all their sides and angles when you already know a few.

Sine Rule or Cosine Rule — well, which is it...?

To decide which of these two rules you need to use, look at how much you already know about the triangle.

SINE RULE

If you know two angles and a side.
(And if you know two angles, you can easily find the other one.)

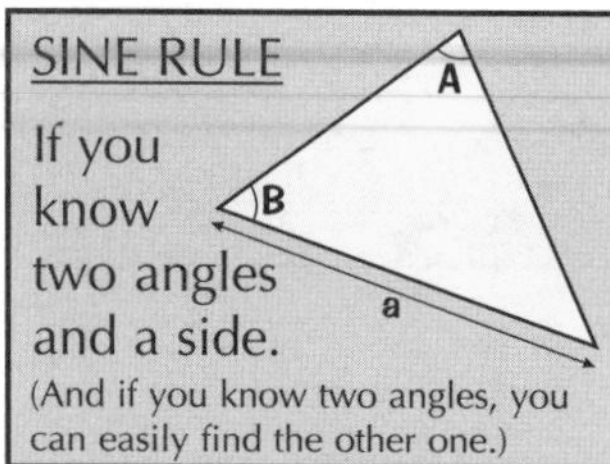

COSINE RULE — THEN SINE RULE

If you know three sides...

...or two sides and the angle between them.

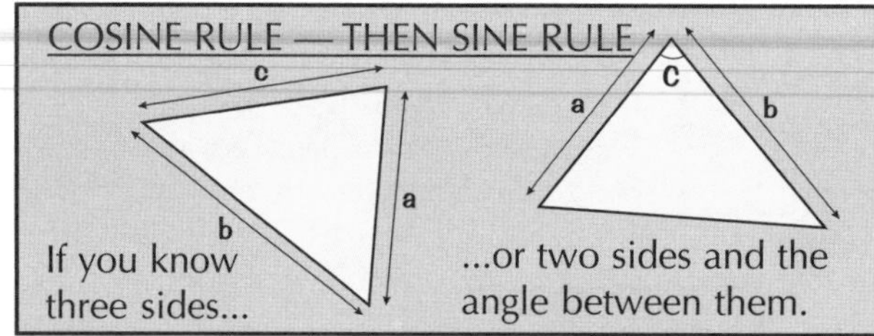

ERR — IT DOESN'T WORK HERE...

If you've got two sides and an angle (but not the angle between them)...

...there are sometimes two possible triangles.

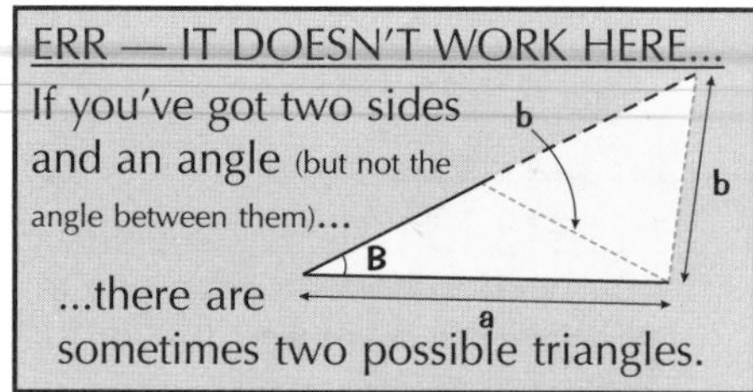

Example: Solve ΔABC, in which $A = 40°$, $a = 27$m, $B = 73°$. Then find its area.

Draw a quick sketch first — don't worry if it's not deadly accurate, though.
You're given two angles and a side, so you need the Sine Rule.

Make sure you put side a opposite angle A.

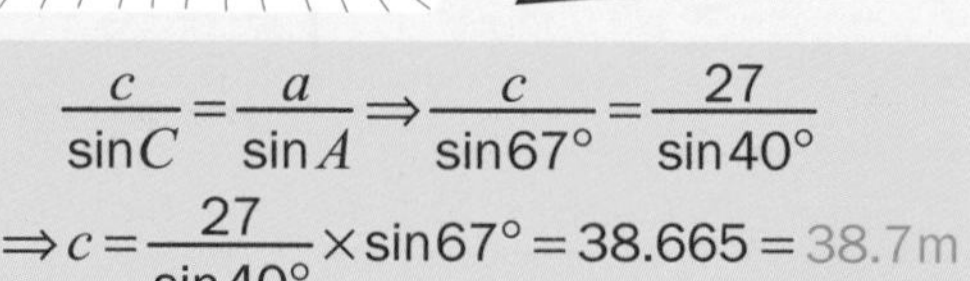

First of all, get the other angle:

$$\angle C = (180 - 40 - 73)° = 67°$$

Then find the other sides, one at a time:

$$\frac{a}{\sin A} = \frac{b}{\sin B} \Rightarrow \frac{27}{\sin 40°} = \frac{b}{\sin 73°}$$
$$\Rightarrow b = \frac{27}{\sin 40°} \times \sin 73° = 40.169 = 40.2\text{m}$$

$$\frac{c}{\sin C} = \frac{a}{\sin A} \Rightarrow \frac{c}{\sin 67°} = \frac{27}{\sin 40°}$$
$$\Rightarrow c = \frac{27}{\sin 40°} \times \sin 67° = 38.665 = 38.7\text{m}$$

Now just use the formula to find its area.

$$\text{Area of } \Delta \text{ ABC} = \tfrac{1}{2}ab\sin C$$
$$= \tfrac{1}{2} \times 27 \times 40.169 \times \sin 67°$$
$$= 499.2\text{m}^2$$

Example: Find X, Y and z.

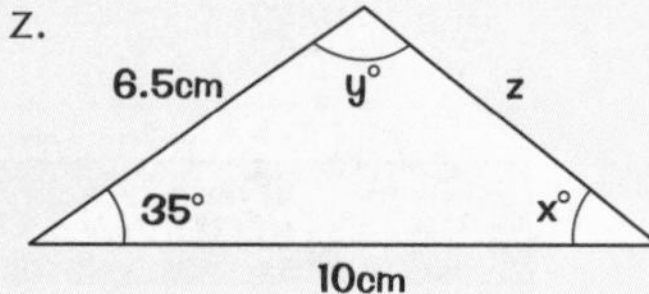

$$a^2 = b^2 + c^2 - 2bc\cos A$$

You've been given 2 sides and the angle between them, so you're going to need the Cosine Rule, then the Sine Rule.

$$z^2 = (6.5)^2 + 10^2 - 2(6.5)(10)\cos 35°$$
$$\Rightarrow z^2 = 142.25 - 130\cos 35°$$
$$\Rightarrow z^2 = 35.7602$$
$$\Rightarrow z = 5.9800 = 5.98\text{cm} \quad \text{(to 2 d.p.)}$$

In this case, angle A is 35°, and side *a* is actually *z*

$$\frac{a}{\sin A} = \frac{b}{\sin B} = \frac{c}{\sin C}$$

When you use an earlier answer, don't use one that's been rounded too much. Use at least 3 or 4 decimal places. So it's a good idea to write down a really exact answer to each part and then round it.

You've got all the sides. Now use the Sine Rule to find another two angles.

The Sine Rule just says: Dividing "length of a side" by "the sine of the opposite angle" gives the same answer, whichever pair you choose.

$$\frac{6.5}{\sin X} = \frac{5.9800}{\sin 35°}$$
$$\Rightarrow \sin X = 0.6234$$
$$\Rightarrow X = \sin^{-1} 0.6234$$
$$\Rightarrow X = 38.6°$$

To get the last angle, just subtract the two angles you know from 180°:

$$35° + 38.6° + Y = 180°$$
$$Y = 180 - 35 - 38.6 = 106.4°$$

...and I feel that love is dead, I'm loving angles instead...

To be honest, this stuff is more likely to come up in a geometry question than a trig question. It won't say "use the cosine rule", you'll just be expected to use it when required as part of a question. So make sure you learn the table showing when to use which rule. And practise lots of questions.

Arc Lengths and Sector Areas

This page is for AQA Core 2, OCR Core 2, Edexcel Core 2

Enough about triangles. What about circles? They're exciting too, and also involve angles. And it's here that radians really come into their own — because this stuff only works if your angle's in radians.

Find Arc Lengths and Sector Areas using Radians

If you have a part of a circle, like a section of pie chart, you can work out the length of the curved side, or the area of the 'slice of pie' — as long as you know the angle at the centre (θ) and the length of the radius (r).

Arc Length:

For any circle with a radius of r, where the angle θ is measured in radians, the arc length of the sector S is given by the formula:

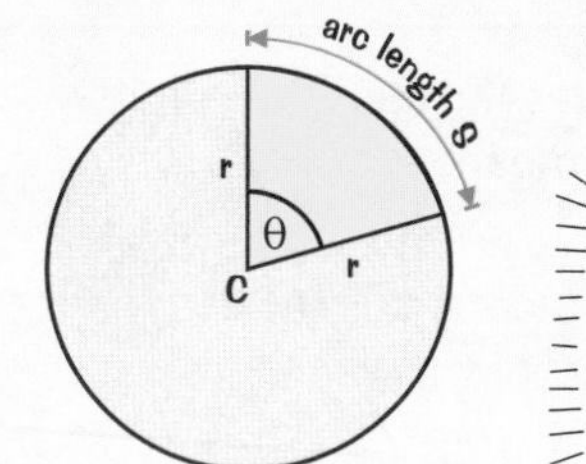

$$S = r\theta$$

Put $\theta = 2\pi$ into $S = r\theta$ and you get:

$S = r \times 2\pi = 2\pi r$.

That's the formula for the circumference of a circle.

And you can work out A, the area of the sector using:

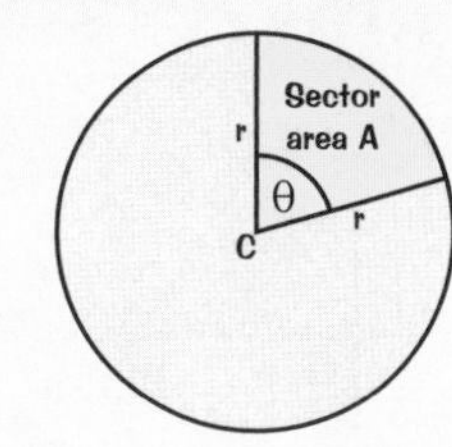

$$A = \frac{1}{2}r^2\theta$$

Put $\theta = 2p$ into $A = \frac{1}{2}r^2\theta$ and you get:

$A = \frac{1}{2}r^2 \times 2\pi = \pi r^2$.

And that's the area of a circle.

You can use these to work out the Areas and Perimeters of Other Shapes

Yep, with a bit of imagination you can work out the areas of loads of other shapes too.

Example: Find a formula for the area and perimeter of the segment A in the diagram below.

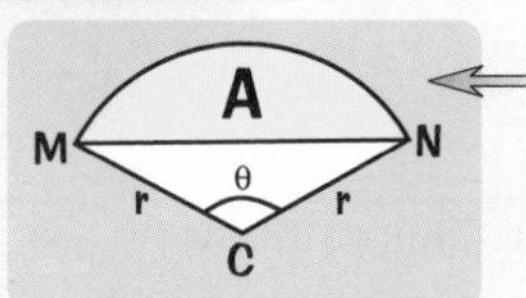

A segment's the 'top bit' of a sector — the yellow bit. You can find its area by subtracting the area of the triangle CMN from the area of the sector.

Area of the whole sector is given by: $A_{sector} = \frac{1}{2}r^2\theta$

Area of the triangle CMN is given by: $A_{triangle} = \frac{1}{2}r^2\sin\theta$

The area of a triangle is $A = \frac{1}{2}ab\sin\theta$, where a and b are the lengths of two of the sides, and θ is the angle between them. Here, a and b both equal r.

See page 33 for more info.

So the formula for area A must be:

$$\begin{aligned} A &= A_{sector} - A_{triangle} \\ &= \frac{1}{2}r^2\theta - \frac{1}{2}r^2\sin\theta \\ &= \frac{1}{2}r^2(\theta - \sin\theta) \end{aligned}$$

The perimeter of the segment is the sum of the curved length S_1 and the straight line S_2.

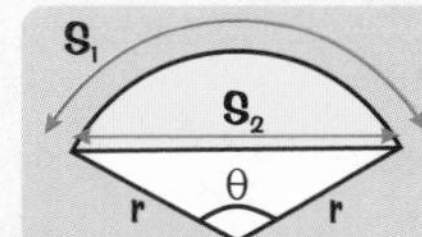

The curved length S_1 is given by: $S_1 = r\theta$.

You can work out the length of the straight line S_2 using the Cosine Rule on the white triangle:

$$\begin{aligned} S_2^2 &= r^2 + r^2 - 2r^2\cos\theta \\ &= 2r^2 - 2r^2\cos\theta \\ S_2 &= \sqrt{2r^2(1-\cos\theta)} \end{aligned}$$

The Cosine Rule: $a^2 = b^2 + c^2 - 2bc\cos A$

See pages 51-52 for more info.

So the total perimeter S is the sum of these two lengths: $S = S_1 + S_2 = r\theta + \sqrt{2r^2(1-\cos\theta)}$

...and through it all she offers me a sector, an arc length and a segment...

The formulas at the top of the page are obvious really, if you think about it. A sector's just a fraction of a circle — and to work out what fraction, take your angle and divide it by 2π. So if your sector had an angle $\pi/2$ at the centre, then you've got $\frac{(\pi/2)}{2\pi} = \frac{1}{4}$ of a circle. Then to get the length of the arc or the area of the sector, just multiply $2\pi r$ or πr^2 by $\frac{1}{4}$.

Graphs of Trig Functions

This page is for AQA Core 2, OCR Core 2, Edexcel Core 2
Before you leave this page, you should be able to close your eyes and picture these three graphs in your head, properly labelled and everything. If you can't, you need to learn them more. I'm not kidding.

sin x and cos x are always in the range –1 to 1

sin x and cos x are similar — they just bob up and down between –1 and 1.

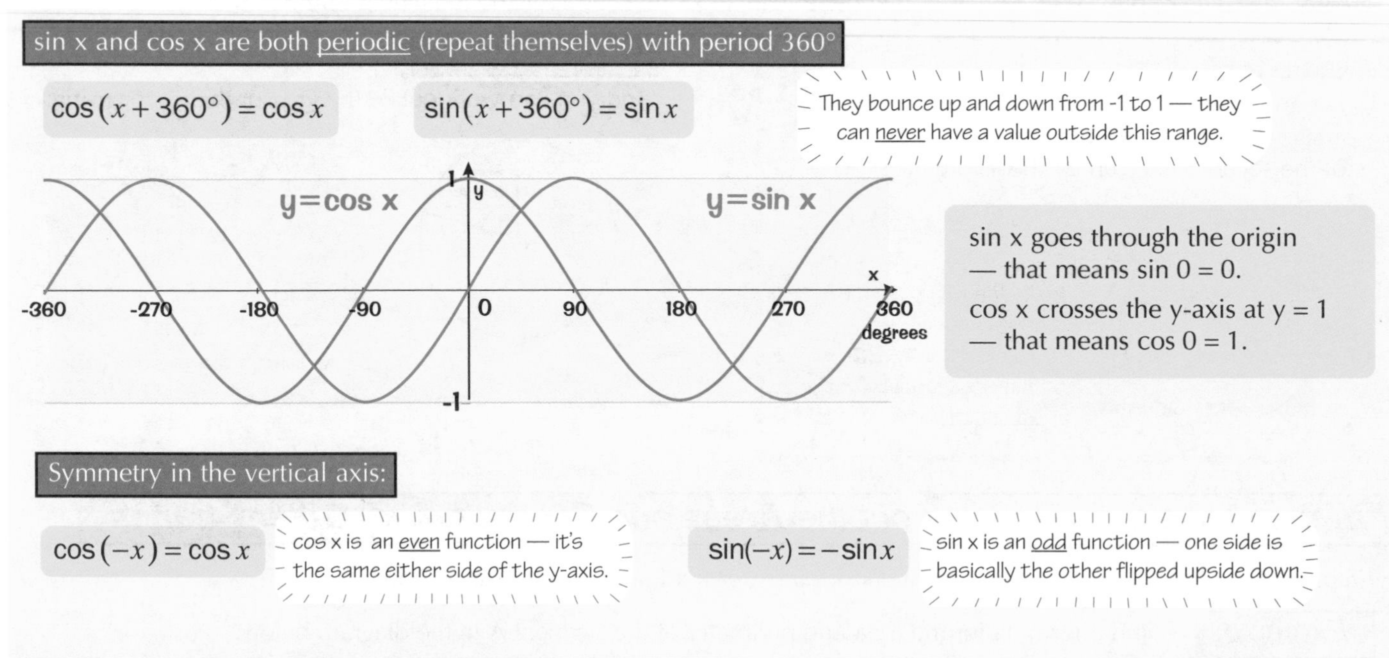

sin x and cos x are both periodic (repeat themselves) with period 360°
$\cos(x+360°) = \cos x$
$\sin(x+360°) = \sin x$
They bounce up and down from -1 to 1 — they can never have a value outside this range.
y=cos x
y=sin x
1
-1
y
x
-360
-270
-180
-90
0
90
180
270
360
degrees
sin x goes through the origin — that means sin 0 = 0.
cos x crosses the y-axis at y = 1 — that means cos 0 = 1.
Symmetry in the vertical axis:
$\cos(-x) = \cos x$
cos x is an even function — it's the same either side of the y-axis.
$\sin(-x) = -\sin x$
sin x is an odd function — one side is basically the other flipped upside down.

tan x can be Any Value at all

tan x is different from sin x and cos x — it doesn't go gently up and down between –1 and 1. It goes between $-\infty$ and $+\infty$.

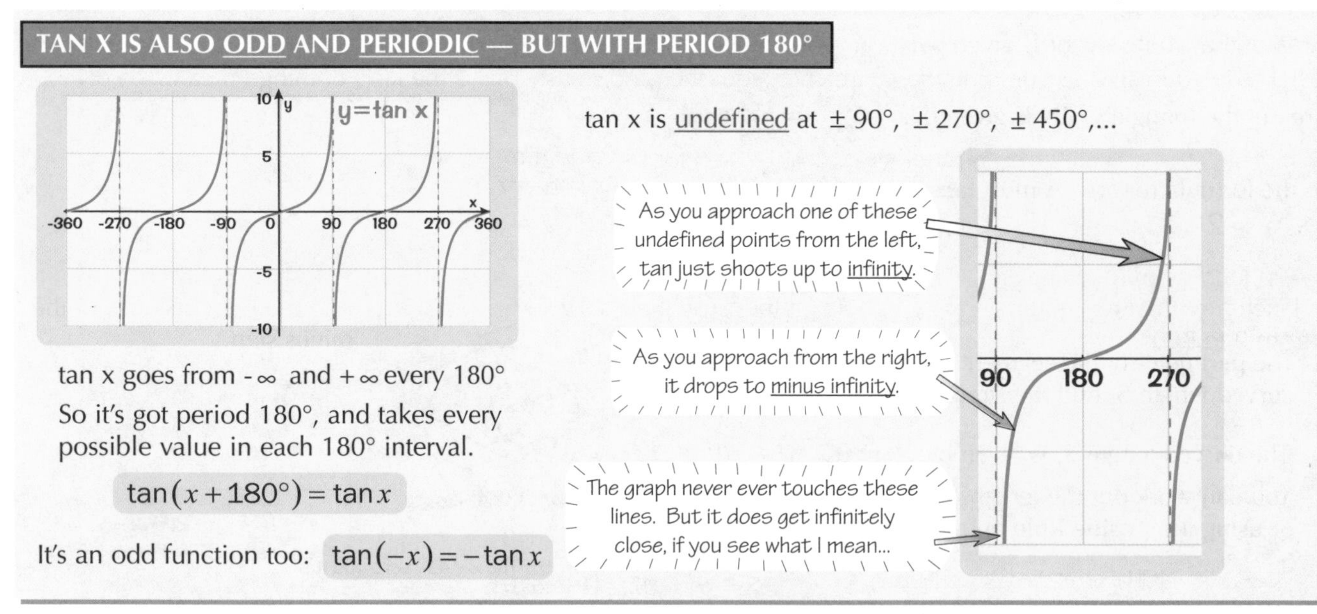

TAN X IS ALSO ODD AND PERIODIC — BUT WITH PERIOD 180°
y=tan x
10
5
-5
-10
y
x
-360
-270
-180
-90
0
90
180
270
360
tan x is undefined at ± 90°, ± 270°, ± 450°,...
As you approach one of these undefined points from the left, tan just shoots up to infinity.
As you approach from the right, it drops to minus infinity.
The graph never ever touches these lines. But it does get infinitely close, if you see what I mean...
90
180
270
tan x goes from $-\infty$ and $+\infty$ every 180°
So it's got period 180°, and takes every possible value in each 180° interval.
$\tan(x+180°) = \tan x$
It's an odd function too: $\tan(-x) = -\tan x$

The easiest way to sketch any of these graphs is to plot the important points which happen every 90° (i.e. –180°, –90°, 0°, 90°, 180°, 270°, 360°...) and then just join the dots up.

Sin and cos can make your life worthwhile — give them a chance...

It's really really really really really really important that you can draw the trig graphs on this page, and get all the labels right. Make sure you know what value sin, cos and tan have at the interesting points — i.e. 0°, 90°, 180°, 270°, 360°. Its easy to remember what the graphs look like, but you've got to know exactly <u>where</u> they're max, min, zero, etc.

Transformed Trig Graphs

This page is for AQA Core 2, OCR Core 2, Edexcel Core 2

Transformed trig graphs look much the same as the bog standard ones, just a little different. There's three types:

There's 3 basic types of ***Transformed Trig Graph...***

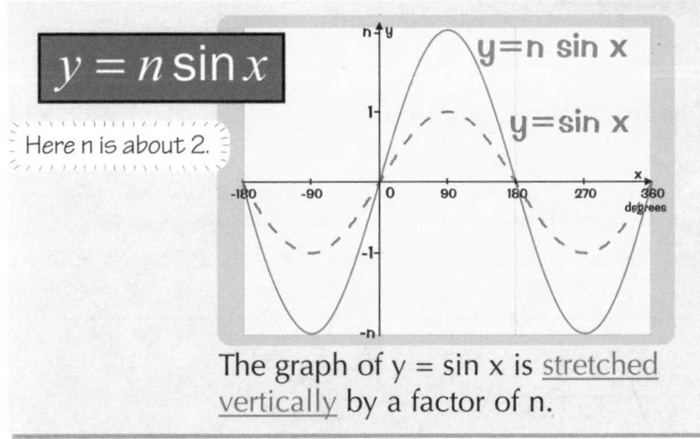

The graph of y = sin x is stretched vertically by a factor of n.

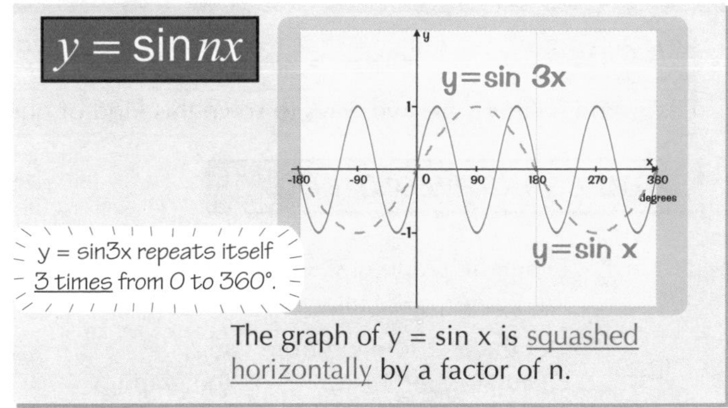

The graph of y = sin x is squashed horizontally by a factor of n.

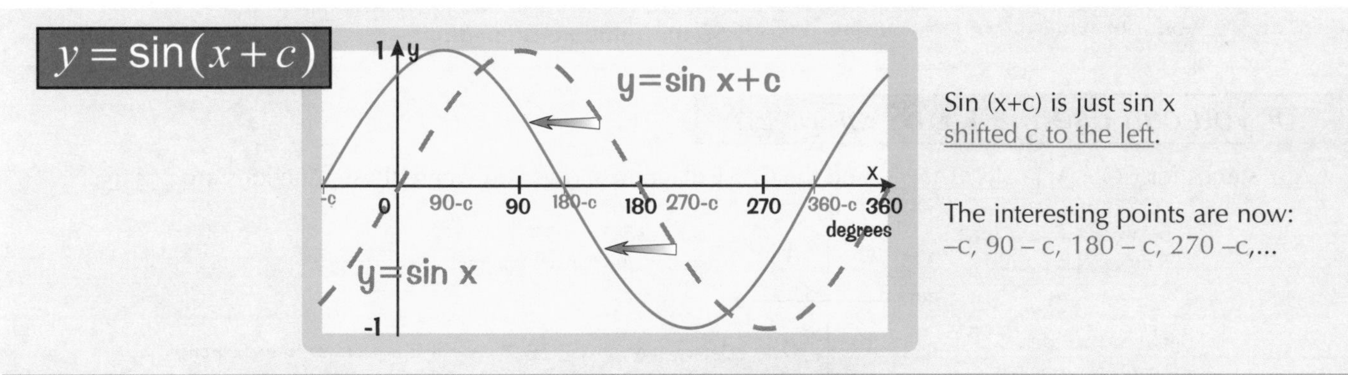

Sin (x+c) is just sin x shifted c to the left.

The interesting points are now: –c, 90 – c, 180 – c, 270 –c,...

...but you can ***Combine*** *the transformations as well*

They're the three basic kinds of transformations you have to know. But it's not over yet — there's more excitement to come, because you can also combine these stretches, squashes and shifts. This is the kind of thing I mean...

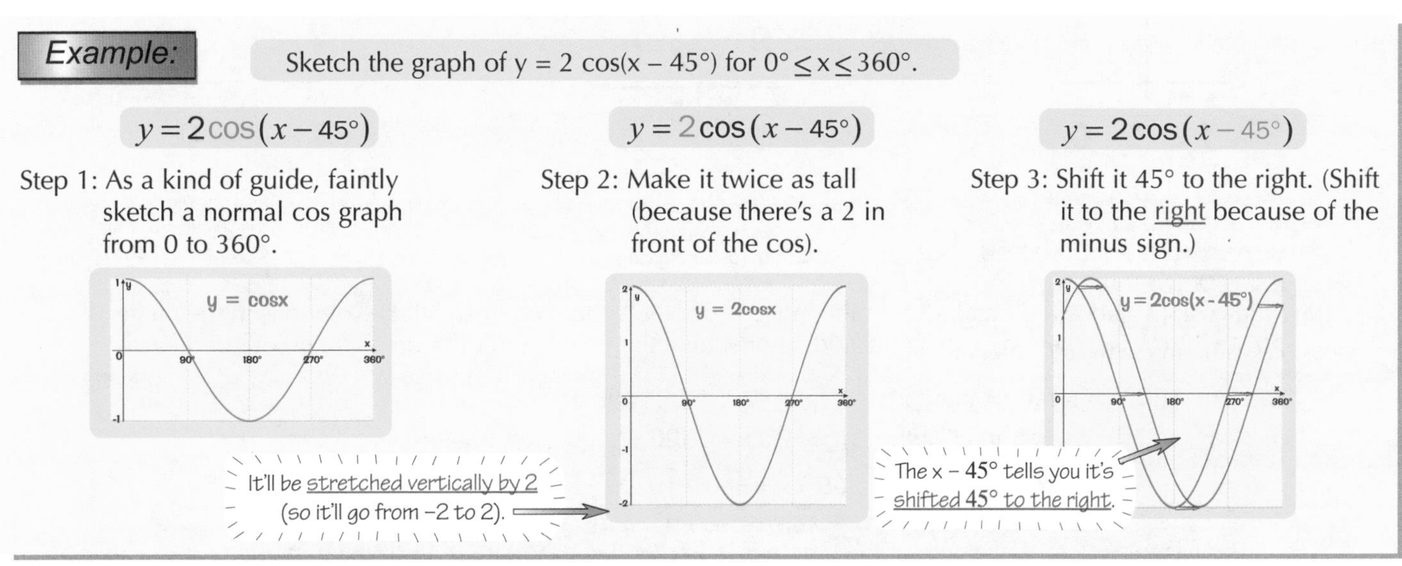

Example: Sketch the graph of y = 2 cos(x – 45°) for $0° \leq x \leq 360°$.

$y = 2\cos(x - 45°)$

Step 1: As a kind of guide, faintly sketch a normal cos graph from 0 to 360°.

$y = 2\cos(x - 45°)$

Step 2: Make it twice as tall (because there's a 2 in front of the cos).

$y = 2\cos(x - 45°)$

Step 3: Shift it 45° to the right. (Shift it to the right because of the minus sign.)

Curling up on the sofa with 2cos x — that's my idea of cosiness ☺

One thing you've really got to be careful about is making sure you move or stretch the graphs in the right direction.
In that last example, the graph moved to the right because the number was negative.
The same sort of thing happens for the stretching and squashing.
If the n in 'y = n sinx' is less than 1, it's not a stretch any more — it's a squash.
And if the n in 'y =sin nx' is less than 1, you have to stretch the graph horizontally, rather than squash it. Clear as mud.

Solving Trig Equations in a Given Interval

This page is for AQA Core 2, OCR Core 2, Edexcel Core 2

I used to really hate trig stuff like this. But once I'd got the hang of it, I just couldn't get enough. I stopped going out, lost interest in the opposite sex — the CAST method became my life. Learn it, but be careful. It's addictive.

There are **Two Ways** to find Solutions in an **Interval**...

Example: Solve $\cos x = \frac{1}{2}$ for $-360° \leq x \leq 720°$.

Like I said — there are two ways to solve this kind of question. Just use the one you prefer...

You can draw a **graph**...

Your calculator gives you a solution of 60° (But see page 33 if you don't know it already). Then you have to work out what the others will be.

The other solutions are 60° either side of the graph's peaks.

1) Draw the graph of y = cosx for the range you're interested in...
2) Get the first solution from your calculator and mark this on the graph,
3) Use the symmetry of the graph to work out what the other solutions are:

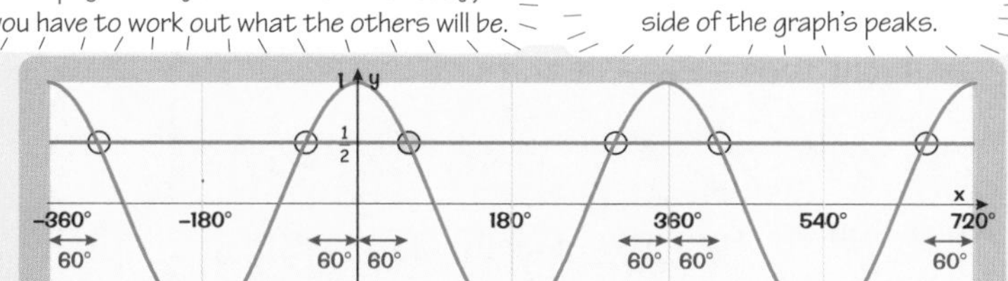

So the solutions are: –300°, –60°, 60°, 300°, 420° and 660°.

...or you can use the **CAST diagram**

CAST stands for COS, ALL, SIN, TAN — and the CAST diagram shows you where these functions are positive:

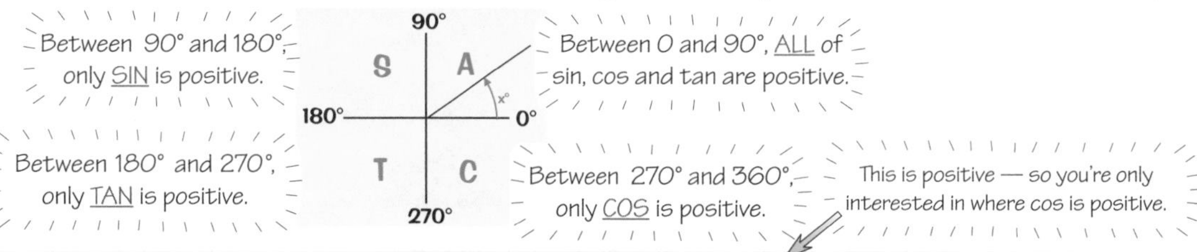

Between 90° and 180°, only SIN is positive.
Between 180° and 270°, only TAN is positive.
90°
180°
0°
270°
S
A
T
C
x°
Between 0 and 90°, ALL of sin, cos and tan are positive.
Between 270° and 360°, only COS is positive.
This is positive — so you're only interested in where cos is positive.

First, to find all the values of x between 0° and 360° where $\cos x = \frac{1}{2}$ — you do this:

Put the first solution onto the CAST diagram.

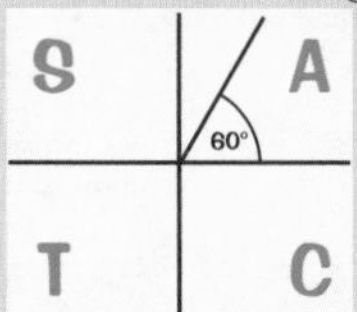

The angle from your calculator goes anticlockwise from the x-axis (unless it's negative — then it would go clockwise into the 4th quadrant).

Find the other angles between 0° and 360° that might be solutions.

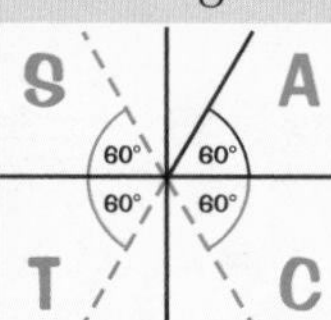

The other solutions come from making the same angle from the horizontal axis into the other 3 quadrants.

Ditch the ones that are the wrong sign.

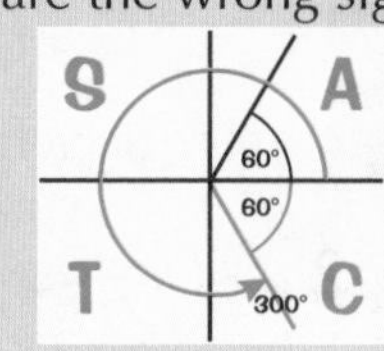

$\cos x = \frac{1}{2}$, which is positive. The CAST diagram tells you cos is positive in the 4th quadrant — but not the 2nd or 3rd — so ditch those two angles.

So you've got solutions 60° and 300° in the range 0° to 360°. But you need all the solutions in the range –360° to 720°. Get these by repeatedly adding or subtracting 360° onto each until you go out of range:

x = 60° ⇒ (adding 360°) x = 420°, 780° (too big)
and (subtracting 360°) x = –300°, –660° (too small)

x = 300° ⇒ (adding 360°) x = 660°, 1020° (too big)
and (subtracting 360°) x = –60, –420° (too small)

So the solutions are: x = –300°, –60°, 60°, 300°, 420° and 660°.

Grow to love the CAST method — but not too much...

Suppose the first solution you get is negative, let's say –d°, then you'd measure it clockwise on the CAST diagram. So it'd be d° in the 4th quadrant. Then you'd work out the other 3 possible solutions in exactly the same way, rejecting the ones which weren't the right sign. Got that? No? Got that? No? Got that? Yes? Good!

Solving Trig Equations in a Given Interval

This page is for AQA Core 2, OCR Core 2, Edexcel Core 2
Sometimes it's a bit more complicated. But only a bit.

Sometimes you end up with **sin kx = number...**

For these, it's definitely easier to draw the graph rather than use the CAST method — that's one reason why being able to sketch these trig graphs properly is so important.

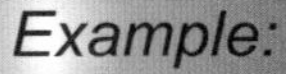

Solve: $\sin 3x = -\frac{1}{\sqrt{2}}$ for $0° \le x \le 360°$.

1) You've got 3x instead of x — so when you draw the graph, make it three times as squashed (the period will 120° instead of 360°).
2) When you use your calculator to get the first solution, it'll probably give you a negative answer (which you don't want).

$$\sin 3x = -\frac{1}{\sqrt{2}}$$
$$\Rightarrow 3x = -45°$$
$$\Rightarrow x = -15°$$

You don't want this solution — so use your graph to work out one that you do want.

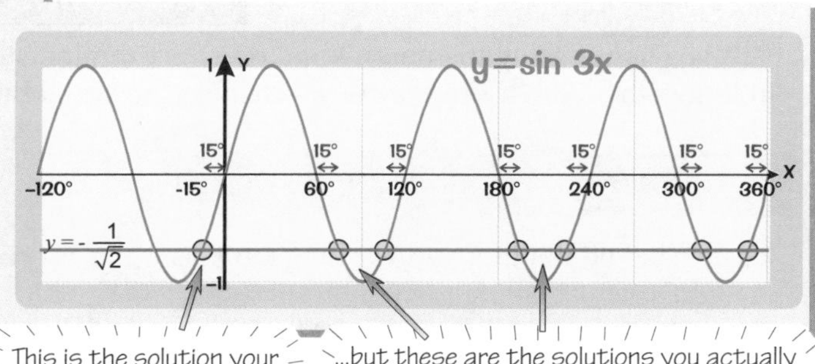

This is the solution your calculator gives you...

...but these are the solutions you actually want — they're all 15° from a point where the graph crosses the x-axis.

3) Using the symmetry of the graph, you can see that the solutions you want are:

   x = 75°, 105°, 195°, 225°, 315° and 345°.

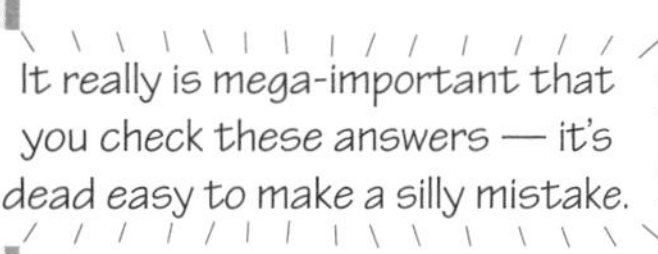

It really is mega-important that you check these answers — it's dead easy to make a silly mistake.

4) Check your answers by putting these values back into your calculator.

...or **sin (x + k) = number**

All the steps in this example are just the same as in the one above.

Example:

Solve $\sin(x + 60°) = \frac{3}{4}$ for $-360° \le x \le 360°$, giving your answers to 2 decimal places.

1) It's easier to find all the values of $(x + 60°)$ first, and then convert them all to x. But first work out the range of $(x + 60°)$.

   If $-360° \le x \le 360°$,

   then $-300° \le x + 60° \le 420°$.
2) Use your calculator to get the first solution...

$$\sin(x + 60°) = \frac{3}{4}$$
$$\Rightarrow x + 60° = 48.59°$$

This is the solution from the calculator.

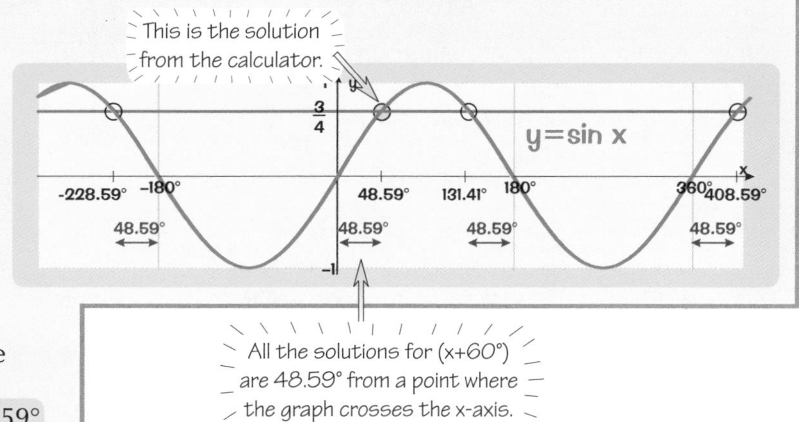

All the solutions for (x+60°) are 48.59° from a point where the graph crosses the x-axis.

3) And again, use the graph's symmetry to get the other three. The solutions for $(x + 60°)$ are:

   $x + 60° = 48.59°, 131.41°, 408.59°$ and $-228.59°$
4) Convert these into solutions for x:

   $x = -11.41°, 71.41°, 348.59°$ and $-228.59°$
5) Check your answers using your calculator.

Live a life of sin (and cos and tan)...

Yep, the examples on this page are pretty fiddly. The most important bit is actually getting the sketch right. If you don't, you're in big trouble. Then you've just got to carefully use the sketch to work out the other solutions. It's tricky, but you'll feel better about yourself when you've got it mastered. Ah you will, you will, you will ...

Solving Trig Equations in a Given Interval

This page is for AQA Core 2, OCR Core 2, Edexcel Core 2

Now for something really exciting — trig identities. Mmm, well, maybe exciting was the wrong word. But they can be dead useful, so here goes...

For equations with **tan x** in, it often helps to use this...

$$\tan x \equiv \frac{\sin x}{\cos x}$$

This is a handy thing to know — and one the examiners love testing. Basically, if you've got a trig equation with a tan in it, together with a sin or a cos — chances are you'll be better off if you rewrite the tan using this formula.

Example: Solve: $3\sin x - \tan x = 0$, for $0° \leq x \leq 360°$.

It's got sin and tan in it — so writing tan x as $\frac{\sin x}{\cos x}$ is probably a good move:

$$3\sin x - \tan x = 0$$

$$\Rightarrow 3\sin x - \frac{\sin x}{\cos x} = 0$$

Get rid of the cos x on the bottom by multiplying the whole equation by cos x.

$$\Rightarrow 3\sin x \cos x - \sin x = 0$$

Now — there's a common factor of sin x. Take that outside a bracket.

$$\Rightarrow \sin x(3\cos x - 1) = 0$$

And now you're almost there. You've got two things multiplying together to make zero. That means either one or both of them is equal to zero themselves.

$$\Rightarrow \sin x = 0 \quad \text{or} \quad 3\cos x - 1 = 0$$

$\sin x = 0$

The first solution is... $\sin^{-1} 0 = 0°$

Now find the other points where sin x is zero in the interval $0° \leq x \leq 360°$.

(Remember the sin graph is zero every *180°*.)

$$\Rightarrow x = 0°, 180°, 360°$$

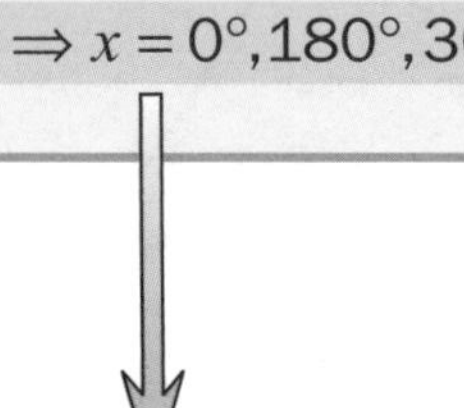

$3\cos x - 1 = 0$

CAST gives any solutions in the interval $0° \leq x \leq 360°$.

Rearrange... $\cos x = \frac{1}{3}$

So the first solution is...

$$\cos^{-1}\tfrac{1}{3} = 70.5°$$

CAST (or the graph of cos x) gives another solution in the 4th quadrant...

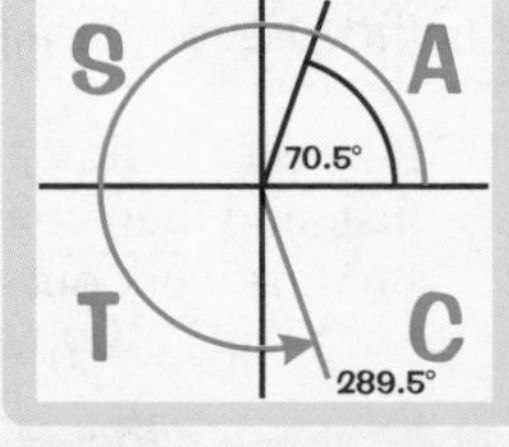

And the two solutions from this part are:

$$\Rightarrow x = 70.5°, 289.5°$$

So altogether you've got five possible solutions:

$$\Rightarrow x = 0°, 180°, 360°, 70.5°, 289.5°$$

Trigonometry is the root of all evil...

What a page — you don't have fun like that every day, do you? No, trig equations are where it's at. This is a really useful trick, though — and can turn a nightmare of an equation into a bit of a pussy-cat. Rewriting stuff using different formulas is always worth trying if it feels like you're getting stuck — even if you're not sure why when you're doing it. You might have a flash of inspiration when you see the new version.

Solving Trig Equations in a Given Interval

This page is for AQA Core 2, OCR Core 2, Edexcel Core 2
Another trig identity — and it's a good 'un — examiners love it. And it's not difficult either.

And if you have a $\sin^2 x$ or a $\cos^2 x$, think of this straight away...

$$\sin^2 x + \cos^2 x \equiv 1 \Rightarrow \begin{cases} \sin^2 x \equiv 1 - \cos^2 x \\ \cos^2 x \equiv 1 - \sin^2 x \end{cases}$$

Use this identity to get rid of a $\sin^2$ or a $\cos^2$ that's making things awkward...

Example: Solve: $2\sin^2 x + 5\cos x = 4$, for $0° \le x \le 360°$.

You can't do much while the equation's got both sin's and cos's in it. So replace the $\sin^2 x$ bit with $1 - \cos^2 x$.

$$2(1-\cos^2 x)+5\cos x=4$$

Multiply out the bracket and rearrange it so that you've got zero on one side — and you get a quadratic in cos x:

$$\Rightarrow 2-2\cos^2 x+5\cos x=4$$
$$\Rightarrow 2\cos^2 x-5\cos x+2=0$$

Now the only trig function is cos.

If you replaced cos x with y, this would be $2y^2 - 5y + 2 = 0$.

This is a quadratic in cos x. It's easier to factorise this if you make the substitution y = cos x.

$$2y^2-5y+2=0$$
$$\Rightarrow (2y-1)(y-2)=0$$
$$\Rightarrow (2\cos x-1)(\cos x-2)=0$$

$2y^2 - 5y + 2 = (2y\ \ ?)(y\ \ ?) = (2y-1)(y-2)$

Now one of the brackets must be 0. So you get 2 equations as usual:

$$(2\cos x-1)=0 \quad or \quad (\cos x-2)=0$$

This is a bit weird. cos x is always between −1 and 1. So you don't get any solutions from this bracket.

$\cos x = \frac{1}{2} \Rightarrow x = 60°$ or $x = 300°$ and $\cos x = 2$ This is impossible — so you get nothing from this bracket.

So at the end of all that, the only solutions you get are x = 60° and x = 300°. How boring.

Use the *Trig Identities* to prove something is the *Same* as something else

Another use for these trig identities is proving that two things are the same.

Example: Show that $\frac{\cos^2\theta}{1+\sin\theta} \equiv 1-\sin\theta$

The identity sign ≡ means that this is true for all θ, rather than just certain values.

Prove things like this by playing about with one side of the equation until it you get the other side.

Left-hand side: $\frac{\cos^2\theta}{1+\sin\theta}$

The only thing I can think of doing here is replacing $\cos^2\theta$ with $1 - \sin^2\theta$. (Which is good because it works.)

$$\equiv \frac{1-\sin^2\theta}{1+\sin\theta}$$

The next trick is the hardest to spot. Look at the top — does that remind you of anything?

The top line is a difference of two squares:

$$\equiv \frac{(1+\sin\theta)(1-\sin\theta)}{1+\sin\theta}$$

$1-a^2=(1+a)(1-a)$
$\Rightarrow 1-\sin^2\theta=(1+\sin\theta)(1-\sin\theta)$

$\equiv 1-\sin\theta$, the right-hand side.

Trig identities — the path to a brighter future...

That was a pretty miserable section. But it's over. These trig identities aren't exactly a barrel of laughs, but they are a definite source of marks — you can bet your last penny they'll be in the exam. That substitution trick to get rid of a $\sin^2$ or a $\cos^2$ and end up with a quadratic in sin x or cos x is a real examiners' favourite. Those identities can be a bit daunting, but it's always worth having a few tricks in the back of your mind — always look for things that factorise, or fractions that can be cancelled down, or ways to use those trig identities. Ah, it's all good clean fun.

Exam Questions

Wow, questions — I love questions, they're like my favourite thing ever...

Exam Questions

1 Find $\cos 315^\circ$, giving your answer in the form $a\sqrt{2}$, where a is an exact fraction.

(5 marks)

2

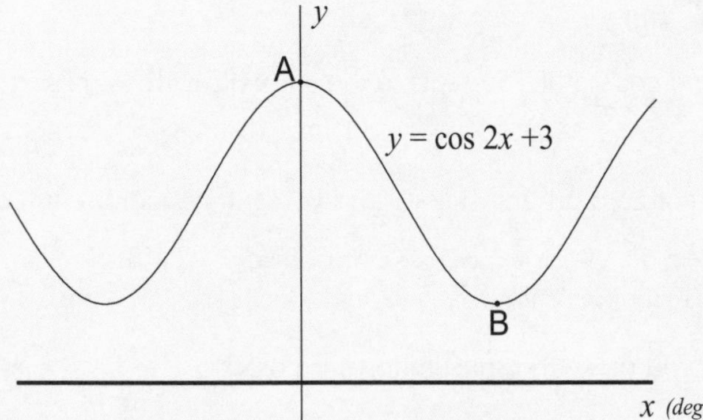

The diagram above shows the curve with equation $y = \cos 2x + 3$.
The curve intersects the y-axis at point A. Point B is a minimum turning point.

(a) What is the y-coordinate at point A?

(1 mark)

(b) What are the coordinates of point B?

(2 marks)

(c) One solution of the equation $\cos 2x + 3 = 3.5$ is $x = 30°$.
Find the other solutions of this equation for $0° \leq x < 360°$.

(8 marks)

3

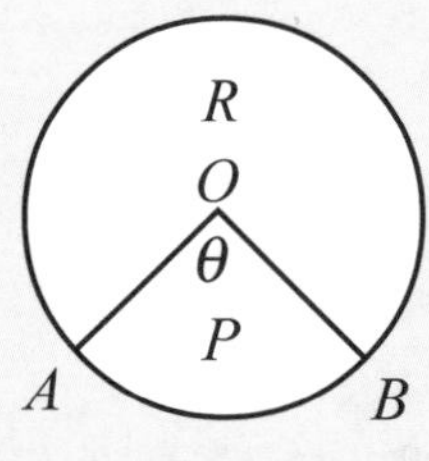

Fig. 1

A symmetrical pendant P is cut from a circular silver disc, leaving the region R, as shown in figure 1.
The radius of the disc is 3 cm, and angle θ equals $\frac{\pi}{4}$ radians.

Find the length of minor arc AB in terms of π.
Find also the perimeter of region R to 2 decimal places.

(4 marks)

4 Find all of the values of θ between 2π and 4π for which $\cos \theta = 0.5$.
Write this angle in radians, giving your answer in terms of π.

(4 marks)

Logs

This page is for AQA Core 2, OCR Core 2, Edexcel Core 2

Don't be put off by your parents or grandparents telling you that logs are hard. Logarithm is just a fancy word for power, and once you know how to use them you can solve all sorts of equations.

You need to be able to Switch between Different Notations

$\log_a b = c$ means the same as $a^c = b$

That means that $\log_a a = 1$ and $\log_a 1 = 0$

The logarithm of 100 to the base 10 is 2, because 10 raised to the power of 2 is 100.

Example:

Index notation: $10^2 = 100$

log notation: $\log_{10} 100 = 2$

The base goes here but it's usually left out if it's 10.

Logs can be to any base. Base 10 is the most common — and the log button on your calculator gives logs to base 10.

Example:

Write down the values of the following:

a) $\log_2 8$ b) $\log_9 3$ c) $\log_5 5$

a) 8 is 2 raised to the power of 3
so $2^3 = 8$ and $\log_2 8 = 3$

b) 3 is the square root of 9, or $9^{1/2} = 3$
so $\log_9 3 = 1/2$

c) anything to the power of 1 is itself
so $\log_5 5 = 1$

Write the following using log notation:

a) $5^3 = 125$ b) $3^0 = 1$

You just need to make sure you get things in the right place.

a) 3 is the power or logarithm that 5 (the base) is raised to to get 125
so $\log_5 125 = 3$

b) you'll need to remember this one:
$\log_3 1 = 0$

The Laws of Logarithms are Unbelievably Useful

Whenever you have to deal with logs, you'll end up using these laws. That means it's no bad idea to learn them off by heart right now.

Laws of Logarithms

$$\log_a x + \log_a y = \log_a (xy)$$

$$\log_a x - \log_a y = \log_a \left(\frac{x}{y}\right)$$

$$\log_a x^k = k \log_a x$$

Here's how to change the base of a log:

Change of Base

$$\log_a x = \frac{\log_b x}{\log_b a}$$

Example: Calculate $\log_7 4$ to 4 decimal places.

$\log_7 4 = \frac{\log_{10} 4}{\log_{10} 7} = 0.7124$. (To check: $7^{0.7124} = 4$)

Logs

This page is for AQA Core 2, OCR Core 2, Edexcel Core 2

Use the Laws to Manipulate Logs

Whenever you get some logs to mess around with, go straight for your log laws. Obvious when you think about it.

Example: Write each expression in the form $\log_a n$, where n is a number.

a) $\log_a 5 + \log_a 4$ b) $\log_a 12 - \log_a 4$ c) $2\log_a 6 - \log_a 9$

a) Use the law of logarithms

$$\log_a x + \log_a y = \log_a (xy)$$

You just have to multiply the numbers together:

$\log_a 5 + \log_a 4 = \log_a (5 \times 4)$

$= \log_a 20$

b) Use the law of logarithms

$$\log_a x - \log_a y = \log_a \left(\frac{x}{y}\right)$$

Divide the numbers:

$\log_a 12 - \log_a 4 = \log_a (12 \div 4)$

$= \log_a 3$

c) Convert $2\log_a 6$ using the law

$$\log_a x^k = k \log_a x$$

$2\log_a 6 = \log_a 6^2 = \log_a 36$

$\log_a 36 - \log_a 9 = \log_a (36 \div 9)$

$= \log_a 4$

Practice Questions

1) *Write down the values of the following*

(a) $\log_3 27$ (b) $\log_3 \left(\frac{1}{27}\right)$ (c) $\log_3 18 - \log_3 2$

2) *Simplify the following*

(a) $\log 3 + 2\log 5$ (b) $\frac{1}{2}\log 36 - \log 3$

3) *Simplify* $\log_b (\chi^2 - 1) - \log_b (\chi - 1)$

4) *Sample exam question:*

a) Write down the value of $\log_3 3$ [1 mark]

b) Given that $\log_a \chi = \log_a 4 + 3\log_a 2$ show that $\chi = 32$ [2 marks]

It's sometimes hard to see the wood for the trees — especially with logs...

Tricky, tricky, tricky... I think of $\log_a b$ as 'the power I have to raise a to if I want to end up with b' — that's all it is. And the log laws make a bit more sense if you think of 'log' as meaning 'power'. For example, you know that $2^a \times 2^b = 2^{a+b}$ — this just says that if you multiply two numbers, you add the powers. Well, the first law of logs is saying the same thing. Any road up, even if you don't really understand why they work, make sure you know the log laws like you know your own navel.

Exponentials

This page is for AQA Core 2, OCR Core 2, Edexcel Core 2

Okay, you've done the theory of logs. Now for a bit about exponentials.

Graphs of a^x Never Reach Zero

All the graphs of $y = a^x$ (where $a > 1$) have the same basic shape.
The graphs for a = 2, a = 3 and a = 4 are shown on the right.

- All the a's are greater than 1 — so y increases as x increases.
- The bigger a is, the quicker the graphs increase.
  The rate at which they increase gets bigger too.
- As x decreases, y decreases at a smaller and smaller rate — y will approach zero, but never actually get there.

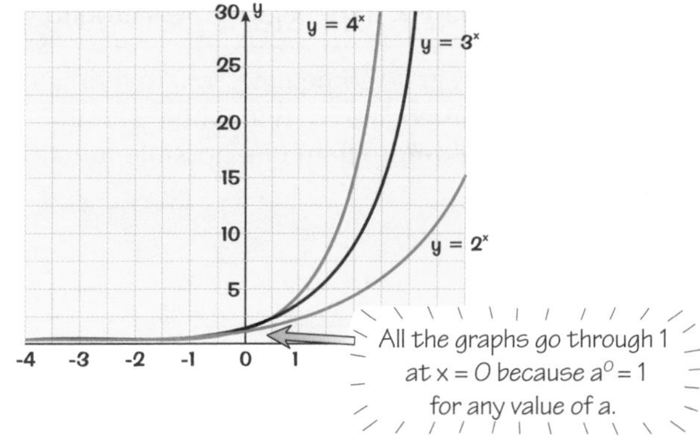

Estimate Roots Using Graphs

Example: Find x to 3 significant figures, where $3^x = 25$.

Your first step is to draw a graph to estimate where the root is. Just use $y = 3^x - 25$, and see where it crosses the axis.

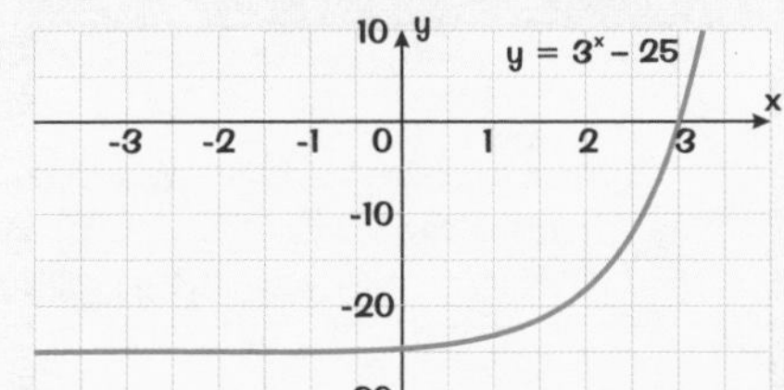

It's obviously between x = 2 and x = 3, nearer to x = 3.

Trying the next decimal point gives x between 2.9 and 3.0 ($3^{2.9} = 24.2$, $3^{3.0} = 27$)

Another decimal point on gives x between 2.92 and 2.93 ($3^{2.92} = 24.7$, $3^{2.93} = 25.001$)

It looks like the answer is x = 2.93 to 3 s.f. — but to make sure, work out: $3^{2.925} = 24.864$

This tells you the answer is definitely between 2.925 (too small) and 2.93 (too big). So $x = 2.93$ to 3 s.f.

If you want more significant figures, you'll have to do more decimal searches.

Use the Calculator Log Button Whenever You Can

Example: Use logarithms to solve the following for x, giving the answers to 4 s.f.
a) $10^x = 170$ b) $10^{3x} = 4000$ c) $7^x = 55$ d) $\log_{10}x = 2.6$ e) $2^{4x} = 80$

You've got the magic buttons on your calculator, but you'd better follow the instructions and show that you know how to use the log rules covered earlier.

a) $10^x = 170$. Taking logs to base 10 of both sides gives $x = \log_{10}170 = 2.230$.

b) $10^{3x} = 4000$. Taking logs of both sides gives $3x = \log_{10} 4000 = 3.602$, so $x = 1.201$.

c) $7^x = 55$. Take logs of both sides, and use the log rules. It doesn't matter what base you use, so why not use base 10? (Make sure you use the same base for each side though.)

$$x \log_{10}7 = \log_{10}55 \quad \text{so} \quad x = \frac{\log_{10} 55}{\log_{10} 7} = 2.059$$

d) $\log_{10}x = 2.6$. You've got to be able to go back the other way, so $x = 10^{2.6} = 398.1$

e) $2^{4x} = 80$. Take logs — again use base 10: $4x(\log_{10} 2) = \log_{10} 80$, so $x = \dfrac{\log_{10} 80}{4\log_{10} 2} = 1.580$

Integration

This page is for AQA Core 1, AQA Core 2, OCR Core 2, Edexcel Core 1, Edexcel Core 2
Integration is the 'opposite' of differentiation — and so if you can differentiate, you can be pretty confident you'll be able to integrate too. There's just one extra thing you have to remember — the constant of integration...

You need the constant because there's More Than One right answer

When you integrate something, you're trying to find a function that returns to what you started with when you differentiate it. And when you add the constant of integration, you're just allowing for the fact that there's more than one possible function that does this...

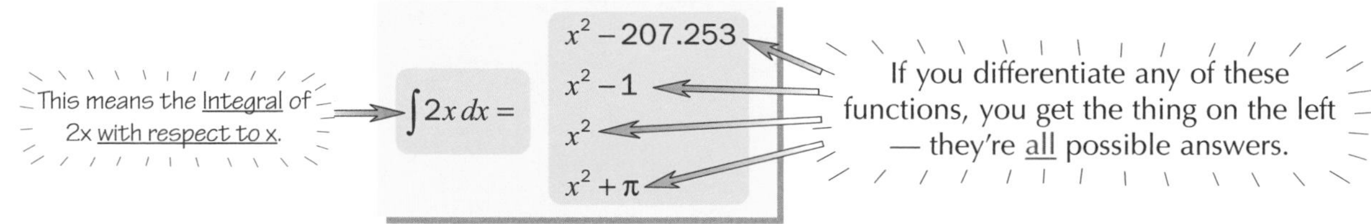

So the answer to this integral is actually... $\int 2x\,dx = x^2 + C$

The 'C' just means 'any number'. This is the constant of integration.

You only need to add a constant of integration to indefinite integrals — these are just integrals without limits (or little numbers) next to the integral sign. (If that doesn't make sense, you'll see what I mean later on.)

Up the power by One — then Divide by it

The formula below tells you how to integrate any power of x (except x^{-1}).

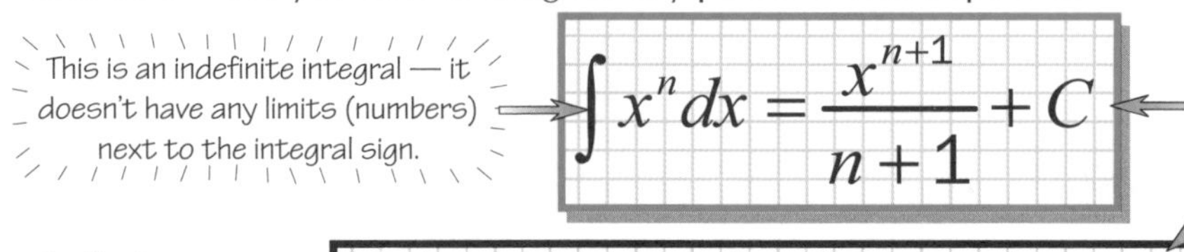

This is an indefinite integral — it doesn't have any limits (numbers) next to the integral sign.

$$\int x^n\,dx = \frac{x^{n+1}}{n+1} + C$$

You can't do this to $\frac{1}{x} = x^{-1}$. When you increase the power by 1 (to get zero) and then divide by zero — you get big problems.

In a nutshell, this says:

To integrate a power of x: (i) Increase the power by one — then divide by it.
and (ii) Stick a constant on the end.

Example: Use the integration formula...

① For 'normal' powers,

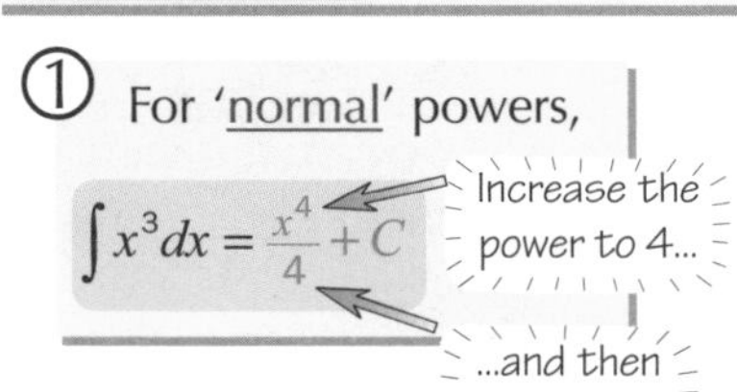

② For negative powers,

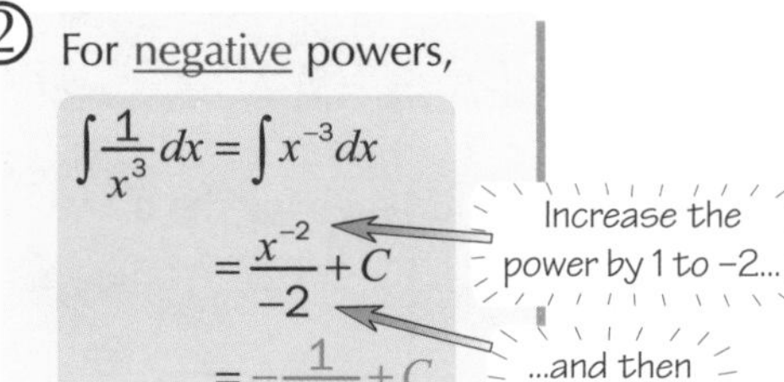

③ For fractional powers,

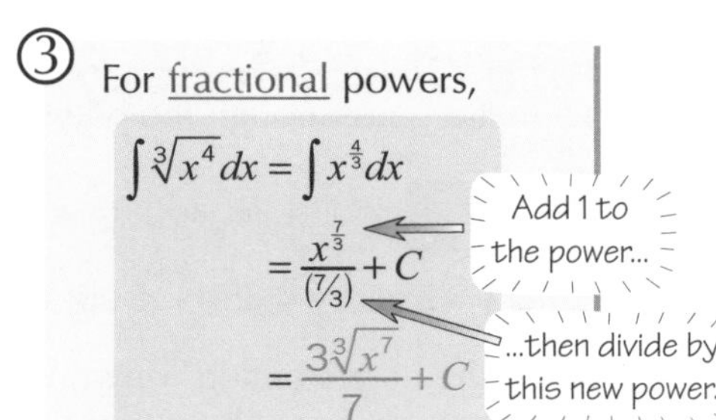

④ And for complicated looking stuff...

$$\int\left(3x^2 - \frac{2}{\sqrt{x}} + \frac{7}{x^2}\right)dx = \int\left(3x^2 - 2x^{-\frac{1}{2}} + 7x^{-2}\right)dx$$

$$= \frac{3x^3}{3} - \frac{2x^{\frac{1}{2}}}{(1/2)} + \frac{7x^{-1}}{-1} + C$$

$$= x^3 - 4\sqrt{x} - \frac{7}{x} + C$$

Do each of these bits separately.

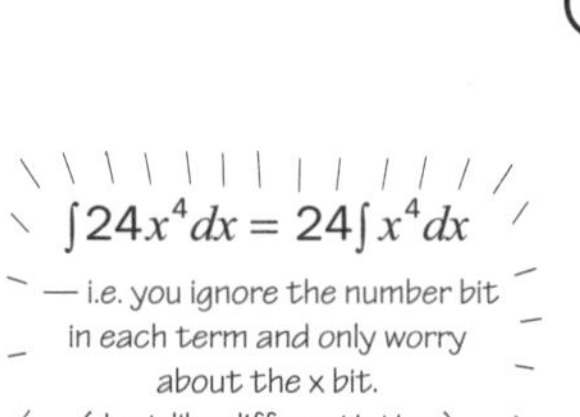

$\int (f(x) + g(x))\,dx = \int f(x)\,dx + \int g(x)\,dx$
— i.e. you can integrate a long expression term by term. (Just like differentiation.)

CHECK YOUR ANSWERS:
You can check you've integrated properly by differentiating the answer — you should end up with the thing you started with.

Indefinite integrals — joy without limits...

This integration lark isn't so bad then — there's only a couple of things to remember and then you can do it no problem. But that constant of integration catches loads of people out — it's so easy to forget — and you'll definitely lose marks if you do forget it. You have been warned. Other than that, there's not much to it. Hurray.

Integration

This page is for AQA Core 1, AQA Core 2, OCR Core 2, Edexcel Core 1

By now, you're probably aware that maths isn't something you do unless you're a bit of a *thrill-seeker.*
You know, sometimes they even ask you to find a curve with a certain derivative that goes through a certain point.

You sometimes need to find the **Value** of the **Constant of Integration**

When they tell you something else about the curve in addition to its derivative, you can work out the value of that constant of integration. Usually the something is the coordinates of one of the points the curve goes through.

Really Important Bit...

When you differentiate y, you get $\frac{dy}{dx}$.
And when you integrate $\frac{dy}{dx}$, you get y*.

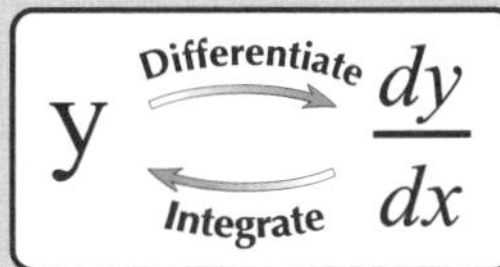

*If you ignore the constant of integration.

Example: Find the equation of the curve through the point (2, 8) with $\frac{dy}{dx} = 6x(x-1)$.

You know the derivative and need to find the function — so integrate.

$$\frac{dy}{dx} = 6x(x-1) = 6x^2 - 6x$$

So integrating both sides gives...

$$y = \int (6x^2 - 6x)\,dx$$
$$\Rightarrow y = \frac{6x^3}{3} - \frac{6x^2}{2} + C$$
$$\Rightarrow y = 2x^3 - 3x^2 + C$$

Don't forget the constant of integration.

Remember:
Even if you don't have any extra information about the curve — you still have to add a constant when you work out an integral without limits.

Check this is correct by differentiating it and making sure you get what you started with.

$$y = 2x^3 - 3x^2 + C = 2x^3 - 3x^2 + Cx^0$$
$$\Rightarrow \frac{dy}{dx} = 2(3x^2) - 3(2x^1) + C(0x^{-1})$$
$$\Rightarrow \frac{dy}{dx} = 6x^2 - 6x$$

So this function's got the correct derivative — but you haven't finished yet.

You now need to find C — and you do this by using the fact that it goes through the point (2, 8).

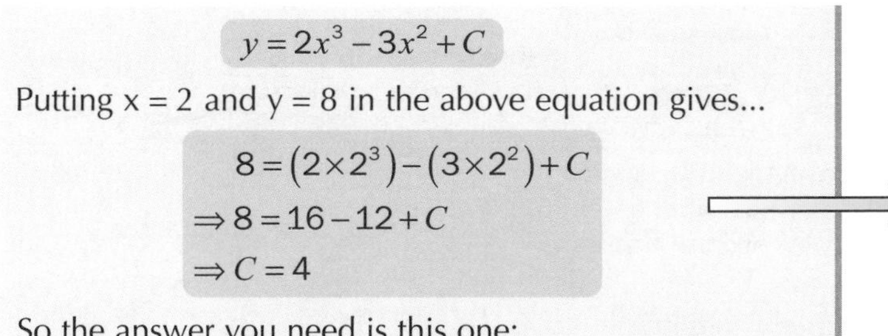

$$y = 2x^3 - 3x^2 + C$$

Putting x = 2 and y = 8 in the above equation gives...

$$8 = (2\times2^3) - (3\times2^2) + C$$
$$\Rightarrow 8 = 16 - 12 + C$$
$$\Rightarrow C = 4$$

So the answer you need is this one:

$$y = 2x^3 - 3x^2 + 4$$

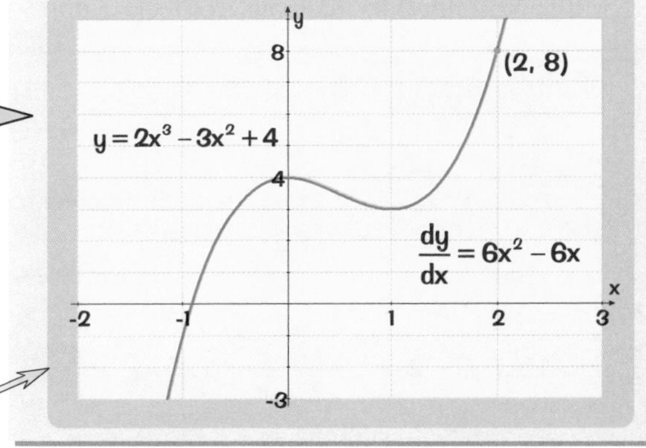

It's a cubic equation — and the graph looks like this...

Maths and alcohol don't mix — so never drink and derive...

That's another page under your belt and — go on, admit it — there was nothing too horrendous on it. If you can do the stuff from the previous page and then substitute some numbers into an equation, you can do everything from this page too. So if you think this is boring, you'd be right. But if you think it's much harder than the stuff before, you'd be wrong.

Integration

This page is for AQA Core 1, AQA Core 2, OCR Core 2, Edexcel Core 2

Some integrals have limits (i.e. little numbers) next to the integral sign. You integrate them in exactly the same way — but you don't need a constant of integration. Much easier. And scrummier and yummier too.

A Definite Integral finds the Area Under a Curve

This definite integral tells you the area between the graph of $y = x^3$ and the x-axis between $x = -2$ and $x = 2$:

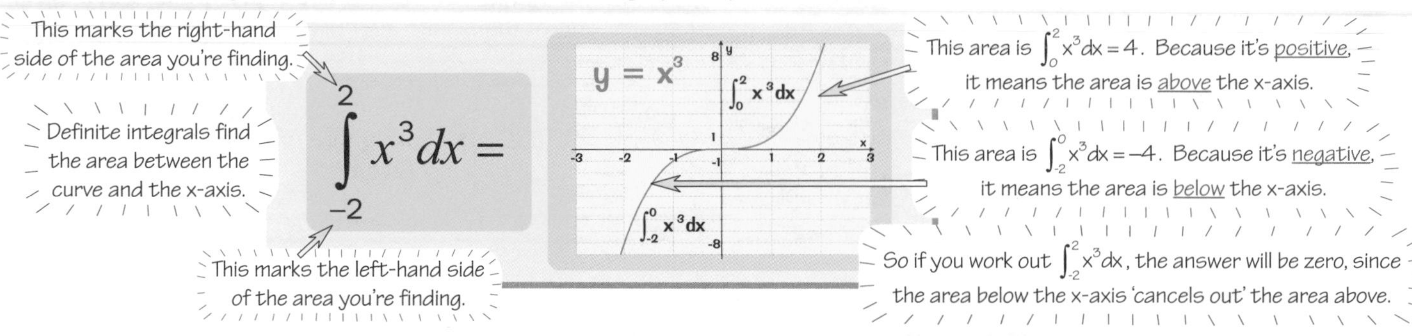

This marks the right-hand side of the area you're finding.
Definite integrals find the area between the curve and the x-axis.
$\int_{-2}^{2} x^3 dx =$
This marks the left-hand side of the area you're finding.
$y = x^3$
$\int_0^2 x^3 dx$
$\int_{-2}^0 x^3 dx$
This area is $\int_0^2 x^3 dx = 4$. Because it's positive, it means the area is above the x-axis.
This area is $\int_{-2}^0 x^3 dx = -4$. Because it's negative, it means the area is below the x-axis.
So if you work out $\int_{-2}^2 x^3 dx$, the answer will be zero, since the area below the x-axis 'cancels out' the area above.

Do the integration in the same way — then use the Limits

Finding a definite integral isn't really any harder than an indefinite one — there's just an extra stage you have to do. After you've integrated the function you have to work out the value of this new function by sticking in the limits.

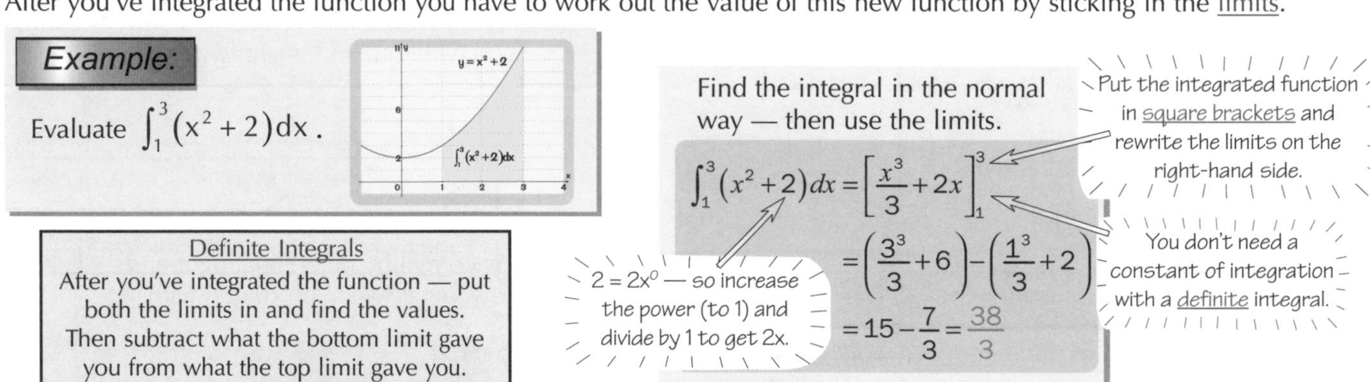

Example:
Evaluate $\int_1^3 (x^2 + 2)dx$.
$y = x^2 + 2$
$\int_1^3 (x^2+2)dx$
Find the integral in the normal way — then use the limits.
$\int_1^3 (x^2+2)dx = \left[\frac{x^3}{3} + 2x\right]_1^3$
$= \left(\frac{3^3}{3} + 6\right) - \left(\frac{1^3}{3} + 2\right)$
$= 15 - \frac{7}{3} = \frac{38}{3}$
Put the integrated function in square brackets and rewrite the limits on the right-hand side.
You don't need a constant of integration with a definite integral.
$2 = 2x^0$ — so increase the power (to 1) and divide by 1 to get $2x$.

Definite Integrals

After you've integrated the function — put both the limits in and find the values. Then subtract what the bottom limit gave you from what the top limit gave you.

Integrate 'to Infinity' with the ∞ (infinity) sign

And you can integrate all the way to infinity as well. Just use the ∞ symbol as your upper limit. Or use $-\infty$ as your lower limit if you want to integrate to 'minus infinity'.

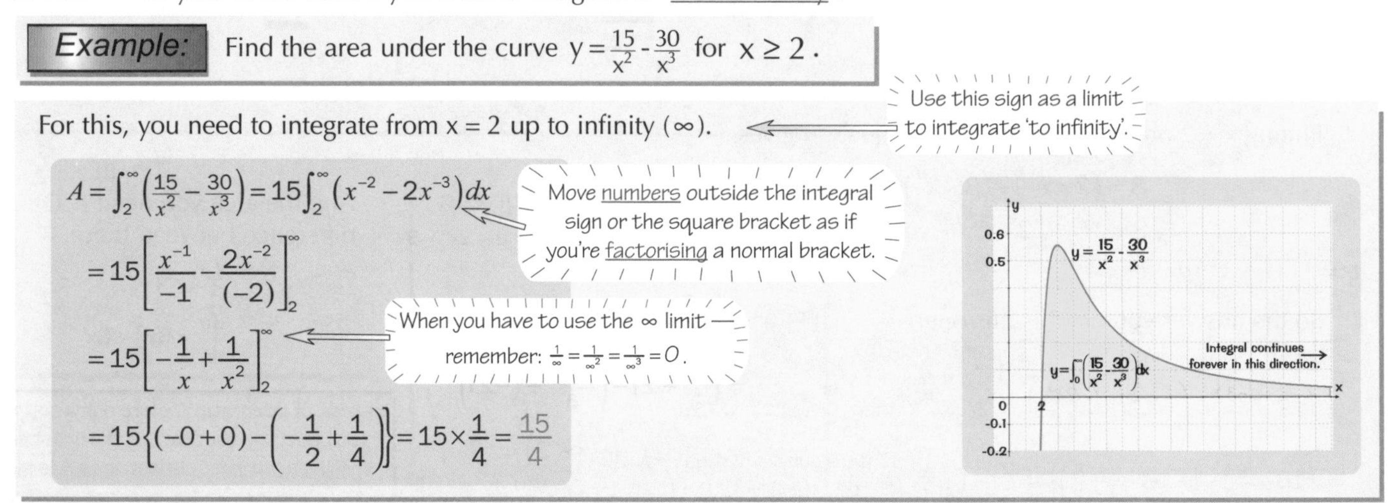

Example: Find the area under the curve $y = \frac{15}{x^2} - \frac{30}{x^3}$ for $x \geq 2$.
For this, you need to integrate from $x = 2$ up to infinity (∞).
Use this sign as a limit to integrate 'to infinity'.
$A = \int_2^\infty \left(\frac{15}{x^2} - \frac{30}{x^3}\right) = 15\int_2^\infty (x^{-2} - 2x^{-3})dx$
$= 15\left[\frac{x^{-1}}{-1} - \frac{2x^{-2}}{(-2)}\right]_2^\infty$
$= 15\left[-\frac{1}{x} + \frac{1}{x^2}\right]_2^\infty$
$= 15\left\{(-0+0) - \left(-\frac{1}{2} + \frac{1}{4}\right)\right\} = 15 \times \frac{1}{4} = \frac{15}{4}$
Move numbers outside the integral sign or the square bracket as if you're factorising a normal bracket.
When you have to use the ∞ limit — remember: $\frac{1}{\infty} = \frac{1}{\infty^2} = \frac{1}{\infty^3} = 0$.
$y = \frac{15}{x^2} - \frac{30}{x^3}$
$y = \int_2^\infty \left(\frac{15}{x^2} - \frac{30}{x^3}\right)dx$
Integral continues forever in this direction.

My hobbies? Well I'm really inte grating. Especially carrots.

It's still integration — but this time you're putting two numbers into an equation afterwards. So although this may not be the wild and crazy fun-packed time your teachers promised you when they were trying to persuade you to take AS maths, you've got to admit that a lot of this stuff is pretty similar — and if you can do one bit, you can use that to do quite a few other bits too. Maths is like that. But I admit it's probably not as much fun as a big banana-and-toffee cake.

Areas Between Curves

This page is for OCR Core 2, Edexcel Core 2

With a bit of thought, you can use integration to find all kinds of areas — even ones that look quite tricky at first. The best way to work out what to do is draw a picture. Then it'll seem easier. I promise you it will.

Sometimes you have to **Add** integrals...

This looks pretty hard — until you draw a picture and see what it's all about.

Example: Find the area enclosed by the curves $y = x^2$, $y = (2 - x)^2$ and the x-axis.

Find out where the curves meet by solving $x^2 = (2-x)^2$. — they meet at x=1.

You have to find area A — but you'll need to split it into two smaller pieces.

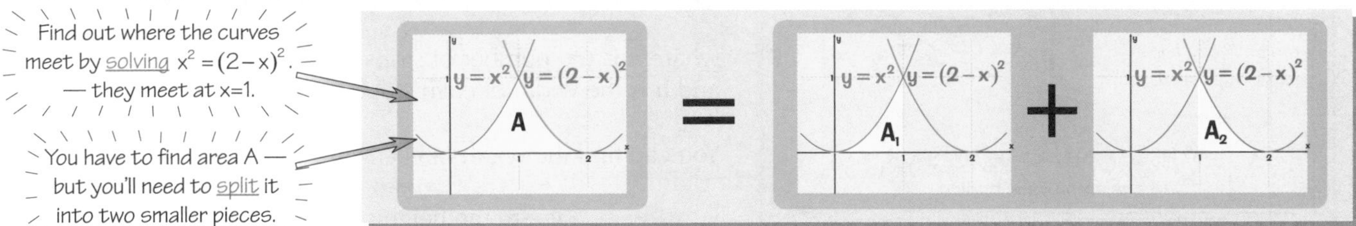

And it's pretty clear from the picture that you'll have to find the area in two lumps, A_1 and A_2.

The first area you need to find is A_1:

$$A_1 = \int_0^1 x^2\,dx = \left[\frac{x^3}{3}\right]_0^1 = \left(\tfrac{1}{3} - 0\right) = \tfrac{1}{3}$$

The other area you need is A_2:

$$A_2 = \int_1^2 (2-x)^2\,dx = \int_1^2 (4 - 4x + x^2)\,dx$$
$$= \left[4x - 2x^2 + \frac{x^3}{3}\right]_1^2$$
$$= \left(8 - 8 + \tfrac{8}{3}\right) - \left(4 - 2 + \tfrac{1}{3}\right)$$
$$= \frac{8}{3} - \frac{7}{3} = \frac{1}{3}$$

And the area the question actually asks for is $A_1 + A_2$. This is

$$A = A_1 + A_2 = \frac{1}{3} + \frac{1}{3} = \frac{2}{3}$$

If you spot that the area A is symmetrical about x = 1, you can save yourself some work by calculating half the area and then doubling it: $A = 2\int_0^1 x^2\,dx$

...sometimes you have to **Subtract** them

Again, it's best to look at the pictures to work out exactly what you need to do.

Example: Find the area enclosed by the curves $y = x^2 + 1$ and $y = 9 - x^2$.

Solve $x^2 + 1 = 9 - x^2$ to find where the curves meet.

$x^2 + 1 = 9 - x^2 \Rightarrow 2x^2 = 8$

$\Rightarrow x^2 = 4$

$\Rightarrow x = \pm 2$

So you'll have to integrate between −2 and 2.

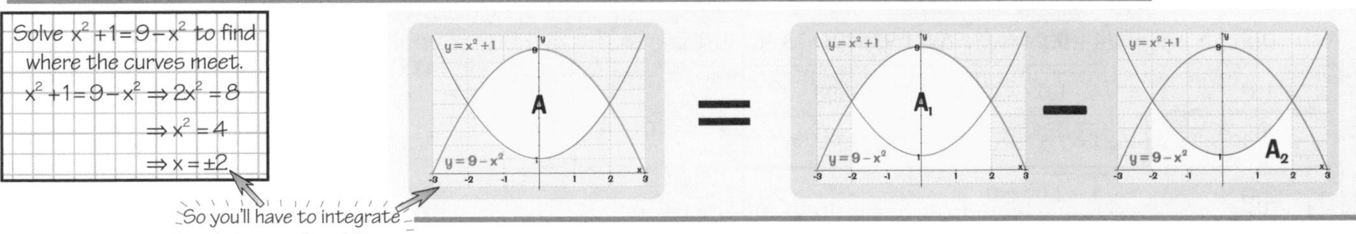

The area under the green curve A_1 is:

$$A_1 = \int_{-2}^{2} (9 - x^2)\,dx = \left[9x - \frac{x^3}{3}\right]_{-2}^{2}$$
$$= \left(18 - \frac{2^3}{3}\right) - \left(-18 - \frac{(-2)^3}{3}\right)$$
$$= \left(18 - \tfrac{8}{3}\right) - \left(-18 - \left(-\tfrac{8}{3}\right)\right) = \frac{46}{3} - \left(-\frac{46}{3}\right) = \frac{92}{3}$$

The area under the red curve is:

$$A_2 = \int_{-2}^{2} (x^2 + 1)\,dx = \left[\frac{x^3}{3} + x\right]_{-2}^{2}$$
$$= \left(\frac{2^3}{3} + 2\right) - \left(\frac{(-2)^3}{3} + (-2)\right)$$
$$= \left(\tfrac{8}{3} + 2\right) - \left(-\tfrac{8}{3} - 2\right) = \frac{28}{3}$$

And the area you need is the difference between these:

$$A = A_1 - A_2 = \frac{92}{3} - \frac{28}{3} = \frac{64}{3}$$

Instead of integrating before subtracting — you could try 'subtracting the lines', and then integrating. This last area A is also: $A = \int_{-2}^{2} \{(9 - x^2) - (x^2 + 1)\}\,dx$

And so, our hero integrates the area between two curves, and saves the day...

That's the basic idea of finding the area enclosed by two curves and lines — draw a picture and then break the area down into smaller, easier chunks. And it's always a good idea to keep an eye out for anything symmetrical that could save you a bit of work — like in the first example. Questions like this aren't hard — but they can sometimes take a long time. Great.

Numerical Integration of Functions

This page is for AQA Core 2, OCR Core 2, Edexcel Core 2

Sometimes integrals can be just too hard to do using the normal methods — then you need to know other ways to solve them. That's where the Trapezium Rule comes in.

The **Trapezium Rule** is Used to Find the **Approximate Area** Under a Curve

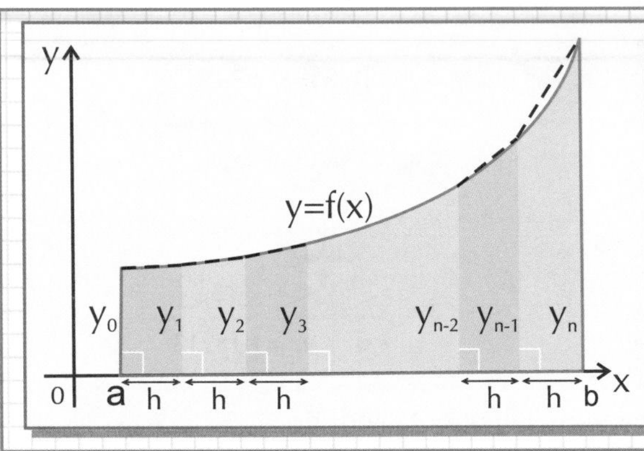

The area represented by $\int_a^b y\,dx$ is approximately:

$$\int_a^b y\,dx \approx \frac{h}{2}[y_0 + 2(y_1 + y_2 + \ldots + y_{n-1}) + y_n]$$

where **n** is the number of strips or intervals and **h** is the width of each strip.

You can find the width of each strip using $h = \frac{(b-a)}{n}$

$\mathbf{y_0, y_1, y_2, \ldots, y_n}$ are the heights of the sides of the trapeziums — you get these by putting the x-values into the curve.

So basically the formula for approximating $\int_a^b y\,dx$ works like this:

'Add the first and last heights $(y_0 + y_n)$ and add this to twice all the other heights added up — then multiply by $\frac{h}{2}$.'

Example: Find an approximate value to $\int_0^2 \sqrt{4-x^2}\,dx$ using 4 strips. Give your answer to 4 s.f.

Start by working out the width of each strip: $h = \frac{(b-a)}{n} = \frac{(2-0)}{4} = 0.5$

This means the x-values are $x_0 = 0$, $x_1 = 0.5$, $x_2 = 1$, $x_3 = 1.5$ and $x_4 = 2$ (the question specifies 4 strips, so n = 4).
Set up a table and work out the y-values or heights using the equation in the integral.

x	$y = \sqrt{4-x^2}$
$x_0 = 0$	$y_0 = \sqrt{4-0^2} = 2$
$x_1 = 0.5$	$y_1 = \sqrt{4-0.5^2} = \sqrt{3.75} = 1.936491673$
$x_2 = 1.0$	$y_2 = \sqrt{4-1.0^2} = \sqrt{3} = 1.732050808$
$x_3 = 1.5$	$y_3 = \sqrt{4-1.5^2} = \sqrt{1.75} = 1.322875656$
$x_4 = 2.0$	$y_4 = \sqrt{4-2.0^2} = 0$

Now put all the y-values into the formula with h and n:

$$\int_a^b y\,dx \approx \frac{0.5}{2}[2 + 2(1.936491673 + 1.732050808 + 1.322875656) + 0]$$
$$\approx 0.25[2 + 2 \times 4.991418137]$$
$$\approx 2.996 \text{ to 4 s.f.}$$

Watch out — if they ask you to work out a question with 5 x-values (or 'ordinates') then this is the same as 4 strips. The x-values usually go up in nice jumps — if they don't then check your calculations carefully.

The Approximation might be an **Overestimate** or an **Underestimate**

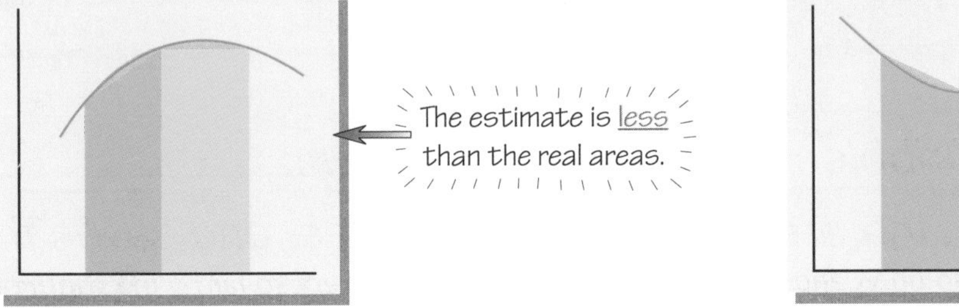

The estimate is less than the real areas.

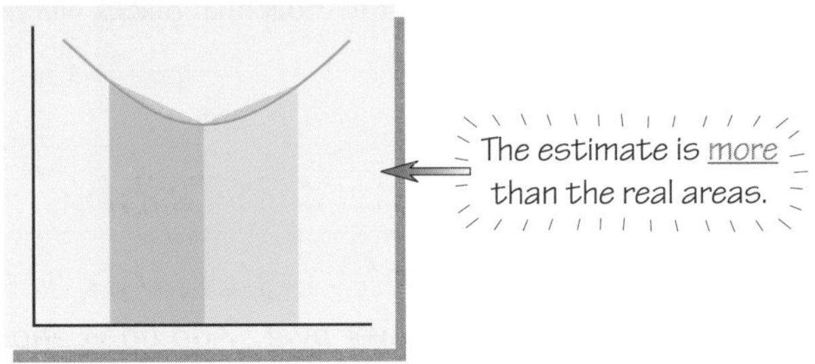

The estimate is more than the real areas.

Numerical Integration of Functions

This page is for AQA Core 2, OCR Core 2, Edexcel -Core 2

These are usually popular questions with examiners — as long as you're careful, there are plenty of marks to be had.

The Trapezium Rule is in the Formula Booklet

...so don't try any heroics — always look it up and use it with these questions.

Example: Use the trapezium rule with 6 ordinates to find an approximation to $\int_{3.5}^{4.5} 2\log_{10} x \, dx$

Remember, 6 ordinates means 5 strips — so n = 5.

Calculate the width of the strips: h = $\frac{(b-a)}{n} = \frac{(4.5-3.5)}{5} = 0.2$

Set up a table and work out the y-values using y = $2\log_{10}$ x:

x	$y = 2\log_{10} x$
$x_0 = 3.5$	$y_0 = 2\log_{10} 3.5 = 1.08814$
$x_1 = 3.7$	$y_1 = 2\log_{10} 3.7 = 1.13640$
$x_2 = 3.9$	$y_2 = 1.18213$
$x_3 = 4.1$	$y_3 = 1.22557$
$x_4 = 4.3$	$y_4 = 1.26694$
$x_5 = 4.5$	$y_5 = 1.30643$

$x_5 = 2\log_{10} b = 1.30643$

Putting all these values in the formula gives:

$$\int_a^b y\,dx \approx \frac{0.2}{2}[1.08814 + 2(1.13640 + 1.18213 + 1.22557 + 1.26694) + 1.30643]$$
$$\approx 0.1[1.08814 + (2 \times 4.81104) + 1.30643]$$
$$\approx 1.201665$$
$$\approx 1.202 \text{ to 3 d.p.}$$

Practice Questions

1) *Use the trapezium rule with n intervals to estimate:*

a) $\int_0^3 (9-x^2)^{\frac{1}{2}} dx$ with $n = 3$

b) $\int_{0.2}^{1.4} x^3 + 4\,dx;\ n = 6$

2) *Sample exam question:*

The following is a table of values for y = 2x² – 2x.

x	1.2	1.4	1.6	1.8	2
y	0.48	1.12	p	2.88	q

a) Find the values of p and q. [2 marks]

b) Use the trapezium rule and all the values of y in the completed table to obtain an estimate of *I* to 3 decimal places where $I = \int_{1.2}^{2} 2x^2 - 2x\,dx$. [4 marks]

Maths rhyming slang #3: Dribble and drool — Trapezium rule...

Take your time with Trapezium Rule questions — it's so easy to make a mistake with all those numbers flying around. Make a nice table showing all your ordinates (careful — this is always one more than the number of strips). Then add up y_1 to y_{n-1} and multiply the answer by 2. Add on y_0 and y_n. Finally, multiply what you've got so far by the width of a strip and divide by 2. It's a good idea to write down what you get after each stage, by the way — then if you press the wrong button (easily done) you'll be able to pick up from where you went wrong. They're not hard — just fiddly.

Exam Questions

Exam Questions

1 Find the x-intercepts of the curve that passes through the origin, and for which $\frac{dy}{dx} = 6(x^2 - 1)$.

(6 marks)

2 The diagram shows the curve with equation $y = x^3 - 10x^2 + 25x$ which is tangential to the x-axis at the point (5, 0).

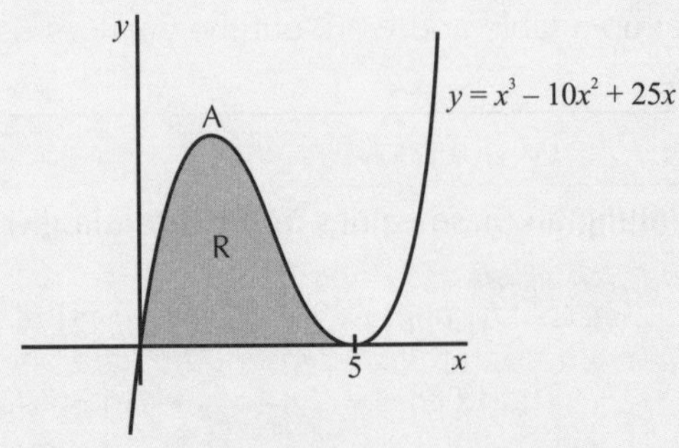

(a) Find the coordinates of the maximum turning point of the curve and prove that it is a maximum.

(6 marks)

(b) Find the area of the shaded region R, giving your answer as an exact fraction.

(5 marks)

3 The diagram shows the curve with equation $y = x(x + 3)$ and the line with equation $y = x + 8$.

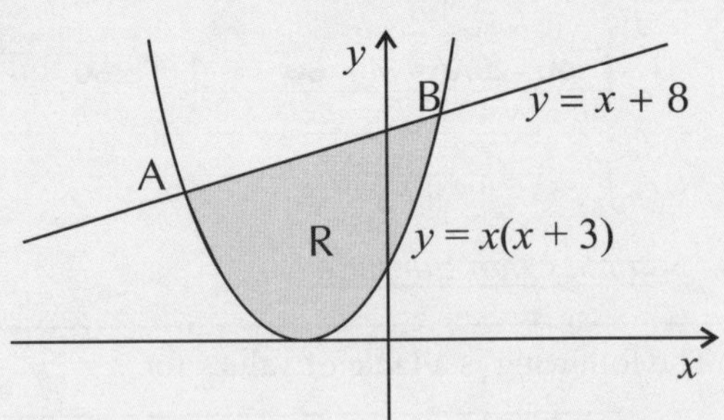

(a) Find the coordinates of points A and B, where the two graphs meet.

(3 marks)

(b) Find the area of the shaded region R.

(8 marks)

4 Use the trapezium rule with 4 intervals to estimate $\int_0^{\frac{\pi}{3}} \cos x \, dx$ to 3 d.p.

(6 marks)

Histograms

This page is for OCR S1, Edexcel S1

Histograms are glorified bar charts. The main difference is that you plot the frequency density (rather than the frequency). Frequency density is easy to find — you just divide the frequency by the width of a class.

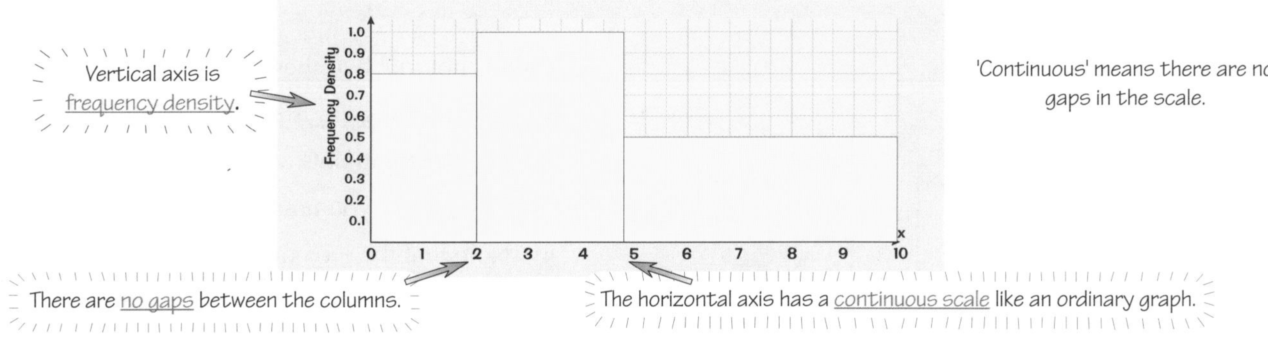
Vertical axis is frequency density.
Frequency Density
1.0
0.9
0.8
0.7
0.6
0.5
0.4
0.3
0.2
0.1
0 1 2 3 4 5 6 7 8 9 10
x
'Continuous' means there are no gaps in the scale.
There are no gaps between the columns.
The horizontal axis has a continuous scale like an ordinary graph.

To Draw a Histogram it's best to Draw a Table First

Getting histograms right depends on finding the right upper and lower bounds for each class.

Example:

Draw a histogram to represent the data below showing the masses of parcels (given to the nearest 100 g).

Mass of parcel (to nearest 100 g)	100 - 200	300 - 400	500 - 700	800 - 1100
Number of parcels	100	250	600	50

First draw a table showing the upper and lower class bounds, plus the frequency density:

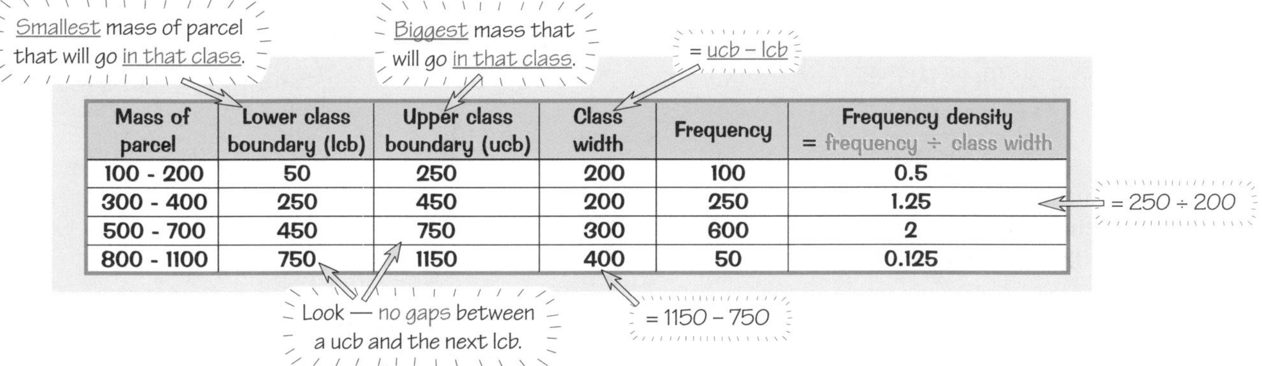

Mass of parcel	Lower class boundary (lcb)	Upper class boundary (ucb)	Class width	Frequency	Frequency density = frequency ÷ class width
100 - 200	50	250	200	100	0.5
300 - 400	250	450	200	250	1.25
500 - 700	450	750	300	600	2
800 - 1100	750	1150	400	50	0.125

Smallest mass of parcel that will go in that class.
Biggest mass that will go in that class.
= ucb – lcb
= 250 ÷ 200
Look — no gaps between a ucb and the next lcb.
= 1150 – 750

Now you can draw the histogram.

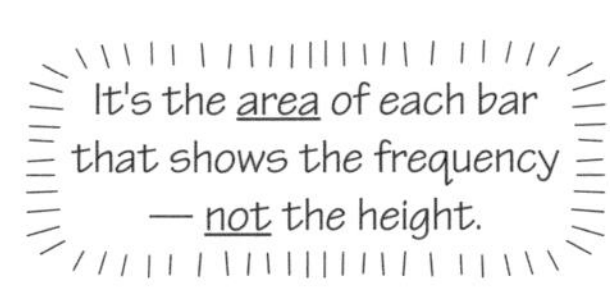
It's the area of each bar that shows the frequency — not the height.

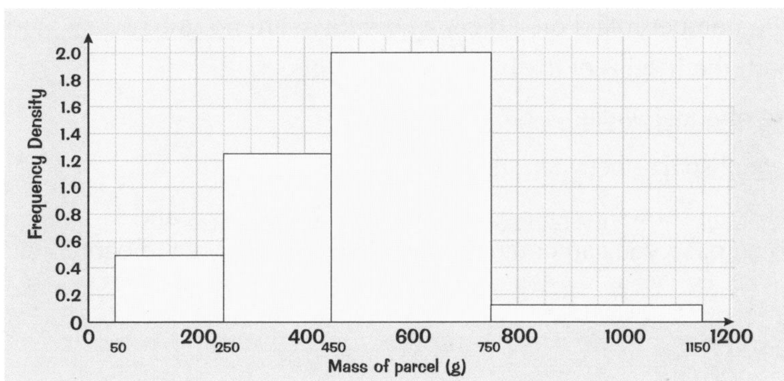
Frequency Density
2.0
1.8
1.6
1.4
1.2
1.0
0.8
0.6
0.4
0.2
0
0 50 200 250 400 450 600 750 800 1000 1150 1200
Mass of parcel (g)

Note: A class with a lower class boundary of 50 g and upper class boundary of 250 g can be written in different ways.

So you might see: "100 - 200 to nearest 100 g"
"50 ≤ mass < 250"
"50–", followed by "250–" for the next class and so on.

They all mean the same — just make sure you know how to spot the lower and upper class boundaries.

Presenting Data

This page is for OCR S1

Join Up the Histogram Bars to Make a Frequency Polygon

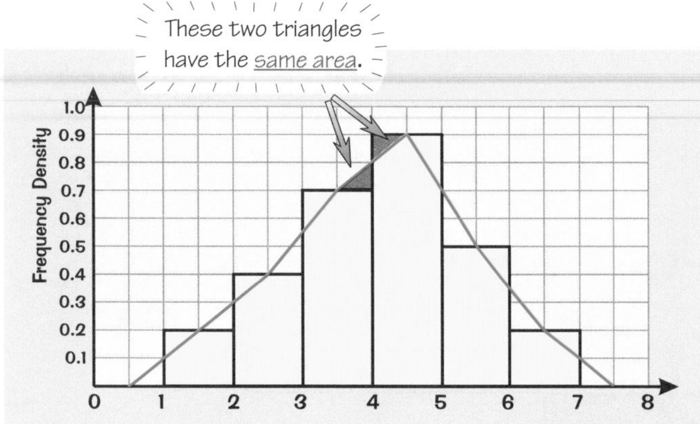

There's not much to these:

1) Take a histogram (see p.71), with bars of equal width.
2) Join the midpoints of the tops of the bars.
3) Hey presto, you now have a frequency polygon.
4) Note that the area under the polygon is the same as the area under the histogram — look at the triangle bits on the diagram.

In a Pie Chart, the Area Represents Frequency

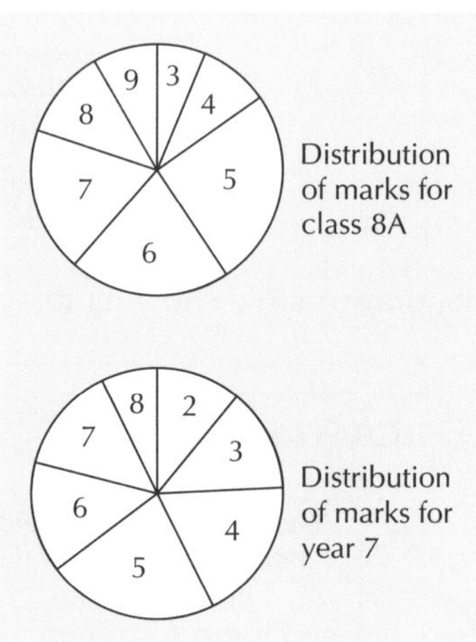

1) If you draw comparative pie charts correctly then their areas will be proportional to the frequency.
2) Look at the pie charts on the right. If you said, "more pupils in class 8A scored 6 marks than those in year 7", you could well be wrong — you don't know how many pupils there are in total in class 8A or year 7. All you can say is "a greater proportion of the pupils in class 8A scored 6 marks than in year 7."
3) Suppose class 8A contains 30 pupils and that year 7 contains 180 pupils. If the radius of the pie-chart for class 8A is 4 cm, then the radius of the year 7 pie chart should be 9.8 cm, as shown:

$$\frac{\text{Area for year 7}}{\text{Area for class}} = \frac{\pi r^2}{\pi 4^2} = \frac{180}{30}, \text{ hence } r = 9.8 \text{ cm}$$

4) When the pie charts are redrawn with these radii, they can now be compared one against the other. The area for 6 marks on the year 7 pie chart will actually be larger than that on the class 8A pie chart.

In a Bar Chart, the Heights of the Bars Represent Frequency

Bar charts are pretty straightforward. The main thing is that they're for discrete data — things like the number rolled on a dice, where there are no in-between values.

1) In a bar chart, the heights of the bars represent frequency.
2) The bars are of equal width.
3) There are equal spaces between the bars.
4) The vertical scale shows frequency. It should be continuous and go down to zero, so you can compare the relative heights of the bars.

Example:

A survey of eye colours gave the following results:

eye colour	blue	brown	green
frequency	2	7	9

...which turns into this bar chart...

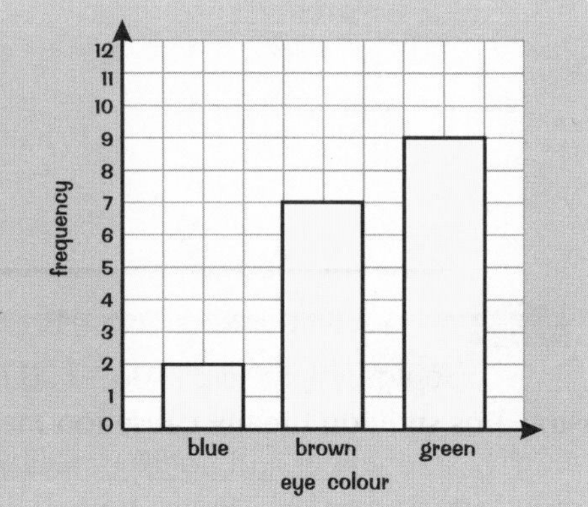

Stem and Leaf Diagrams

This page is for OCR S1, Edexcel S1

Stem and Leaf Diagrams *look nothing like stems or leaves*

They're just an easy way to represent your data. And they come in two flavours — plain or back-to-back.

Example: The lengths in metres of cars in a car park were measured to the nearest 10 cm. Draw a stem and leaf diagram to show the following data: 2.9, 3.5, 4.0, 2.8, 4.1, 3.7, 3.1, 3.6, 3.8, 3.7

It's best to do a rough version first, and then put the 'leaves' in order afterwards.

My 'stems' are the numbers before the decimal point, and my 'leaves' are the numbers after.

It's a good idea to cross out the numbers (in pencil) as you add them to your diagram.

```
2 | 9, 8
3 | 5, 7, 1, 6, 8, 7
4 | 0, 1
```

Digits after the decimal point — this row represents 4.0 m and 4.1 m.

Put the digits after the decimal point in order

```
2 | 8, 9
3 | 1, 5, 6, 7, 7, 8
4 | 0, 1
```

Key 2|9 means 2.9 m

Always give a key.

Example: The heights of boys and girls in a year 11 class are given to the nearest cm in a back-to-back stem and leaf diagram below. Write out the data in full.

First boy, 8|16|, has height 168 cm. The boys are read backwards.

Boys		Girls
	15	9
8	16	1, 5, 9
9, 8, 1	17	0, 2, 3, 5
5, 2	18	0
1	19	

First girl, |15|9, has height 159 cm.

Key 8|16|5 means Boys 168 cm and girls 165 cm

Boys: 168, 171, 178, 179, 182, 185, 191

Girls: 159, 161, 165, 169, 170, 172, 173, 175, 180

Practice Questions

1) *The stem and leaf diagram on the right represents the lengths (in cm) of 15 bananas. Write down the original data as a list.*

```
12 | 8
13 | 2, 5
14 | 3, 3, 6, 8
15 | 2, 9
16 | 1, 1, 2, 3
17 | 0, 2
```

Key 12|8 means 12.8 cm

2) *Construct a back-to-back stem and leaf diagram to represent the following data:*
Boys' test marks 34, 27, 15, 39, 20, 26, 32, 37, 19, 22
Girls' test marks 21, 38, 37, 12, 27, 28, 39, 29, 25, 24, 31, 36

3) *Twenty phone calls were made by a householder one evening. The lengths of the calls (in minutes to the nearest minute) are recorded below. Draw a histogram of the data.*

Length of call	**0 - 2**	**3 - 5**	**6 - 8**	**9 - 15**
Number of calls	**10**	**6**	**3**	**1**

Sample exam question:

4) The profits of 100 businesses are given in the table.

Profit, £x million.	Number of businesses
$4.5 \leqslant x < 5.0$	24
$5.0 \leqslant x < 5.5$	26
$5.5 \leqslant x < 6.0$	21
$6.0 \leqslant x < 6.5$	19
$6.5 \leqslant x < 8.0$	10

(a) Represent the data in a histogram. [3 marks]

(b) Comment on the distribution of the profits of the businesses. [2 marks]

First things first: remember — there are lies, damned lies and statistics...

Histograms shouldn't really cause too many problems — this is quite a friendly topic really. The main things to remember are to work out the lower and upper boundaries of each class properly, and then make sure you use frequency density (rather than just the frequency). Stem and leaf diagrams — hah, they're easy, I do them in my sleep. Make sure you can too.

Mean, Median, Mode and Range

This page is for AQA S1, OCR S1, Edexcel S1

The **Definitions** are really GCSE stuff

You more than likely already know these definitions. But if you don't, learn them now — you'll be needing them loads.

Mean = $\bar{x} = \frac{\sum x}{n}$ or $\frac{\sum fx}{n}$

The Σ (sigma) just means you add stuff up — so Σx means you add up all the values of x.

where each x is a data value, f is the frequency of each x (the number of times it occurs), and n is the total number of data values.

Median = middle data value when all the data values are placed in order of size.

Mode = most frequently occurring data value.

Range = highest value – lowest value

This will be the $\left(\frac{n+1}{2}\right)$th value in the ordered list.

Example: Find the mean, median, mode and range of the following data: 2, 3, 6, 2, 5, 9, 3, 8, 7, 2

Put in order first: 2, 2, 2, 3, 3, 5, 6, 7, 8, 9

Mean $= \frac{2+2+2+3+3+5+6+7+8+9}{10} = \underline{4.7}$

Median = average of 5th and 6th values = average of 3 and 5 = $\underline{4}$

Since $\frac{n+1}{2} = 5.5$

Mode = $\underline{2}$

Range = 9 – 2 = $\underline{7}$

Use a **Table** when there are a lot of **Numbers**

Example:

The number of letters received one day in 100 houses was recorded. Find the mean, median, mode and range of the number of letters.

Number of letters	Number of houses
0	11
1	25
2	27
3	21
4	9
5	7

The number of letters received by each house is measured in discrete quantities (e.g. 3 letters). There isn't a continuous set of possible values between getting 3 and 4 letters (e.g. 3.45 letters).

The first thing to do is make a table like this one:

Number of letters x	Number of houses f		fx
0	11	(11)	0
1	25	(36)	25
2	27	(63)	54
3	21		63
4	9		36
5	7		35
totals	100		213

Multiply x by f to get this column.

Put the running total in brackets — it's handy when you're finding the median. (But you can stop when you get past halfway.)

$n = \sum f = 100$

$\sum fx = 213$

① The mean is easy — just divide the total of the fx-column by the total of the f-column (= n).

Mean $= \frac{213}{100} =$ **2.13 letters**

② To find the position of the median, add 1 to the total frequency (= n) and then divide by 2.
Here the median is in position: (100 + 1) ÷ 2 = 50.5.

So the median is halfway between the 50th and 51st data values.

Using your running total of f, you can see that the data values in positions 37 to 63 are all 2s. This means the data values at positions 50 and 51 are both 2 — so: **Median = 2 letters**

③ The highest frequency is for 2 letters — so: **Mode = 2 letters**

④ Range = highest data value – lowest data value. So: **Range = 5 – 0 = 5 letters**

Mean, Median, Mode and Range

This page is for AQA S1, OCR S1, Edexcel S1

*If the data's **Grouped** you'll have to **Estimate***

If the data's grouped, you can only estimate the mean, median and mode.

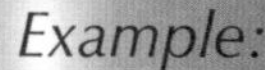

The height of a number of trees was recorded. The data collected is shown in this table:

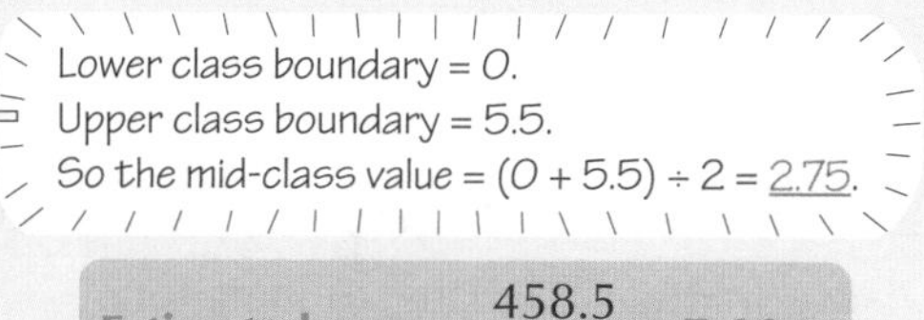

Height of tree to nearest m	0 - 5	6 - 10	11 - 15	16 - 20
Number of trees	26	17	11	6

Find an estimate of the mean height of the trees.

Here, you assume that every reading in a class takes the mid-class value (which you find by adding the lower class boundary to the upper class boundary and dividing by 2). It's best to make another table...

Height of tree to nearest m	mid-class value x	Number of trees f	fx
0 - 5	2.75	26 (26)	71.5
6 - 10	8	17 (43)	136
11 - 15	13	11	143
16 - 20	18	6	108
	Totals	60 (= n)	458.5(= Σfx)

Lower class boundary = 0.
Upper class boundary = 5.5.
So the mid-class value = (0 + 5.5) ÷ 2 = 2.75.

Estimated mean = $\frac{458.5}{60}$ = **7.64 m**

*Estimate the **Median** by assuming the values are **Evenly Spread***

The median position here is (60 + 1) ÷ 2 = 30.5, so the median is the 30.5th reading (halfway between the 30th and 31st). Your 'running total' tells you the median must be in the '6 - 10' class.
Now you have to assume that all the readings in this class are evenly spread.

There are 26 trees before class 6 - 10, so the 30.5th tree is the 4.5th value of this class.

Divide the class into 17 equally wide parts (as there are 17 readings) and assume there's a reading at each new point.

Then you want the reading that's 4.5 parts along.

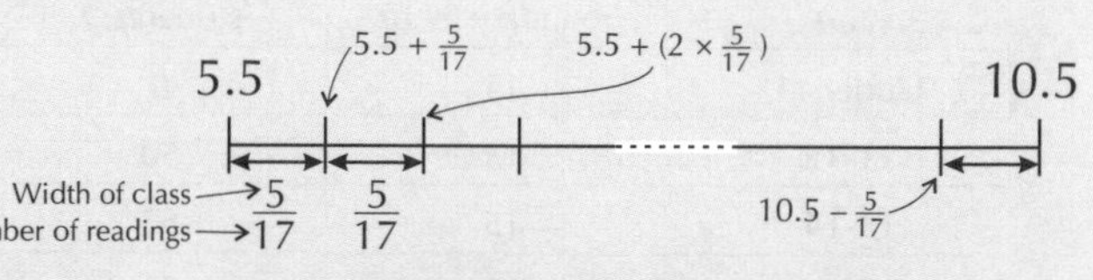

So the median = lower class boundary + (4.5 × width of each 'bit')

$= 5.5 + \left(4.5 \times \frac{5}{17}\right) = 6.8 \text{ m}$

The modal class is the class with most readings in it. In this example the modal class is 0 - 5 m.

Practice Questions

1) *Calculate the mean, median and mode of the data in the table on the right.*

x	0	1	2	3	4
f	5	4	4	2	1

2) *The speeds of 60 cars travelling in a 40 mph speed limit area were measured to the nearest mph. The data is summarised in the table. Calculate estimates of the mean and median, and state the modal class.*

Speed (mph)	30 - 34	35 - 39	40 - 44	45 - 50
Frequency	12	37	9	2

Sample Exam Question:

3) The stem and leaf diagram shows the test marks for 30 male students and 16 female students.

(a) Find the median test mark of the male students. [1 mark]

(b) Compare the distribution of the male and female marks. [2 marks]

Male students		Female students
8, 3, 3	4	
8, 7, 7, 7, 5, 3, 2	5	5, 6, 7
9, 7, 6, 6, 5, 5, 2, 2, 1, 1, 0	6	1, 2, 3, 3, 4, 5, 6, 7, 9
9, 9, 8, 5, 4, 3, 1, 0, 0	7	2, 4, 8, 9

Key 5|6|2 means Male student test mark 65 and Female student test mark 62

I can't deny it — this page really is kind of average...

Doing all this stuff isn't that hard — it's remembering all the different names that gives me a headache. But it's all made easier if you learn that the MeDian is the one in the MiDdle, while the MOde is the one that there's MOst of. The mean, well, that's just your common or garden 'average' that you learnt about while you were still in short trousers.

Cumulative Frequency Diagrams

This page is for AQA S1, OCR S1, Edexcel S1

Quartiles divide the data into Four

The median divides the data into two — the quartiles divide the data into four.

Example: Find the median and quartiles of the following data: 2, 5, 3, 11, 6, 7, 1

First put the list in order: 1 2 3 5 6 7 11

There are 7 numbers, so the median is in position 4 (i.e. you take the fourth number along): Median = 5

The middle of the set of numbers below the median is the lower quartile (Q_1) and the middle of the set above the median is the upper quartile (Q_3).

1 2 3 5 6 7 11 (Q_1 = 2, Q_3 = 7)

About 25% of the readings are less than the lower quartile. About 75% are less than the upper quartile.

Lower quartile = 2 Upper quartile = 7

Use Cumulative Frequency Graphs to find the Median and Quartiles

Cumulative frequency means 'running total'. Cumulative frequency diagrams make medians and quartiles easy to find...

Example: The ages of 200 students in a school are recorded in the table below.

Draw a cumulative frequency graph and use it to estimate the median age and the interquartile range. Also estimate how many students are older than 18.

Age in completed years	11 - 12	13 - 14	15 - 16	17 - 18
Number of students	50	65	58	27

(1) First draw a table showing the upper class boundaries and the cumulative frequency:

Age in completed years	Upper class boundary (ucb)	Number of students, f	Cumulative frequency (cf)
Under 11	11	0	0
11-12	13	50	50
13-14	15	65	115
15-16	17	58	173
17-18	19	27	200

The first reading in a cumulative frequency table must be zero — so add this extra row to show the number of students with age less than 11 is 0.

CF is the number of students with age less than the ucb — it's the same thing as your running total from the last two pages.

The last number in the CF column should always be the total number of readings.

People say they're '18' right up until their 19th birthday — so the ucb of class 17-18 is 19.

Next draw the axes — cumulative frequency always goes on the vertical axis. Here, age goes on the other axis. Then plot the upper class boundaries against the cumulative frequencies, and join the points.

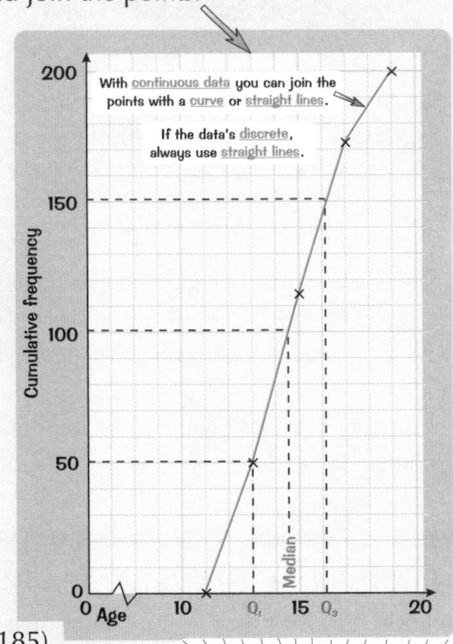

Always plot the upper class boundary of each class.

(2) To find the median from a graph, go to the median position on the vertical scale and read off the value from the horizontal axis.

Median position = $\frac{1}{2}(200+1) = 100.5$ so Median = 14.5 years

Then you can find the **quartiles** in the same way. Find their positions first:

Q_1 position $= \frac{1}{4} \times (200+1) = 50.25$ (i.e. between the 50th and 51st readings)

Q_3 position $= \frac{3}{4} \times (200+1) = 150.75$ (i.e. between the 150th and 151st readings)

Lower quartile, Q_1 = 13 years Upper quartile, Q_3 = 16.2 years

The **interquartile range** (IQR) = $Q_3 - Q_1$. It measures spread. The smaller it is the less spread the data is.

IQR = $Q_3 - Q_1$ = 16.2 – 13 = 3.2 years

(3) To find how many students are older than 18, first go up from 18 on the horizontal axis, and read off the number of students younger than 18 (= 185).

Then the number of students older than 18 is just 200 – 185 = 15 (approximately)

Cumulative Frequency Diagrams

This page is for AQA S1, OCR S1, Edexcel S1

Percentiles divide the data into 100

Percentiles divide the data into 100 — the median is the 50th percentile and Q_1 is the 25th percentile, etc.

Example:

The position of the 11th percentile (P_{11}) is $\frac{11}{100} \times$ (total frequency +1) = $\frac{11}{100} \times 201 = 22.11$

Or by going $\frac{11}{100}$ up the graph, you can see the 11th percentile is about 12.

You find interpercentile ranges by subtracting two percentiles, e.g. the middle 60% of the readings = $P_{80} - P_{20}$.

Box and Whisker Diagrams are useful for comparing distributions

Box and whisker plots show the median and quartiles in an easy-to-look-at kind of way...
These are sometimes called box plots.

They look like this:

ALWAYS DRAW A SCALE

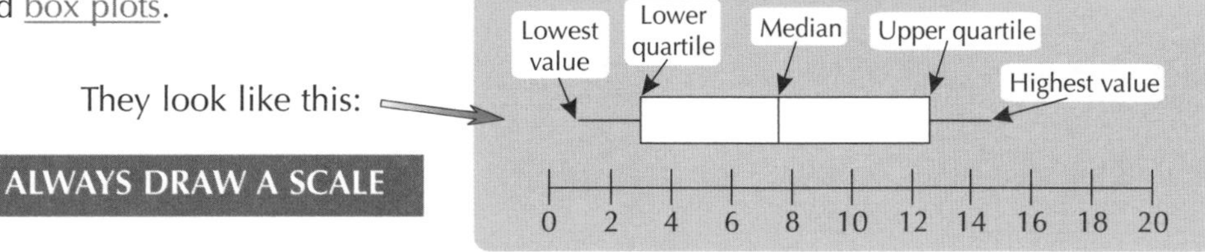

Practice Questions

1) *Draw a cumulative frequency diagram of the data given in the table. Use your diagram to estimate the median and interquartile range.*

Distance walked (km)	0 - 2	2 - 4	4 - 6	6 - 8
Number of walkers	10	5	3	2

Sample Exam questions:

2) A shopkeeper records the age, to the nearest year, of the customers that enter his shop before 9:00 am one morning.

Age of customer to nearest year	5 - 10	11 - 15	16 - 20	21 - 30	31 - 40	41 - 70
Number of customers	2	3	10	2	2	1

(a) On graph paper, draw a cumulative frequency diagram. [4 marks]

(b) From your graph estimate

(i) the median age of the customers. [1 mark]

(ii) the number of customers at least 12 years old. [2 marks]

3) Two workers iron clothes. Each irons 10 items, and records the time it takes them for each, to the nearest minute.

Worker A: 3 5 2 7 10 4 5 5 4 12

Worker B: 3 4 8 6 7 8 9 10 11 9

(a) For worker A's times. Find:

(i) the median, [1 mark]

(ii) the lower and upper quartiles. [2 marks]

(b) On graph paper draw, using the same scale, two box plots to represent the times of each worker. [6 marks]

(c) Make one statement comparing the two sets of data. [1 mark]

(d) Which worker would be best to employ? Give a reason for you answer. [1 mark]

WG Grace — an old-time box and whisker man...

Cumulative frequency sounds a bit scarier than running total — but if you remember that they're the same thing, that'll help. You might get a question on quartiles... to find these, you have to find the median of the bottom half and the median of the top half of the data — not including the middle value in either half if you have an odd number of values.

Variance and Standard Deviation

This page is for AQA S1, OCR S1, Edexcel S1

Standard deviation and variance both measure how spread out the data is from the mean — the bigger the variance, the more spread out your readings are.

The **Formulas** look pretty **Tricky**

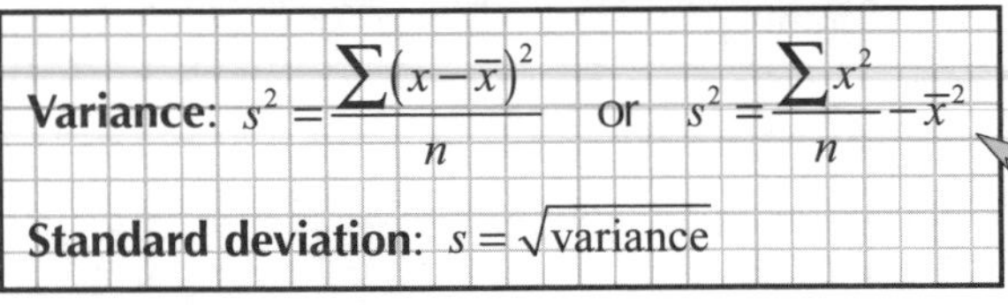

Variance: $s^2 = \frac{\sum(x-\bar{x})^2}{n}$ or $s^2 = \frac{\sum x^2}{n} - \bar{x}^2$

Standard deviation: $s = \sqrt{\text{variance}}$

The x-values are the data, $\bar{x}$ is the mean, and n is the number of data values.

The second formula is easier to use.

Example: Find the mean and standard deviation of the following numbers: 2, 3, 4, 4, 6, 11, 12

1) Find the total of the numbers first: $\sum x = 2+3+4+4+6+11+12 = 42$
2) Then the mean is easy: $\text{Mean} = \bar{x} = \frac{\sum x}{n} = \frac{42}{7} = 6$
3) Next find the sum of the squares: $\sum x^2 = 4+9+16+16+36+121+144 = 346$
4) Use this to find the variance: $\text{Variance, } s^2 = \frac{\sum x^2}{n} - \bar{x}^2 = \frac{346}{7} - 6^2 = 49.43 - 36 = 13.43$
5) And take the square root to find the standard deviation: $\text{Standard deviation} = \sqrt{13.43} = 3.66$ to 3 sig. fig.

Questions about **Standard Deviation** can look a bit **Weird**

They can ask questions about standard deviation in different ways. But you just need to use the same old formulas.

Example:

The mean of 10 boys' heights is 180 cm, and the standard deviation is 10 cm. The mean for 9 girls is 165 cm, and the standard deviation is 8 cm. Find the mean and standard deviation of the whole group of 19 girls and boys.

(1) Let the boys' heights be x and the girls' heights be y.

Write down the formula for the mean and put the numbers in for the boys: $\bar{x} = \frac{\sum x}{n} \Rightarrow 180 = \frac{\sum x}{10} \Rightarrow \sum x = 1800$

Do the same for the girls: $165 = \frac{\sum y}{9} \Rightarrow \sum y = 1485$

So the sum of the heights for the boys and the girls = $\sum x + \sum y = 1800 + 1485 = 3285$

And the mean height of the boys and the girls is: $\frac{3285}{19} = 172.9 \text{ cm}$

Round the fraction to 1 dp to give your answer. But if you need to use the mean in more calculations, use the fraction (or your calculator's memory) so you don't lose accuracy.

(2) Now for the variance — write down the formula for the boys first: $s_x^2 = \frac{\sum x^2}{n} - \bar{x}^2 \Rightarrow 10^2 = \frac{\sum x^2}{10} - 180^2 \Rightarrow \sum x^2 = 10 \times (100 + 32400) = 325000$

Do the same for the girls: $s_y^2 = \frac{\sum y^2}{n} - \bar{y}^2 \Rightarrow 8^2 = \frac{\sum y^2}{9} - 165^2 \Rightarrow \sum y^2 = 9 \times (64 + 27225) = 245601$

Okay, so the sum of the squares of the heights of the boys and the girls is: $\sum x^2 + \sum y^2 = 325000 + 245601 = 570601$

Which means the variance of all the heights is: $s^2 = \frac{570601}{19} - \left(\frac{3285}{19}\right)^2 = 139.0 \text{ cm}^2$

Don't use the rounded mean (172.9) — you'll lose accuracy.

And finally the standard deviation of the boys and the girls is: $s = \sqrt{139.0} = 11.8 \text{ cm}$

Phew.

Variance and Standard Deviation

This page is for AQA S1, OCR S1, Edexcel S1

Use Mid-Class Values if your data's in a Table

With grouped data, assume every reading takes the mid-class value. Then use the frequencies to find $\sum x$ and $\sum x^2$.

Example: The heights of sunflowers in a garden were measured and recorded in the table below. Estimate the mean height and the standard deviation.

Height of sunflower	$150 \le x < 170$	$170 \le x < 190$	$190 \le x < 210$	$210 \le x < 230$
Number of sunflowers	5	10	12	3

Draw up another table, and include columns for the mid-class values x, as well as fx and fx^2:

Height of sunflower	Mid-class value, x	x^2	f	fx	fx^2
$150 \le x < 170$	160	25600	5	800	128000
$170 \le x < 190$	180	32400	10	1800	324000
$190 \le x < 210$	200	40000	12	2400	480000
$210 \le x < 230$	220	48400	3	660	145200
		Totals	30 ($= n$)	5660 ($= \Sigma x$)	1077200 ($= \Sigma x^2$)

fx^2 means $f \times (x^2)$ — not $(fx)^2$.

Now you've got the totals in the table, you can calculate the mean and variance:

$$\text{Mean} = \bar{x} = \frac{\sum x}{n} = \frac{5660}{30} = 189 \text{ to 3 sig. fig.}$$

$$\text{Variance} = s^2 = \frac{\sum x^2}{n} - \bar{x}^2 = \frac{1077200}{30} - \left(\frac{5660}{30}\right)^2 = 312 \text{ to 3 sig. fig.}$$

$$\text{Standard deviation} = \sqrt{312} = 17.7 \text{ to 3 sig. fig.}$$

Practice Questions

1) *Find the mean and standard deviation of the following numbers: 11, 12, 14, 17, 21, 23, 27.*

2) *The scores in an IQ test for 50 people are recorded in the table below.*

Score	100 - 106	107 - 113	114 - 120	121 - 127	128 - 134
Frequency	6	11	22	9	2

Calculate the mean and variance of the distribution.

Sample Exam question:

3) In a supermarket two types of chocolate drops were compared.
The weights (in grams) of 20 chocolate drops of brand A are summarised by:

$$\sum A = 60.3 \text{ g} \qquad \sum A^2 = 219 \text{ g}^2$$

The mean weight of 30 chocolate drops of brand B was 2.95 g, and the standard deviation was 1 g.

(a) Find the mean weight of a brand A chocolate drop. [1 mark]

(b) Find the standard deviation of the weight of the brand A chocolate drops. [3 marks]

(c) Compare brands A and B. [2 marks]

(d) Find the standard deviation of the weight of all 50 chocolate drops. [4 marks]

People who enjoy this stuff are standard deviants...

The formula for the variance looks pretty scary, what with the S's and $\bar{x}$'s floating about. But it comes down to 'the mean of the squares minus the square of the mean'. That's how I remember it anyway — and my memory's rubbish.
Ooh, while I remember... don't forget to work out mid-class values carefully, using the upper and lower class boundaries.

Coding

This page is for AQA S1, OCR S1, Edexcel S1

Coding means doing something to every reading (like adding a number, or multiplying by a number) to make life easier.

Coding can make the Numbers much Easier

Finding the mean of 1001, 1002 and 1006 looks hard(ish). But take 1000 off each number and finding the mean of what's left (1, 2 and 6) is much easier — it's 3. So the mean of the original numbers must be 1003. That's coding.

You usually change your original variable x to an easier one to work with y (so here, if $x = 1001$, then $y = 1$.)

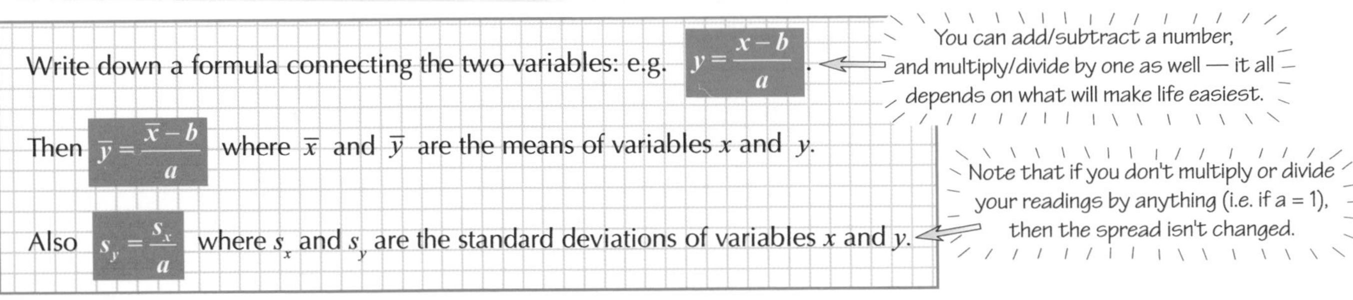

Write down a formula connecting the two variables: e.g. $y = \frac{x-b}{a}$.
You can add/subtract a number, and multiply/divide by one as well — it all depends on what will make life easiest.
Then $\bar{y} = \frac{\bar{x}-b}{a}$ where $\bar{x}$ and $\bar{y}$ are the means of variables x and y.
Also $s_y = \frac{s_x}{a}$ where s_x and s_y are the standard deviations of variables x and y.
Note that if you don't multiply or divide your readings by anything (i.e. if a = 1), then the spread isn't changed.

Example: Find the mean and standard deviation of: 1 000 020, 1 000 040, 1 000 010 and 1 000 050.

The obvious thing to do is subtract a million from every reading to leave 20, 40, 10 and 50.
Then make life even simpler by dividing by 10 — giving 2, 4, 1 and 5.

(1) So use the coding: $y = \frac{x - 1000000}{10}$. Then $\bar{y} = \frac{\bar{x} - 1000000}{10}$ and $s_y = \frac{s_x}{10}$.

(2) Find the mean and standard deviation of the y values: $\bar{y} = \frac{2+4+1+5}{4} = \underline{3}$ $\quad s_y = \frac{2^2+4^2+1^1+5^2}{4} - 3^2 = \frac{46}{4} - 9 = \underline{2.5}$

(3) Then use the formulas to find the mean and standard deviation of the original values:

$\bar{x} = 10\bar{y} + 1000000 = (10 \times 3) + 1000000 = \underline{1000030}$ $\quad s_x = 10 s_y = 10 \times 2.5 = \underline{25}$

You can use coding with Summarised Data

This kind of question looks tricky at first — but use the same old formulas and it's a piece of cake.

Example: A set of 10 numbers, x, can be summarised as shown: $\sum(x-10) = 15$ and $\sum(x-10)^2 = 100$
Find the mean and standard deviation of x.

(1) Okay, the obvious first thing to try is: $y = x - 10$

That means: $\sum y = 15$ and $\sum y^2 = 100$

(2) Work out $\bar{y}$ and s_y^2 using the normal formulas: $\bar{y} = \frac{\sum y}{n} = \frac{15}{10} = 1.5$

$s_y^2 = \frac{\sum y^2}{10} - \bar{y}^2 = \frac{100}{10} - 1.5^2 = 10 - 2.25 = 7.75$
so $s_y = 2.78$ to 3 sig. fig.

(3) Then finding the mean and standard deviation of the x-values is easy: $\bar{x} = \bar{y} + 10 = 1.5 + 10 = \underline{11.5}$

The spread of x is the same as the spread of y since you've only subtracted 10 from every number. $\rightarrow$ $s_x = s_y = \underline{2.78}$ to 3 sig. fig.

Coding

This page is for AQA S1, OCR S1, Edexcel S1

Sensible Coding can make life Much Easier

Find the mean and standard deviation of the data in this table:

Class	10 - 19	20 - 29	30 - 39
f	2	5	3

It's grouped data, so use mid-class values — these are $x = 14.5$, 24.5 and 34.5.

Now let $y = \frac{x - 24.5}{10}$ and then draw up another table:

This coding will make all the numbers in the table dead easy.

Class	Mid-class x	$y = \frac{x-24.5}{10}$	f	fy	fy^2
10 - 19	14.5	-1	2	-2	2
20 - 29	24.5	0	5	0	0
30 - 39	34.5	1	3	3	3
		Totals	10 (= n)	1 (= Σy)	5 (= Σy^2)

$\bar{y} = \frac{\sum y}{n} = \frac{1}{10} = 0.1$ and $s_y^2 = \frac{\sum y^2}{10} - \bar{y}^2 = \frac{5}{10} - 0.1^2 = 0.5 - 0.01 = 0.49$

So $s_y = 0.7$

But $y = \frac{x - 24.5}{10}$ so $\bar{y} = \frac{\bar{x} - 24.5}{10}$, which means $\bar{x} = 10\bar{y} + 24.5 = (10 \times 0.1) + 24.5 = 25.5$

Since everything has been divided by 10, the spread of y is not the same as the spread of x.

In fact, $s_y = \frac{s_x}{10}$ so $s_x = 10s_y = 7$

Practice Questions

1) *For a set of data, n = 100,* $\sum(x-20) = 125$, *and* $\sum(x-20)^2 = 221$.
   Find the mean and standard deviation of x.

2) *The time taken (to the nearest minute) for a commuter to travel to work on 20 consecutive days is recorded in the table. Use coding to find the mean and standard deviation of the times.*

Time to nearest minute	**30 - 33**	**34 - 37**	**38 - 41**	**42 - 45**
Frequency	**3**	**6**	**7**	**4**

Sample Exam question:

3) A group of 19 people played a game. The scores, x, that the people achieved are summarised by:

$$\sum(x-30) = 228 \text{ and } \sum(x-30)^2 = 3040$$

(a) Calculate the mean and the standard deviation of the 19 scores. [3 marks]

(b) Show that $\sum x = 798$ and $\sum x^2 = 33820$. [3 marks]

(c) Another student played the game. Her score was 32.
Find the new mean and standard deviation of all 20 scores. [4 marks]

I thought the coding page would be a little more... well, James Bond...

Coding data isn't hard — the only tricky thing can be to work out how best to code it, although there will usually be some pretty hefty clues in the question if you care to look. But remember that adding/subtracting a number from every reading won't change the spread (the variance or standard deviation), but multiplying/dividing readings by something will.

Skewness and Outliers

This page is for Edexcel S1

Skewness tells you whether your data is symmetrical — or kind of lopsided.

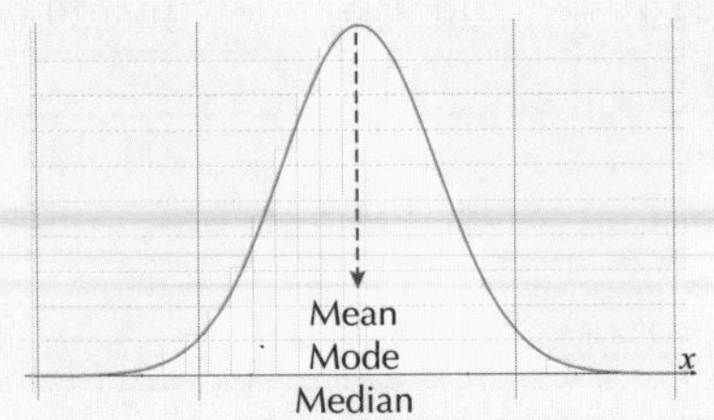

This is a typical symmetrical distribution.

Notice: mean = median = mode

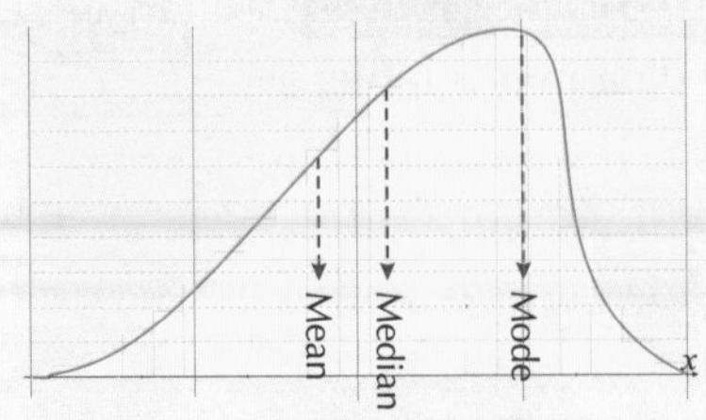

A negatively skewed distribution has a tail on the left. Most data values are on the higher side.

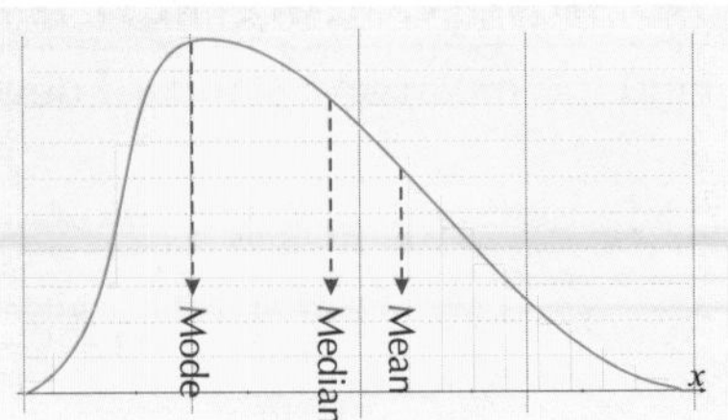

A positively skewed distribution has a tail on the right. Most data values are on the lower side.

For all distributions: **mean – mode = 3 × (mean – median)** — approximately.

Measure skewness using **Pearson's Coefficient of Skewness...**

A coefficient of skewness measures how 'all-up-one-end' your data is. You need to know a couple of formulas...

$$\text{Pearson's coefficient of skewness} = \frac{\text{mean} - \text{mode}}{\text{standard deviation}} = \frac{3(\text{mean} - \text{median})}{\text{standard deviation}}$$

This usually lies between -3 and +3.
So if Pearson's coefficient of skewness is -0.1, then the distribution is slightly negatively skewed.

...or the **Quartile Coefficient of Skewness**

Remember that Q_1 is the lower quartile, Q_3 is the upper quartile, and the median is Q_2.

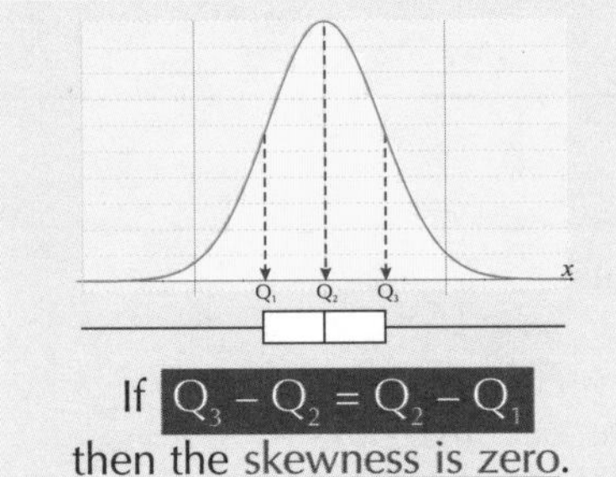

If $Q_3 - Q_2 = Q_2 - Q_1$ then the skewness is zero.

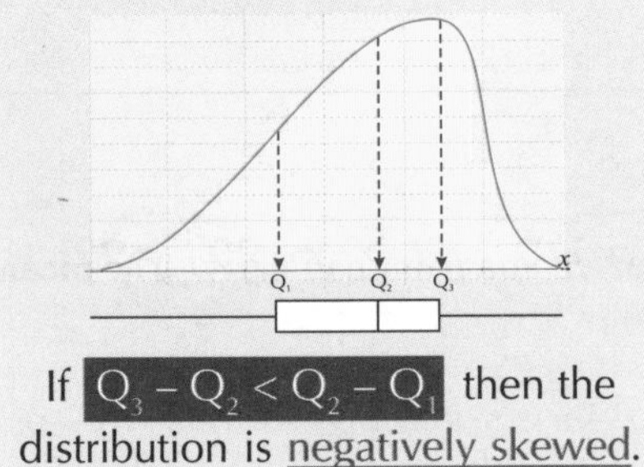

If $Q_3 - Q_2 < Q_2 - Q_1$ then the distribution is negatively skewed.

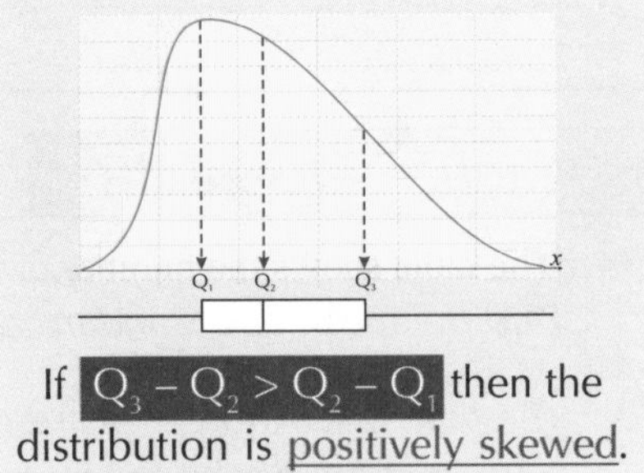

If $Q_3 - Q_2 > Q_2 - Q_1$ then the distribution is positively skewed.

$$\text{Quartile coefficient of skewness} = \frac{(Q_3 - Q_2) - (Q_2 - Q_1)}{Q_3 - Q_1} = \frac{Q_3 - 2Q_2 + Q_1}{Q_3 - Q_1}$$

Example: This table summarises the marks obtained in Maths 'calculator' and 'non-calculator' papers.

Calculate the Pearson's and Quartile coefficient of skewness for each paper. Comment on the distributions.

Calculator Paper		Non-calculator paper
40	Lower quartile, Q_1	35
58	Median, Q_2	42
70	Upper quartile, Q_3	56
55	Mean	46.1
21.2	Standard deviation	17.8

Calculator Paper		Non-calculator Paper
$\frac{3\times(55.0-58)}{21.2} = \frac{-9}{21.2} = -0.425$	Pearson's coefficient of skewness	$\frac{3\times(46.1-42)}{17.8} = \frac{12.3}{17.8} = 0.691$
$\frac{70-2\times58+40}{70-40} = \frac{-6}{30} = -0.2$	Quartile coefficient of skewness	$\frac{56-2\times42+35}{56-35} = \frac{7}{21} = 0.333$

The quartile coefficient of skewness tells you that the calculator paper scores are slightly negatively skewed and that the non-calculator paper scores are positively skewed.

Generally students have done better (compared to the mean) on the calculator paper.
Pearson's coefficient of skewness confirms these results.

Skewness and Outliers

This page is for Edexcel S1

An outlier is a freak piece of data that lies a long way from the rest of the readings.
To find whether a reading is an outlier you have to measure how far away from the rest of the data it is.

Outliers fall Outside Fences

There are various ways to decide if a reading is an outlier — the method you should use is always described in the question.

Example: A data value is considered to be an outlier if it is more than 3 times the IQR above the upper quartile or more than 3 times the IQR below the lower quartile.

The lower and upper quartiles of a data set are 70 and 100. Decide whether the data values 20 and 210 are outliers.

First you need the IQR: $Q_3 - Q_1 = 100 - 70 = 30$

Then it's a piece of cake to find where your fences are.
Lower fence first: $Q_1 - (3 \times \text{IQR}) = 70 - (3 \times 30) = -20$

And the upper fence: $Q_3 + (3 \times \text{IQR}) = 100 + (3 \times 30) = 190$

-20 and 190 are called fences. Any reading lying outside the fences is considered an outlier.

20 is inside the lower fence, so it is not an outlier. 210 is outside the upper fence, so it is an outlier.

Practice Questions

1) *A data value is considered an outlier if it's more than 3 times the IQR above the upper quartile or more than 3 times the IQR below the lower quartile. If the lower and upper quartiles of a data set are 62 and 88, decide which of the following data are outliers: a) 161, b) 176, c) 0*

2) *Find the median and quartiles of the data below. Draw a box and whisker diagram, and comment on any skewness.*
Amount of pocket money (in £) received per week by twenty 15-year-olds:
10, 5, 20, 50, 5, 1, 6, 5, 15, 20, 5, 7, 5, 10, 12, 4, 8, 6, 7, 30.

3) *A set of data has a mean of 10.3, a mode of 10 and a standard deviation of 1.5.*
Calculate Pearson's coefficient of skewness, and draw a possible sketch of the distribution.

Practice Exam Questions:

4) The table shows the number of hits received at a paint ball party. The mean is 16.4 hits.

Age	12	13	14	15	16	17	18	19	20	21	22	23	24	25
Frequency	2	4	6	7	6	4	4	2	1	1	0	0	0	1

(a) Find the median and mode number of hits. [3 marks]

(b) An outlier is a data value which is greater than $3(Q_3 - Q_1)$ above Q_3 or below Q_1.
Is 25 an outlier? Show your working. [2 marks]

(c) Sketch the shape of the distribution and comment on any skewness. [2 marks]

(d) How would the shape of the distribution be affected if the value of 25 was removed? [1 mark]

5) The data in the table shows the number of mm of rain that fell on 30 days on a tropical island.

mm of rain	5 - 10	10 - 15	15 - 20	20 - 25	25 - 30	30 - 35
No. of days	2	3	5	7	10	3

(a) Draw a cumulative frequency diagram of the data. [3 marks]

(b) Using your diagram estimate the median and quartiles. [3 marks]

(c) Calculate the quartile coefficient of skewness and describe the shape. [2 marks]

'Outlier' is the name I give to something that my theory can't explain...

Those definitions of positive and negative skew aren't the most obvious in the world — and it's easy to get them mixed up. Remember that negative skew involves a tail on the left, which means that a lot of your readings are on the high side. Positive skew is the opposite — a tail on the right, and a bunch of readings that are a little on the low side.

Populations and Sampling

This page is for AQA S1

In a random sample each item from the population must have an equal chance of being selected.

In a **Random Sample** *everything has an* **Equal Chance** *of being picked*

1) With simple random sampling, every person or thing in a population has an equal chance of being in the sample.
2) To get a truly random sample, you need a complete list of the population, and it must include every last person or thing — this isn't always easy to get.
3) To choose the sample, give every member of the population a number and then use a calculator or random number tables to pick the ones to include in your sample.

Example: Use the random number table below to select a sample of 3 people from a population of 80.

8330	3992	1840	0330	1290	3237	9165	4815	0766
2508	9927	6948	8532	1646	1931	8502	8636	2296
9310	0572	1826	3667	6848	3169	6858	9349	4586

1) Give each person in the population a two-digit number from 01 to 80.
2) Roll a dice (or choose some other method) to find a place to start in the random number tables. So if you roll a three, start at the 3rd digit.
3) The first two-digit number is 30, so include item 30 in your sample. The next is 39, so include item 39, and so on.
4) The next two-digit number is 92, but this is no good because it's bigger than 80. So forget 92 and use the next number instead, which is 18.

So the sample of three would be the 30th, 39th and 18th items.

The Ran# button on your calculator can be used in a similar way. Read the instructions to find out more.

You can sample **With Replacement** *or* **Without Replacement**

SAMPLING WITH REPLACEMENT is when an item may be chosen for a sample more than once.
SAMPLING WITHOUT REPLACEMENT is when an item may not be chosen more than once.

Example: Find a sample of size 6 from the population of 80 above: a) with replacement, b) without replacement.

Starting at the 3rd digit, the first few numbers would be: 30, 39, 92, 18, 40, 03, 30, 12, 90, 32...
You have to get rid of 92 and 90, making your sample list: 30, 39, 18, 40, 03, 30, 12, 32...

a) If you're sampling with replacement then the sample would be: 30, 39, 18, 40, 03, 30
So choose the items with these numbers. Note that the 30th item has been chosen twice.
b) If you're sampling without replacement then you can't use the 30th item twice.
You'd have to get rid of the second 30. The final sample would be: 30, 39, 18, 40, 03, 12

Sampling without replacement leads to results that are more representative of the population as a whole, i.e. sampling without replacement is more precise than sampling with replacement.

You can **Reallocate** *numbers rather than* **Getting Rid** *of them*

This method avoids having to get rid of numbers greater than the size of your population (like in the example above, when we got rid of 90 and 92, since the population only had 80 things in it).

Example: Choose a sample, without replacement, from a population of 100.

1) This time you could give each member of the population a three-digit number — from 001 to 100.
2) Instead of disregarding numbers greater than 100, you could use the following rules:
   001-100 leave as they are,
   101-200 subtract 100,
   210-300 subtract 200 etc.,
3) So 239 would become 039, and 184 would become 084, etc.

Populations and Sampling

This page is for AQA S1

Sampling is used to find out about a Population

Sampling is used to estimate things about a group of people or items by just looking at a few of them.
The group of people or items you wish to know about is called the population.

This could be: all the students at a university,
all the trees in a forest,
all the bags of sugar produced by a particular company.

If you want to find out about a population, then you could question every person or examine every item.
However, this would usually take too much time and effort, so instead you look at a sample of the population.

Sampling is the process of picking the people or things to examine.
There are many different methods of sampling — you just need to know about simple random sampling.

Using a sample to estimate the parameters of a population

Once you've selected your sample, you use it to find out information about the population it's drawn from.
The information you want is often the population mean (μ) or variance (σ^2) — parameters of the population.
These parameters are estimated from the sample data.
A statistic used to estimate the value of a parameter is called an estimator.
For a set of data, the numerical value taken by an estimator is called an estimate.

Parameter vs. Statistic
A numerical property, for example the mean, calculated using all the population data is a "parameter". If you calculate such a value from sample data, it's called a statistic.

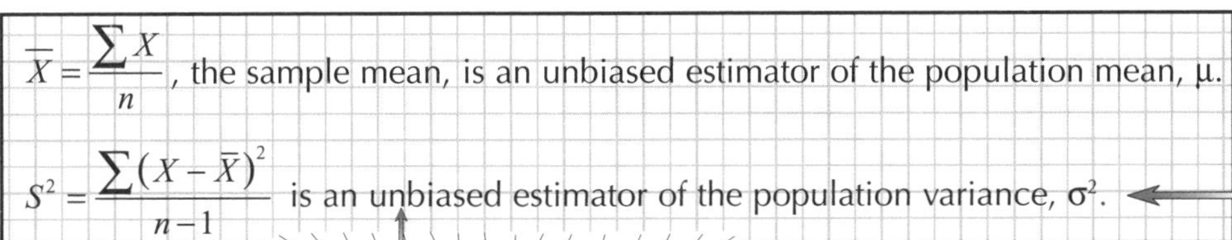

$\overline{X} = \frac{\sum X}{n}$, the sample mean, is an unbiased estimator of the population mean, μ.

$S^2 = \frac{\sum (X - \overline{X})^2}{n-1}$ is an unbiased estimator of the population variance, σ^2. ← $\overline{X}$ is the estimate of the population mean.

Here unbiased means it is the most accurate estimate possible.

Example:

You choose a simple random sample of 6 trees from a forest and record the height of each in metres.

10 13 11 15 15 14

Use this data to estimate the average height and variance of the trees in the forest.

1) Calculate the sample mean: $\bar{x} = \frac{\sum x}{n} = \frac{78}{6} = 13$ ← When calculating the numerical value taken by an estimator (the estimate) use lower case letters.

So an estimate of the average height of the trees in the forest is 13 metres.

2) Calculate $s^2 = \frac{\sum (x - \bar{x})^2}{n-1} = \frac{9+0+4+4+4+1}{5} = \frac{22}{5} = 4.4$

So an estimate of the variance of the heights of the trees in the forest is 4.4 metres.

Practice Questions

1) A sample of 100 items from a population of 1000 items is needed.
Explain how you would obtain a simple random sample without replacement.

Sample Exam question:

2) 15 of the 50 people in an office were randomly selected and asked how often they shopped at a supermarket:

Number of times the supermarket was visited in a month														
10	12	4	5	4	2	1	8	9	10	4	4	2	6	9

Use this sample to estimate the population mean and variance. [4 marks]

Statistically — surely I should have taken part in an opinion poll by now...

The important thing here is to make sure you read the questions carefully. Sounds simple, but it's easy to get mixed up.
If you're asked to calculate the variance of a set of numbers, then use the formula on page 78. If you're asked to estimate the variance of a population from a sample of numbers, then use the formula given here.

Random Events and Their Probability

This page is for AQA S1, OCR S1, Edexcel S1

Random events happen by chance. Probability is a measure of how likely they are. It can be a chancy business.

A Random Event has ***Various Outcomes***

1) In a trial (or experiment) the things that can happen are called outcomes (so if I time how long it takes to eat my dinner, 63 seconds is a possible outcome).
2) Events are 'groups' of one or more outcomes (so an event might be 'it takes me less than a minute to eat my dinner every day one week').
3) When all outcomes are equally likely, you can work out the probability of an event by counting the outcomes.

$$P(\text{event}) = \frac{\text{Number of outcomes where event happens}}{\text{Total number of possible outcomes}}$$

Example: Suppose I've got a bag with 15 balls in — 5 red, 6 blue and 4 yellow.

If I take a ball out without looking, then any ball is equally likely — there are 15 possible outcomes. Of these 15 outcomes, 5 are red, 6 are blue and 4 are yellow. And so...

$P(\text{red ball}) = \frac{5}{15} = \frac{1}{3}$ $\quad P(\text{blue ball}) = \frac{6}{15} = \frac{2}{5}$ $\quad P(\text{yellow ball}) = \frac{4}{15}$

You can find the probability of either a red or a yellow ball in a similar way...

$P(\text{red or yellow ball}) = \frac{9}{15} = \frac{3}{5}$

The ***Sample Space*** is the Set of ***All Possible Outcomes***

Drawing the sample space (called S) helps you count the outcomes you're interested in.

Example: The classic probability machine is a dice. If you roll it twice, you can record all the possible outcomes in a 6 × 6 table (a possible diagram of the sample space).

There are 36 outcomes in total. You can find probabilities by counting the ones you're interested in (and using the above formula). For example:

(i) The probability of an odd number and then a '1'. There are 3 outcomes that make up this event, so the probability is: $\frac{3}{36} = \frac{1}{12}$

(ii) The probability of the total being 7. There are 6 outcomes that correspond to this event, giving a probability of: $\frac{6}{36} = \frac{1}{6}$

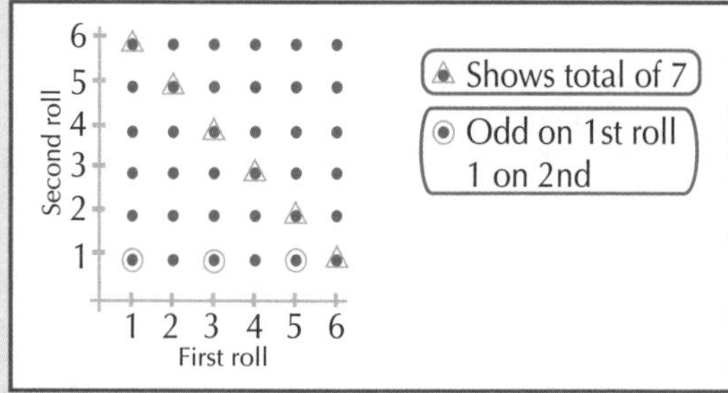

Venn Diagrams show which ***Outcomes*** correspond to which ***Events***

Say you've got 2 events, A and B — a Venn diagram shows which outcomes satisfy event A, which satisfy B, which satisfy both, and which satisfy neither.

(i) All outcomes satisfying event A go in one part of the diagram, and all outcomes satisfying event B go in another bit.

(ii) If they satisfy 'both A and B', they go in the dark green middle bit, written $A \cap B$ (and called the intersection of A and B).

(iii) The whole of the green area is written $A \cup B$ — it means 'either A or B' (and is called the union of A and B).

A B

Again, you can work out probabilities of events by counting outcomes and using the formula above.

You can also get a nice formula linking $P(A \cap B)$ and $P(A \cup B)$.

$$P(A \cup B) = P(A) + P(B) - P(A \cap B)$$

If you just add up the outcomes in A and B, you end up counting $A \cap B$ twice — that's why you have to subtract it.

Example: If you roll a dice, event A could be 'I get an even number', and B 'I get a number bigger than 4'. The Venn diagram would be:

$P(A) = \frac{3}{6} = \frac{1}{2}$ $\quad P(B) = \frac{2}{6} = \frac{1}{3}$ $\quad P(A \cap B) = \frac{1}{6}$ $\quad P(A \cup B) = \frac{4}{6} = \frac{2}{3}$

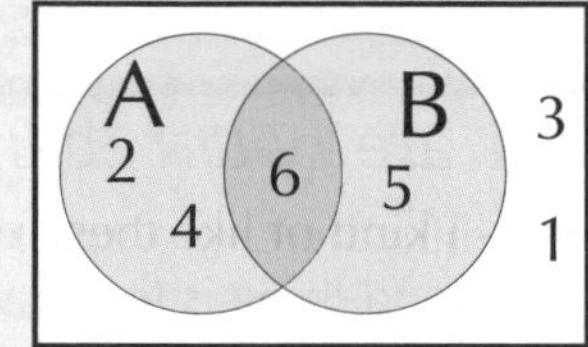

Here, I've just counted outcomes — but I could have used the formula.

Random Events and Their Probability

This page is for AQA S1, OCR S1, Edexcel S1

Venn Diagrams make it easy to get your head round **Tricky Things**

Example: A survey was carried out to find what pets people like.

The probability they like dogs is 0.6. The probability they like cats is 0.5. The probability they like gerbils is 0.4. The probability they like dogs and cats is 0.4. The probability they like cats and gerbils is 0.1, and the probability they like gerbils and dogs is 0.2. Finally, the probability they like all three kinds of animal is 0.1.

You can draw all this in a Venn diagram. (Here I've used C for 'likes cats', D for 'likes dogs' and G for 'likes gerbils'.)

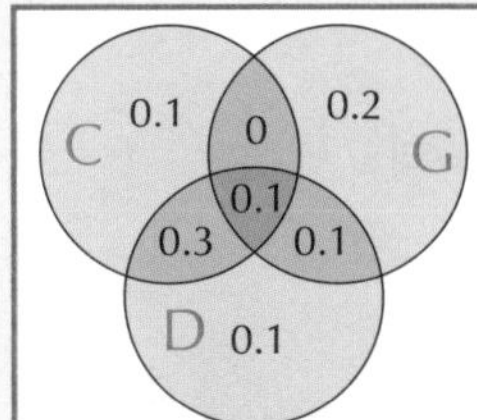

1) Stick in the middle one first — 'likes all 3 animals' (i.e. $C \cap D \cap G$).
2) Then do the 'likes 2 animals' probabilities by taking 0.1 from each of the given 'likes 2 animals' probabilities. (If they like 3 animals, they'll also be in the 'likes 2 animals' bits.)
3) Finally, do the 'likes 1 kind of animal' probabilities, by making sure the total probability in each circle adds up to the probability in the question.

① From the Venn diagram, the probability that someone likes either dogs or cats is 0.7.

② The probability that someone likes gerbils but not dogs is 0.2.

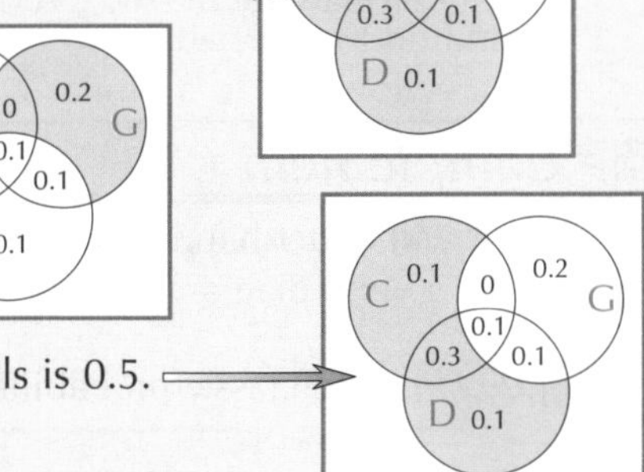

③ The probability that someone likes cats and dogs, but not gerbils is 0.5.

Practice Questions

1. ***A dice and a coin are thrown and the outcomes recorded.***
If a head is thrown, the score on the dice is doubled. If a tail is thrown, 4 is added to the score on the dice.
a) Represent this by means of a sample space diagram.
b) What is the probability that you score more than 5?
c) If you throw a tail, what is the probability that you get an even score?

2. ***Half the students in a sixth form college eat sausages for dinner and 20% eat chips.***
10% of those who eat chips also eat sausages. By use of a Venn diagram or otherwise, find:
a) the percentage of students who eat both chips and sausages,
b) the percentage of students who eat chips but not sausages,
c) the percentage of students who eat either chips or sausages but not both.

Sample Exam question:

3. A soap company asked 120 people about the types of soap (from Brands A, B and C) they bought. Brand A was bought by 40 people, Brand B by 30 people and Brand C by 25. Both Brands A and B (and possibly C as well) were bought by 8 people, B and C (and maybe A) were bought by 10 people, and A and C (and maybe B) by 7 people. All three brands were bought by 3 people.

(a) Represent this information in a Venn diagram. [5 marks]

(b) If a person is selected at random, find the probability that:

(i) they buy at least one of the soaps. [2 marks]

(ii) they buy least two of the soaps, [2 marks]

(iii) they buy soap B, given that they buy only one type of soap. [3 marks]

Two heads are better than one — though only half as likely using two coins...

I must admit — I kind of like these pages. This stuff isn't too hard, and it's really useful for answering loads of questions. And one other good thing is that Venn diagrams look, well, nice somehow. But more importantly, when you're filling one in, the thing to remember is that you usually need to 'start from the inside and work out'.

Probability

This page is for AQA S1, OCR S1, Edexcel S1

So far so good. But I can see you want more.

Mutually Exclusive *Events Have* ***No Overlap***

If two events can't both happen at the same time (i.e. $P(A \cap B) = 0$) they're called mutually exclusive (or just 'exclusive').

If A and B are exclusive, then the probability of A or B is: $P(A \cup B) = P(A) + P(B)$. ← Use the formula from page 86, but put $P(A \cap B) = 0$.

More generally,

For n exclusive events (i.e. only one of them can happen at a time):

$$P(A_1 \cup A_2 \cup ... \cup A_n) = P(A_1) + P(A_2) + ... + P(A_n)$$

Example: Find the probability that a card pulled at random from a pack of cards (no jokers) is either a picture card (a Jack, Queen or King) or the 7, 8 or 9 of clubs.

Call event A — 'I get a picture card', and event B — 'I get the 7, 8 or 9 of clubs'.

Events A and B are mutually exclusive — they can't both happen. Also, $P(A) = \frac{12}{52} = \frac{3}{13}$ and $P(B) = \frac{3}{52}$.

So the probability of either A or B is: $P(A \cup B) = P(A) + P(B) = \frac{12}{52} + \frac{3}{52} = \frac{15}{52}$

The ***Complement*** *of 'Event A' is* ***'Not Event A'***

An event A will either happen or not happen. The event 'A doesn't happen' is called the complement of A (or A'). On a Venn diagram, it would look like this (because $A \cup A' = S$, the sample space):

S
A'
A

At least one of A and A' has to happen, so...

$$P(A) + P(A') = 1 \quad or \quad P(A') = 1 - P(A)$$

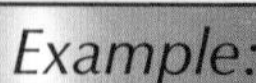

Example: A teacher keeps socks loose in a box. One morning, he picks out a sock. He quickly calculates that the probability of then picking out a matching sock is 0.56. What is the probability of him not picking a matching sock?

Call event A 'picks a matching sock'. Then A' is 'doesn't pick a matching sock'. Now A and A' are complementary events (and P(A) = 0.56), so P(A) + P(A') = 1, and therefore P(A') = 1 – 0.56 = 0.44

Tree Diagrams *Show Probabilities for* ***Two or More*** *Events*

Each 'chunk' of a tree diagram is a trial, and each branch of that chunk is a possible outcome. Multiplying probabilities along the branches gives you the probability of a series of outcomes.

Example: If Susan plays tennis one day, the probability that she'll play the next day is 0.2. If she doesn't play tennis, the probability that she'll play the next day is 0.6. She plays tennis on Monday. What is the probability she plays tennis:

(i) on both the Tuesday and Wednesday of that week?
(ii) on the Wednesday of the same week?

Let T mean 'plays tennis' (and then T′ means 'doesn't play tennis').

Notice that these add up to 1.

(i) Then the probability that she plays on Tuesday and Wednesday is P(T and T) = 0.2 × 0.2 = 0.04 (multiply probabilities since you need a series of outcomes — T and then T).

(ii) Now you're interested in either P(T and T) or P(T' and T). To find the probability of one event or another happening, you have to add probabilities: P(plays on Wednesday) = 0.04 + 0.48 = 0.52.

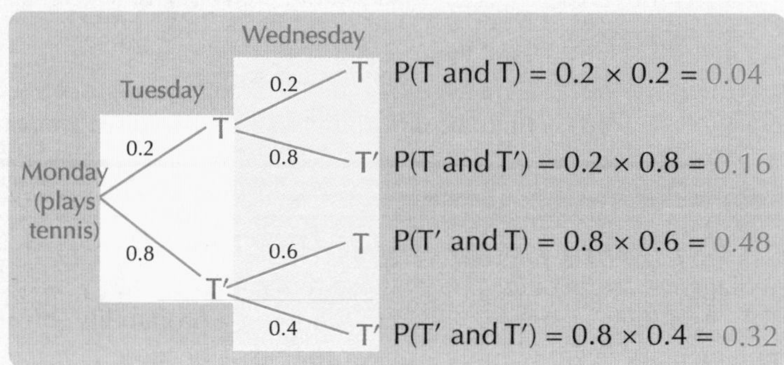

Probability

This page is for AQA S1, OCR S1, Edexcel S1

Sometimes a Branch is Missing

Example:

A box of biscuits contains 5 chocolate biscuits and 1 lemon biscuit.
George takes out three biscuits at random, one at a time, and eats them.

a) Find the probability that he eats 3 chocolate biscuits.

b) Find the probability the last biscuit is chocolate.

Let C mean 'picks a chocolate biscuit' and L mean 'picks the lemon biscuit'

After the lemon biscuit there are only chocolate biscuits left, so the tree diagram doesn't 'branch' after an 'L'.

a) Three chocolate biscuits is shown by only one 'path' along the branches.

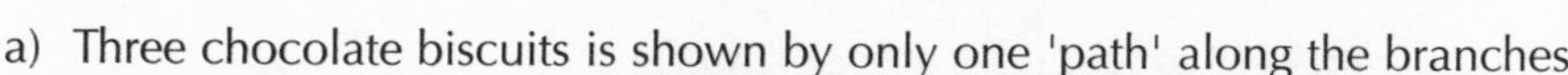

$$P(\text{C and C and C}) = \frac{5}{6}\times\frac{4}{5}\times\frac{3}{4} = \frac{60}{120} = \frac{1}{2}$$

b) The third biscuit being chocolate is shown by 3 'paths' along the branches — so you can add up the probabilities:

$$P(\text{third biscuit is chocolate}) = \left(\frac{5}{6}\times\frac{4}{5}\times\frac{3}{4}\right)+\left(\frac{5}{6}\times\frac{1}{5}\times1\right)+\left(\frac{1}{6}\times1\times1\right) = \frac{1}{2}+\frac{1}{6}+\frac{1}{6} = \frac{5}{6}$$

There's a quicker way to do this, since there's only one outcome where the chocolate isn't picked last:

$$P(\text{third biscuit is \underline{not} chocolate}) = \frac{5}{6}\times\frac{4}{5}\times\frac{1}{4} = \frac{1}{6}, \text{ so } P(\text{third biscuit is chocolate}) = 1-\frac{1}{6} = \frac{5}{6}$$

Working out the probability of the complement of the event you're interested in is sometimes easier.

Practice Questions

1. Arabella rolls two dice and adds the two results together.

a) What is the probability that she scores a prime number?

b) What is the probability that she scores a square number?

c) What is the probability that she scores a number that is either a prime number or a square number?

2. In a school orchestra (made up of pupils in either the upper or lower school), 40% of the musicians are boys. Of the boys, 30% are in the upper school. Of the girls in the orchestra, 50% are in the upper school.

a) Draw a tree diagram to show the various probabilities.

b) Find the probability that a musician chosen at random is in the upper school.

Sample Exam question:

3. A box contains counters of various colours. There are 3 red counters, 4 white counters and 5 green counters. Two random counters are removed from the jar one at a time. Once removed, the colour of the counter is noted. The first counter is not replaced before the second one is drawn.

(a) Draw a tree diagram to show the probabilities of the various outcomes. [4 marks]

(b) Find the probability that the second disc is green. [2 marks]

(c) Find the probability that both the discs are red. [2 marks]

(d) Find the probability that the two discs are not both the same colour. [3 marks]

Useful quotes: I can live for two months on a good compliment*...

Tree diagrams are another one of those things that are fairly easy to get your head round, but at the same time, are incredibly useful. And if you get stuck trying to work out a probability, it's worth checking to see if the probability of the complementary event would be easier to find — because if you can find one, then you can easily work out the other.

* Mark Twain

Conditional Probability

This page is for AQA S1, OCR S1, Edexcel S1

Examiners love conditional probability — they can't get enough of it. So learn this well...

*P(B|A) means **Probability of B**, given that **A has Already Happened***

Conditional probability means the probability of something, given that something else has already happened. For example, P(B|A) means the probability of B, given that A has already happened. More tree diagrams...

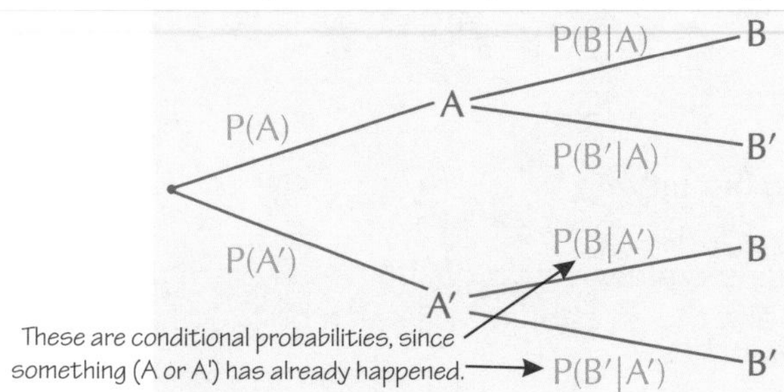

If you multiply probabilities along the branches, you get:

i.e. P(A and B) $\Rightarrow P(A \cap B) = P(A) \times P(B \mid A)$

You can rewrite this as:

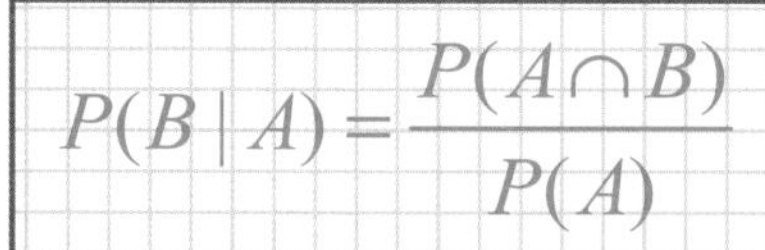

$$P(B \mid A) = \frac{P(A \cap B)}{P(A)}$$

Example: Horace either walks (W) or runs (R) to the bus stop.
If he walks he catches (C) the bus with a probability of 0.3. If he runs he catches it with a probability of 0.7. He walks to the bus stop with a probability of 0.4.
Find the probability that Horace catches the bus.

P(C) = P(C∩W) + P(C∩R)
= P(W)P(C|W) + P(R)P(C|R)
= (0.4 × 0.3) + (0.6 × 0.7) = 0.12 + 0.42 = 0.54

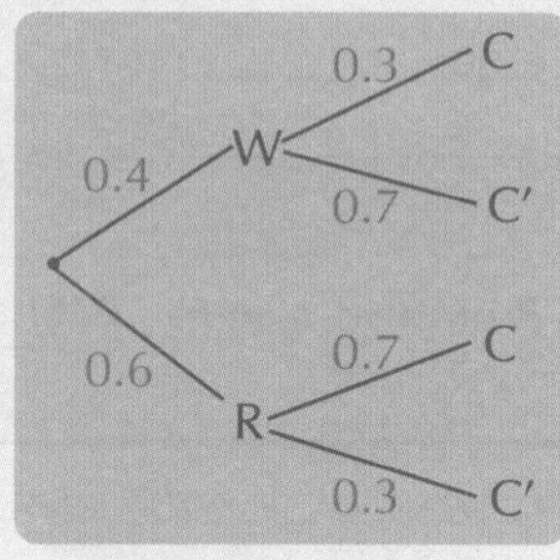

This is easier to follow if you match each part of this working to the probabilities in the tree diagram.

*If **B is Conditional** on A then **A is Conditional** on B*

If B depends on A then A depends on B — and it doesn't matter which event happens first.

Example: Horace turns up at school either late (L) or on time (L'). He is then either shouted at (S) or not (S'). The probability that he turns up late is 0.4. If he turns up late the probability that he is shouted at is 0.7. If he turns up on time the probability that he is shouted at is 0.2.
If you hear Horace being shouted at, what is the probability that he turned up late?

1) The probability you want is P(L|S). — Get this the right way round — he's already being shouted at.

2) Use the conditional probability formula: $P(L \mid S) = \dfrac{P(L \cap S)}{P(S)}$

3) The best way to find $P(L \cap S)$ and $P(S)$ is with a tree diagram.

Be careful with questions like this — the information in the question tells you what you need to know to draw the tree diagram with L (or L') considered first. But you need P(L|S) — where S is considered first. So don't just rush in.

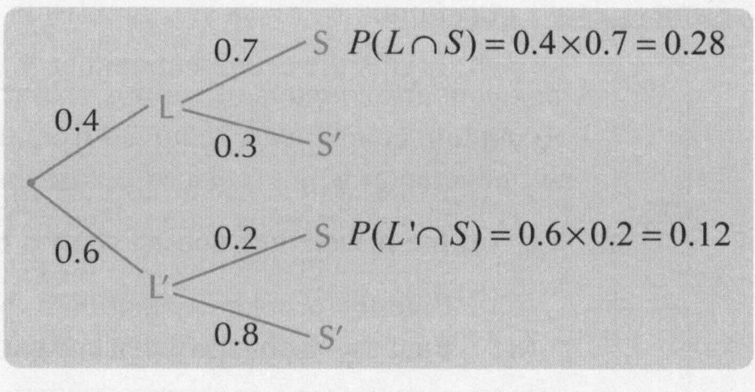

$$P(L \cap S) = 0.4 \times 0.7 = 0.28$$
$$P(S) = P(L \cap S) + P(L' \cap S) = 0.28 + 0.12 = 0.40$$

4) Put these in your conditional probability formula to get:

$$P(L \mid S) = \frac{0.28}{0.4} = 0.7$$

Conditional Probability

This page is for AQA S1, OCR S1, Edexcel S1

Independent Events Have No Effect on Each Other

If the probability of B happening doesn't depend on whether or not A has happened, then A and B are independent.

1) If A and B are independent, P(A|B) = P(A).
2) If you put this in the conditional probability formula, you get: $P(A|B)=P(A)=\frac{P(A\cap B)}{P(B)}$

Or, to put that another way: For independent events: $P(A\cap B)=P(A)P(B)$

Example: You are exposed to two infectious diseases — one after the other. The probability you catch the first (A) is 0.25, the probability you catch the second (B) is 0.5, and the probability you catch both of them is 0.2. Are catching the two diseases independent events?

1) You need to compare P(A|B) and P(A) — if they're different, the events aren't independent.

$$P(A|B)=\frac{P(A\cap B)}{P(B)}=\frac{0.2}{0.5}=0.4 \qquad P(A)=0.25$$

2) P(A|B) and P(A) are different, so they are not independent.

Practice Questions

1. A and B are two events. P(A) = 0.4, P(B) = 0.3, P(A∩B) = 0.1.
   a) Find P(B|A).
   b) Find P(B|A').
   c) Say whether or not A and B are independent.

2. Albert eats a limited choice of lunch. He eats either chicken or beef for his main course, and either chocolate pudding or ice cream for dessert. The probability that he eats chicken is 1/3, the probability that he eats ice cream given that he has chicken is 2/5, and the probability that he has ice cream given that he has beef is 3/4.
   a) Find the probability he has either chicken or ice cream — but not both.
   b) Find the probability that he eats ice cream.
   c) Find the probability that he had chicken given that you see him eating ice cream.

Sample Exam questions:

3. V and W are independent events, where P(V) = 0.2 and P(W) = 0.6.
   (a) Find: (i) P(V ∩ W), and (ii) P(V ∪ W). [3 marks]
   (b) If U is the event that neither V or W occurs, find P(U|V'). [3 marks]

4. For a particular biased dice, the event 'throw a 6' is called event B. P(B) = 0.2.
   This biased dice and a fair dice are rolled together. Find the probability that:
   (a) the biased dice doesn't show a 6, [1 mark]
   (b) at least one of the dice shows a 6, [2 marks]
   (c) exactly one of the dice shows a 6, given that at least one of them shows a 6. [3 marks]

Statisticians say: P(Having cake ∩ Eating it) = 0...

It's very easy to assume that events are independent and use the P(A∩B) formula in the red box on this page — when in fact they're not independent at all and you should be using the formula in the blue box on page 90. I admit that using the wrong formula may make the calculations slightly easier — but think how virtuous you'll feel if you do things the hard way.

Permutations and Combinations

This page is for OCR S1

*In **Permutations**, the **Order Matters***

Permutations and combinations are all about the numbers of ways you can arrange things.
In a permutation, the order matters — so '123' is a different permutation from '321'.

The number of permutations of n distinct objects is n!, where n! = n × (n – 1) × (n – 2) × ... × 2 × 1.

Example: How many different permutations are there of the numbers 5, 7 and 9?

1) The way to do these is to think: 'How many choices do I have for the first number?', 'How many choices do I then have for the second number?' and so on.
2) You can put either 5, 7 or 9 in the first position, so you have 3 choices — I'll choose 7.
3) Then you can put either 5 or 9 in the second position, so you have 2 choices — I'll choose 9.
4) Then I have only 1 choice for what goes in the third position — I have to choose 5.
   So there are 3 × 2 × 1 = 6 permutations of 5, 7 and 9.

Sometimes you only want to arrange a few of the available objects.

The number of permutations of r objects selected from n distinct objects is $\frac{n!}{(n-r)!}$.

This is sometimes written ^nP_r.

Example: How many permutations of 5 letters can you make from the letters of the alphabet (if you don't use any letter twice)?

You have 26 choices for the first letter, 25 for the second, then 24, 23 and 22 for the third, fourth and fifth.
This gives a total of 26 × 25 × 24 × 23 × 22 = 7 893 600 permutations.

This is $\frac{26!}{21!}$.

*The number of permutations is **Less** if some objects are the **Same***

Say you have *n* objects, but they're not all different — there's a group of *r* that are identical, and another group of *s* that are identical. Then the number of permutations is:

$$\frac{n!}{r!\times s!}$$

Example: How many different arrangements of the letters in the word MISSISSIPPI are there?

There are 11 letters in total — including 4 S's, 4 I's and 2 P's.

So the number of arrangements is: $\frac{11!}{4!\times 4!\times 2!} = 34650$

*When the **Order Doesn't Matter** it's a **Combination***

In a combination, the order of the objects doesn't matter — so '123' is the same combination as '231'.

The number of combinations of r objects selected from n distinct objects is ^nC_r or $\binom{n}{r}$, where $^nC_r = \binom{n}{r} = \frac{n!}{r!(n-r)!}$

Example: How many ways are there to choose a team of 11 players from a squad of 16?

Easy — just use the formula: $\binom{16}{11} = \frac{16!}{5!\times 11!} = 4368$

Notice that $\binom{16}{11} = \binom{16}{5}$.

Example: How many ways are there to choose 6 Lotto numbers from a possible 49?

Use the formula again: $\binom{49}{6} = \frac{49!}{6!\times 43!} = 13\,983\,816$

Cancel the 43! with the rest of the 49! to get $\frac{49!}{6!43!} = \frac{49\times 48\times 47\times 46\times 45\times 44}{6\times 5\times 4\times 3\times 2\times 1}$

Permutations and Combinations

This page is for OCR S1

Use **Permutations** and **Combinations** to Find Probabilities

This is a bit like the stuff on page 88 — you count the outcomes that you're interested in, and divide by the total number of possible outcomes. Use the 'permutations and combinations' formulas to do the counting.

Example: I've bought a Lotto ticket — what's the probability that I win the jackpot?

I'm only interested in 1 outcome (the one where my ticket wins). But there are 13983816 possibles (see p92).

So the probability that I win is $\frac{1}{13983816}$.

Example: Six people are to be sat in a row of seats at random. Three of the people are the Bills family. What's the probability that the Bills family are sat together?

1) This is 'number of ways the Bills family are together ÷ the total number of seating arrangements'.
2) The total number of seating arrangements is 6! = 720.
3) Now assume that the Bills are together by pretending they're one person — now you have only 4 people to arrange.
4) These 4 people can be arranged in 4! = 24 ways.
5) However for each of these 24 ways, there are 3! = 6 ways to arrange the Bills family. So the number of arrangements with the Bills together is actually 24 × 6 = 144.
6) Therefore the probability that the Bills are together is: $\frac{144}{720} = 0.2$

Practice Questions

1. ***How many arrangements of the letters in the word statistics are there?***

2. ***Three married couples are about to sit on the 6 back seats of a coach.***
   a) ***How many different arrangements of the 6 people are possible?***
   b) ***In how many of these is Mr Brown sitting next to his wife?***
   c) ***What is the probability that, from a randomly chosen arrangement, Mr Brown is not sitting next to his wife?***

3. ***The letters of the alphabet are written on 26 identical discs and placed in a bag. Six discs are then taken at random.***
   a) ***How many different sets of 6 are possible?***
   b) ***What is the probability that 5 vowels are taken?***
   c) ***What is the probability that at least 1 vowel is taken?***
   d) ***If you select the letters A, B, R, I, O and Z, and arrange 4 of them randomly, what is the probability that your arrangement starts with a vowel?***

Sample Exam question:

4. A group of 5 men and 5 women stand in a line to have their photos taken.
   (a) In how many different ways can they stand in line? [1 mark]
   (b) What is the probability that no two men and no two women stand next to each other? [3 marks]
   (c) Six people from the group are to pose for an additional photograph. At least 2 of the people chosen must be men. In how many different ways can the 6 people be chosen? [3 marks]

Maths proverb: Absence of stats makes the heart grow fonder...

Now you have to promise not to say 'combination' when you really mean 'permutation', okay. The "*How many choices?*" approach to finding the number of permutations is a good one to remember. But it's definitely worth committing the ^nP_r and ^nC_r formulas to memory as well — they can save you having to actually think. At the first sight of a "*Choose r from n*" question, you just need to decide whether the order of the things matters, and then bung the numbers in the right formula.

Probability Distributions

This page is for OCR S1, Edexcel S1

This stuff isn't hard — but it can seem a bit weird at times.

*Getting your head round this **Boring Stuff** will help a bit*

This first bit isn't particularly interesting. But understanding the difference between X and x (bear with me) might make the later stuff a bit less confusing. Might.

1) **X (upper case) is just the** name **of a** random variable. So X could be 'score on a dice' — it's just a name.
2) **A** random variable **doesn't have a** fixed **value.** Like with a dice score — the value on any 'roll' is all down to chance.
3) **x (lower case) is a** particular value **that X can take.** So for one roll of a dice, x could be 1, 2, 3, 4, 5 or 6.

4) Discrete **random variables only have a** certain number **of possible values.** Often these values are whole numbers, but **they don't have to be. Usually there are only a few possible values** (e.g. the possible scores with one roll of a dice).
5) **The** probability density function **(pdf) is a** list of the possible values **of x,** plus the probability for each one.

All the Probabilities Add up to 1

For a discrete random variable X:

$$\sum_{\text{all } x} P(X=x)=1$$

This says that if you add up the probabilities of all the possible values of X, you get 1.

Example: The random variable X has pdf $P(X = x) = kx$ for $x = 1, 2, 3$. Find the value of k.

So X has three possible values (x = 1, 2 and 3), and the probability of each is kx (where you need to find k).
It's easier to understand with a table:

x	1	2	3
$P(X = x)$	$k \times 1 = k$	$k \times 2 = 2k$	$k \times 3 = 3k$

Now just use the formula: $\sum_{all\ x} P(X=x)=1$ Here, this means: $k+2k+3k=6k=1$

i.e. $k=\frac{1}{6}$

Piece of cake.

The **mode** is the most likely value — so it's the value with the biggest probability.

Example: The discrete random variable X has pdf as shown in the table below.

x	0	1	2	3	4
$P(X = x)$	0.1	0.2	0.3	0.2	a

Find: (i) the value of a, (ii) $P(2 \leq X < 4)$, (iii) the mode

(i) Use the formula $\sum_{all\ x} P(X=x)=1$ again.

From the table: $0.1 + 0.2 + 0.3 + 0.2 + a = 1$
$0.8 + a = 1$
$a = 0.2$

Careful with the inequality signs — you need to include $x = 2$ but not $x = 4$.

(ii) This is asking for the probability that 'X is greater than or equal to 2, but less than 4'. Easy — just add up the probabilities.

$P(2 \leq X < 4) = P(X = 2) + P(X = 3) = 0.3 + 0.2 = 0.5$

(iii) The mode is the value of x with the biggest probability — so mode = 2.

Probability Distributions

This page is for OCR S1, Edexcel S1

Do Complicated questions **Bit by Bit**

Example: A game involves rolling two fair dice. If the sum of the scores is greater than 10 then the player wins 50p. If the sum is between 8 and 10 (inclusive) then he wins 20p. Otherwise he gets nothing. If X is the random variable "amount player wins", find the pdf of X.

There are 3 possible values for X (0, 20 and 50) and you need the probability of each. To work these out, you need the probability of getting various totals on the dice.

① You need to know P(8 ≤ score ≤ 10) — the probability that the score is between 8 and 10 inclusive (i.e. including 8 and 10) and P(11 ≤ score ≤ 12) — the probability that the score is greater than 10. This means working out: P(score = 8), P(score = 9), P(score = 10), P(score = 11) and P(score = 12). Use a table...

②

Score on dice 2 \ Score on dice 1 (+)	1	2	3	4	5	6
1	2	3	4	5	6	7
2	3	4	5	6	7	8
3	4	5	6	7	8	9
4	5	6	7	8	9	10
5	6	7	8	9	10	11
6	7	8	9	10	11	12

There are 36 possible outcomes...

...5 of these have a total of 8 — so the probability of scoring 8 is $\frac{5}{36}$,

...4 have a total of 9 — so the probability of scoring 9 is $\frac{4}{36}$,

...the probability of scoring 10 is $\frac{3}{36}$

...the probability of scoring 11 is $\frac{2}{36}$

...the probability of scoring 12 is $\frac{1}{36}$

③ To find the probabilities you need, you just add the right bits together:

$P(X = 20p) = P(8 \leq \text{score} \leq 10) = \frac{5}{36}+\frac{4}{36}+\frac{3}{36}=\frac{12}{36}=\frac{1}{3}$

$P(X = 50p) = P(11 \leq \text{score} \leq 12) = \frac{2}{36}+\frac{1}{36}=\frac{3}{36}=\frac{1}{12}$

To find P(X = 0) just take the total of the two probabilities above from 1 (since X = 0 is the only other possibility).

$P(X = 0) = 1-\left(\frac{12}{36}+\frac{3}{36}\right)=1-\frac{15}{36}=\frac{21}{36}=\frac{7}{12}$

④ Now just stick all this info in a table (and check that the probabilities all add up to 1):

x	0	20	50
$P(X = x)$	$\frac{7}{12}$	$\frac{1}{3}$	$\frac{1}{12}$

Practice Questions

1) The probability density function of Y is shown in the table.
(a) Find the value of k. (b) Find P(Y<2).

y	0	1	2	3
P(Y = y)	0.5	k	k	3k

2) An unbiased six-sided dice has faces marked 1, 1, 1, 2, 2, 3.
The dice is rolled twice. Let X be the random variable "sum of the two scores on the dice".
Show that P(X = 4) = $\frac{5}{18}$. Find the probability density function of X.

Sample exam question:

3) In a game a player tosses three fair coins. If three heads occur then the player gets 20p; if two heads occur then the player gets 10p; otherwise the player gets nothing.

(a) If X is the random variable 'amount received' tabulate the probability density function of X. [4 marks]

The player pays 10p to play one game.

(b) Use the probability density function to find the probability that the player wins (i.e. gets more money than he pays to play) in one game. [2 marks]

Useful quotes: All you need in life is ignorance and confidence, then success is sure...*

Remember I said on page 86 that the 'counting the outcomes' approach was useful — well there you go. And if you remember how to do that, then you can work out a pdf. And if you can work out a pdf, then you can often begin to unravel even fairly daunting-looking questions. But most of all, REMEMBER THAT ALL THE PROBABILITIES ADD UP TO 1. (Ahem.)

* Mark Twain

The Distribution Function

This page is for Edexcel S1

The pdf gives the probability that X will equal this or equal that. The distribution function tells you something else.

'Distribution Function' is the same as 'Cumulative Distribution Function'

The (cumulative) distribution function F(x) gives the probability that X will be less than or equal to a particular value.

$$F(x_0) = P(X \le x_0) = \sum_{x \le x_0} p(x)$$

Example: The probability density function of the discrete random variable H is shown in the table. Find the cumulative distribution function F(H).

h	0.1	0.2	0.3	0.4
$P(H = h)$	$\frac{1}{4}$	$\frac{1}{4}$	$\frac{1}{3}$	$\frac{1}{6}$

There are 4 values of h, so you have to find the probability that H is less than or equal to each of them in turn. It sounds trickier than it actually is — you only have to add up a few probabilities...

$F(0.1) = P(H \le 0.1)$ — this is the same as $P(H = 0.1)$, since H can't be less than 0.1. So $F(0.1) = \frac{1}{4}$.

$F(0.2) = P(H \le 0.2)$ — this is the probability that H = 0.1 or H = 0.2. So $F(0.2) = P(H = 0.1) + P(H = 0.2) = \frac{1}{4} + \frac{1}{4} = \frac{1}{2}$.

$F(0.3) = P(H \le 0.3) = P(H = 0.1) + P(H = 0.2) + P(H = 0.3) = \frac{1}{4} + \frac{1}{4} + \frac{1}{3} = \frac{5}{6}$.

$F(0.4) = P(H \le 0.4) = P(H = 0.1) + P(H = 0.2) + P(H = 0.3) + P(H = 0.4) = \frac{1}{4} + \frac{1}{4} + \frac{1}{3} + \frac{1}{6} = 1$.

$P(X \le \text{largest value of } x)$ is always 1.

Finally, put these values in a table, and you're done...

h	0.1	0.2	0.3	0.4
$F(H) = P(H \le h)$	$\frac{1}{4}$	$\frac{1}{2}$	$\frac{5}{6}$	1

Sometimes they ask you to work backwards...

Example: The formula below gives the cumulative distribution function F(X) for a discrete random variable X. Find k, and the probability density function.
$F(X) = kx$, for $x = 1, 2, 3$ and 4.

(1) First find k. You know that X has to be 4 or less — so $P(X \le 4) = 1$.

Put x = 4 into the cumulative distribution function: $F(4) = P(X \le 4) = 4k = 1$, so $k = \frac{1}{4}$.

(2) Now you can work out the probabilities of X being less than or equal to 1, 2, 3 and 4.

$F(1) = P(X \le 1) = 1 \times k = \frac{1}{4}$, $F(2) = P(X \le 2) = 2 \times k = \frac{1}{2}$, $F(3) = P(X \le 3) = 3 \times k = \frac{3}{4}$, $F(4) = P(X \le 4) = 1$

(3) This is the clever bit... $P(X = 4) = P(X \le 4) - P(X \le 3) = 1 - \frac{3}{4} = \frac{1}{4}$

Think about it... ...if it's less than or equal to 4, ...but it's not less than or equal to 3, ...then it has to be 4.

$P(X = 3) = P(X \le 3) - P(X \le 2) = \frac{3}{4} - \frac{1}{2} = \frac{1}{4}$

$P(X = 2) = P(X \le 2) - P(X \le 1) = \frac{1}{2} - \frac{1}{4} = \frac{1}{4}$

$P(X = 1) = P(X \le 1) = \frac{1}{4}$ — Because x doesn't take any values less than 1.

(4) Finish it all off by making a table. The pdf of X is:

x	1	2	3	4
$P(X = x)$	$\frac{1}{4}$	$\frac{1}{4}$	$\frac{1}{4}$	$\frac{1}{4}$

Or you could write it as a formula: $P(X = x) = \frac{1}{4}$ for $x = 1, 2, 3, 4$

Discrete Uniform Distributions

This page is for Edexcel S1

When every value of X is equally likely, you've got a uniform distribution. For example, rolling an unbiased dice gives you a discrete uniform distribution. (It's 'discrete' because there are only a few possible outcomes.)

In a Discrete Uniform Distribution the Probabilities are Equal

The pdf of a discrete uniform distribution looks like this — in this version there are only 4 possible values:

For a discrete uniform distribution X which can take consecutive whole number values a, a+1, a+2,...,b, the mean (or expected value) and variance are easy to work out.

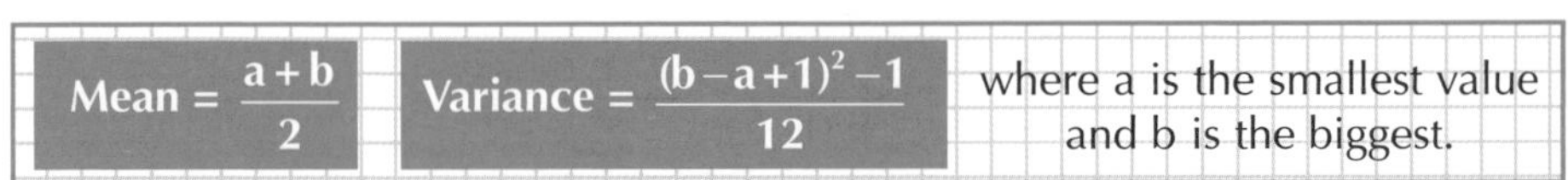

Mean = $\frac{a+b}{2}$ Variance = $\frac{(b-a+1)^2-1}{12}$ where a is the smallest value and b is the biggest.

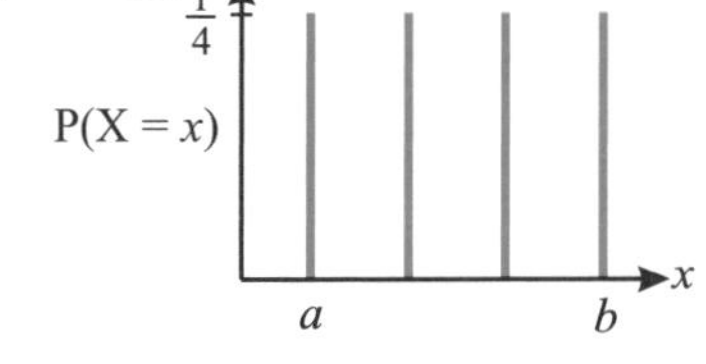

See page 98 for more info about the expected value and variance of a random variable.

Example: Find the mean and variance of the score on an unbiased six-sided dice.

If X is the random variable 'score on a dice', then X has a discrete uniform distribution — like in this table:

x	1	2	3	4	5	6
$P(X = x)$	$\frac{1}{6}$	$\frac{1}{6}$	$\frac{1}{6}$	$\frac{1}{6}$	$\frac{1}{6}$	$\frac{1}{6}$

The symmetry of the distribution should tell you where the mean is — it has to be halfway between 1 and 6.

The smallest value of x is 1 and the biggest is 6 — so a = 1 and b = 6.
Now just stick the numbers in the formulas:

Mean = $\frac{a+b}{2} = \frac{1+6}{2} = \frac{7}{2} = 3.5$

Variance = $\frac{(b-a+1)^2-1}{12} = \frac{(6-1+1)^2-1}{12} = \frac{35}{12} = 2.92$ to 3 sig. fig.

Practice Questions

1) The probability density function for the random variable W is given in the table. Find the cumulative distribution function.

w	0.2	0.3	0.4	0.5
$P(W = w)$	0.2	0.2	0.3	0.3

2) The cumulative distribution function for a random variable R is given in the table. Calculate the probability density function for R. Find P(0 ≤ R ≤ 1).

r	0	1	2
$F(r) = P(R \leq r)$	0.1	0.5	1

3) The discrete random variable X has a uniform distribution, P(X = x) = k for x = 0, 1, 2, 3 and 4. Find the value of k, and then find the mean and variance of X.

Sample exam questions:

4) A discrete random variable X can only take values 0, 1, 2 and 3. Its pdf is shown in the table.

x	0	1	2	3
$P(X = x)$	$2k$	$3k$	k	k

(a) Find the value of k. [1 mark]
(b) Calculate the distribution function for X. [4 marks]
(c) Calculate P(X > 2). [1 mark]

5) The random variable X takes the values 0, 1, 2, 3, 4, 5, 6, 7, 8 and 9 with equal probability.
(a) Write down the pdf of X. [1 mark]
(b) Find the mean and variance of X. [3 marks]
(c) Calculate the probability that X is less than the mean. [2 marks]

Discreet distributions are more 'British' than those lurid, gaudy ones...

If you've got a pdf, then you can easily work out the distribution function. And if you've got a distribution function, you can work out the pdf, as long as you remember the clever wee trick on page 96. The mean of a discrete uniform distribution is weird. I mean (no stats-pun intended), if you've got a dice, then the expected value is 3½. But I went to a casino recently and bet on 3½ every time at the 'Guess the Dice' table. But not once did the dice ever land on 3½. Not once. I lost loads.

Expected Values, Mean and Variance

This page is for OCR S1, Edexcel S1

This is all about the mean and variance of random variables — not a load of data. It's a tricky concept, but bear with it.

The *Mean* of a random variable is the same as the *Expected Value*

You can work out the expected value (or the mean) E(X) of a random variable X.

The expected value (a kind of 'theoretical mean') is what you'd expect the mean of X to be if you took loads of readings. In practice, the mean of your results is unlikely to match the theoretical mean exactly, but it should be pretty near.

If the possible values of X are x_1, x_2, x_3,... then the expected value of X is:

$$\text{Mean} = \text{Expected Value } E(X) = \sum x_i P(X = x_i) = \sum x_i p_i$$

$p_i = P(X = x_i)$

Example: The probability distribution of X, the number of daughters in a family of 3 children, is shown in the table. Find the expected number of daughters.

x_i	0	1	2	3
p_i	$\frac{1}{8}$	$\frac{3}{8}$	$\frac{3}{8}$	$\frac{1}{8}$

$$\text{Mean} = \sum x_i p_i = \left(0\times\tfrac{1}{8}\right)+\left(1\times\tfrac{3}{8}\right)+\left(2\times\tfrac{3}{8}\right)+\left(3\times\tfrac{1}{8}\right) = 0+\tfrac{3}{8}+\tfrac{6}{8}+\tfrac{3}{8} = \tfrac{12}{8} = 1.5$$

So the expected number of daughters is 1.5 — which sounds a bit weird.
But all it means is that if you check a large number of 3-child families, the mean will be close to 1.5.

The *Variance* measures how *Spread Out* the distribution is

You can also find the variance of a random variable. It's the 'expected variance' of a large number of readings.

$$\text{Var}(X) = E(X^2) - [E(X)]^2 = \sum x_i^2 p_i - \left(\sum x_i p_i\right)^2$$

This formula needs $E(X^2) = \sum x_i^2 p_i$ — take each possible value of x, square it, multiply it by its probability and then add up all the results.

Example: Work out the variance for the '3 daughters' example above:

First work out $E(X^2)$: $E(X^2) = \sum x_i^2 p_i = \left(0^2\times\tfrac{1}{8}\right)+\left(1^2\times\tfrac{3}{8}\right)+\left(2^2\times\tfrac{3}{8}\right)+\left(3^2\times\tfrac{1}{8}\right)$
$= 0+\tfrac{3}{8}+\tfrac{12}{8}+\tfrac{9}{8} = \tfrac{24}{8} = 3$

Now you take away the mean squared: $\text{Var}(X) = E(X^2) - [E(X)]^2 = 3 - 1.5^2 = 3 - 2.25 = 0.75$

Example: X has the probability density function $P(X = x) = k(x + 1)$ for $x = 0, 1, 2, 3, 4$.
Find the mean and variance of X.

1. First you need to find k — work out all the probabilities and make sure they add up to 1.
   $P(X = 0) = k \times (0 + 1) = k$. Similarly, $P(X = 1) = 2k$, $P(X = 2) = 3k$, $P(X = 3) = 4k$, $P(X = 4) = 5k$.

   So $k + 2k + 3k + 4k + 5k = 1$, i.e. $15k = 1$, and so $k = \frac{1}{15}$

   Now you can work out p_1, p_2, p_3,... where $p_1 = P(X = 1)$ etc.

2. Now use the formulas — find the mean E(X) first:

$$E(X) = \sum x_i p_i = \left(0\times\tfrac{1}{15}\right)+\left(1\times\tfrac{2}{15}\right)+\left(2\times\tfrac{3}{15}\right)+\left(3\times\tfrac{4}{15}\right)+\left(4\times\tfrac{5}{15}\right) = \tfrac{40}{15} = \tfrac{8}{3}$$

For the variance you need $E(X^2)$:

$$E(X^2) = \sum x_i^2 p_i = \left(0^2\times\tfrac{1}{15}\right)+\left(1^2\times\tfrac{2}{15}\right)+\left(2^2\times\tfrac{3}{15}\right)+\left(3^2\times\tfrac{4}{15}\right)+\left(4^2\times\tfrac{5}{15}\right) = \tfrac{130}{15} = \tfrac{26}{3}$$

And finally: $Var(X) = E(X^2) - [E(X)]^2 = \frac{26}{3} - \left(\frac{8}{3}\right)^2 = \frac{14}{9}$

Expected Values, Mean and Variance

This page is for OCR S1, Edexcel S1

You can use the **Expected Value** and **Variance** formulas for **Functions**

$E(aX + b) = aE(X) + b$ $\quad$ $Var(aX + b) = a^2Var(X)$

Here a and b are any numbers.

Example: If E(X) = 3 and Var(X) = 7, find E(2X+5) and Var(2X+5).

Easy. $E(2X + 5) = 2E(X) + 5 = (2 \times 3) + 5 = 11$

$Var(2X + 5) = 2^2Var(X) = 4 \times 7 = 28$

Example: The discrete random variable X has the following probability distribution:

x	2	3	4	5	6
$P(X = x)$	0.1	0.2	0.3	0.2	k

Find: a) k, b) E(X), c) Var(X), d) E(3X – 1), e) Var(3X – 1)

Slowly, slowly — one bit at a time...

a) Remember the probabilities add up to 1 — $0.1 + 0.2 + 0.3 + 0.2 + k = 1$, and so $k = 0.2$

b) Now you can use the formula to find E(X): $E(X) = \sum x_i p_i = (2\times0.1)+(3\times0.2)+(4\times0.3)+(5\times0.2)+(6\times0.2) = 4.2$

c) Next work out $E(X^2)$: $E(X^2) = \sum x_i^2 p_i = (2^2\times0.1)+(3^2\times0.2)+(4^2\times0.3)+(5^2\times0.2)+(6^2\times0.2) = 19.2$

and then the variance is easy: $Var(X) = E(X^2) - [E(X)]^2 = 19.2 - 4.2^2 = 1.56$

d) You'd expect the question to get harder but it doesn't: $E(3X - 1) = 3E(X) - 1 = 3 \times 4.2 - 1 = 11.6$

e) And finally: $Var(3X - 1) = 3^2Var(X) = 9 \times 1.56 = 14.04$

Practice Questions

1) A discrete random variable X has the probability distribution shown in the table, where k is a constant.

x_i	1	2	3	4
p_i	$\frac{1}{8}$	$\frac{1}{2}$	k	$\frac{5}{24}$

a) Find the value of k.
b) Find E(X) and show Var(X) = 63/64
c) Find E(2X – 1) and Var(2X – 1)

Sample exam question:

2) A discrete random variable X has the pdf $P(X = x) = ax$ for $x = 1, 2, 3$, where a is a constant.

(a) Show $a = \frac{1}{6}$. [1 mark]

(b) Find E(X). [2 marks]

(c) If $Var(X) = \frac{5}{9}$ find $E(X^2)$. [2 marks]

(d) Find E(3X + 4) and Var(3X + 4). [3 marks]

Statisticians say: E(Bird in hand) = E(2 Birds in bush)...

The mean and variance here are theoretical values — don't get them confused with the mean and variance of a load of practical observations. This 'theoretical' variance has a similar formula to the variance formula on p78, though — it's just "E(X-squared) minus E(X)-squared". And you can still take the square root of the variance to get the standard deviation.

The Binomial Distribution

This page is for AQA S1, OCR S1

With the binomial distribution (as with so much in Stats), the notation and jargon can make it seem really hard... ...but once you've practised a few questions, you realise it's about as hard as a really squishy sponge.

The *Binomial Distribution* is all about *Success* and *Failure*

Binomial distributions are for finding out the probability of a certain number of 'successes' in a fixed number of 'trials'.

To use the Binomial distribution, look for the following:

1) **A fixed number of trials (n), e.g. 12 eggs in a box, 10 rolls of a dice...**
2) **Only two outcomes of each trial.** These are called "success" or "failure", but don't necessarily sound like success and failure, e.g. success = the egg is cracked, failure = the egg is not cracked; success = a six is rolled, failure = a six is not rolled.
3) **Each trial is independent of all others, e.g. the first egg being broken does not affect the other eggs being broken. The probability of success (p) is constant for each trial.**

If the random variable X has a Binomial distribution, then we write $X \sim B(n, p)$

Sometimes you need to use the *Probability Formula*

The probability that X takes a particular value x is:

$$P(X = x) = \binom{n}{x} p^x q^{n-x}, \text{ where } q = 1 - p \quad \text{and} \quad \binom{n}{x} = \frac{n!}{x!(n-x)!} \text{ is a binomial coefficient.}$$

Use a calculator's ^nC_r button to work these out, e.g. 5 ^nC_r 3 finds $\binom{5}{3}$.

Example: Eggs are packed in boxes of 12. The probability that each egg is broken is 0.35. Find the probability that in a random box of eggs:
(i) there are 4 broken eggs, (ii) there are less than 3 broken eggs.

Let X be the random variable "number of broken eggs". Here n = 12 (there are 12 eggs in the box, so 12 trials).
Also p = 0.35 (and so q = 1 – 0.35 = 0.65) which means that $X \sim B(12, 0.35)$

(i) First you need the probability of 4 'successes', i.e. X = 4. Just use the formula:

Here, 'success' means the egg is broken and 'failure' means the egg is intact.

$$P(X=4) = \binom{12}{4} \times 0.35^4 \times 0.65^8 = 495 \times 0.35^4 \times 0.65^8 = 0.237 \text{ to 3 significant figures}$$

(ii) Then you need the probability that X is less than 3 — so X could be 0, 1 or 2. Add up the separate probabilities.

$$P(X<3) = P(X=0) + P(X=1) + P(X=2)$$
$$= \binom{12}{0} \times 0.35^0 \times 0.65^{12} + \binom{12}{1} \times 0.35^1 \times 0.65^{11} + \binom{12}{2} \times 0.35^2 \times 0.65^{10}$$
$$= (1 \times 1 \times 0.005688) + (12 \times 0.35 \times 0.008751) + (66 \times 0.1225 \times 0.01346) = 0.151 \text{ to 3 sig. fig.}$$

For a Binomial Distribution, *Mean = np, Variance = npq, Standard Deviation =* $\sqrt{npq}$

Yep, I'll say that again...

If $X \sim B(n, p)$,
Mean or Expected Value $E(X) = np$
Variance, $Var(X) = np(1-p) = npq$
Standard deviation $= \sqrt{npq}$

Example: What is the mean and variance of X, if X is the number of sixes when I throw a fair dice 100 times?

There are a fixed number of independent trials (= 100) and the probability of success each time is the same (= $\frac{1}{6}$).
So this is a binomial distribution — $X \sim B(100, \frac{1}{6})$

Mean = Expected value E(X) = $np = 100 \times \frac{1}{6} = 16.7$

Variance = Var(X) = $np(1-p) = 100 \times \frac{1}{6} \times \frac{5}{6} = 13.9$

The Binomial Distribution

This page is for AQA S1, OCR S1

It's Quicker to use **Tables** when you can

Example: A fair dice is rolled 10 times. Find the probability that: a) there are less than 5 sixes rolled,
b) there are 3 sixes rolled,
c) there are more than 5 sixes rolled.

Let X be the random variable "Number of sixes rolled". Then $X \sim B\left(10, \frac{1}{6}\right)$.

a) Look in your book of tables for the cumulative binomial tables.
These give probabilities for $X \leq x$ (less than or equal to). You need $P(X<5) = P(X = 0,1,2,3 \text{ or } 4) = P(X \leq 4)$.
Find the table for n = 10. Look down the left-hand side until you reach x = 4.
Go across the row of figures until you're in the column with p = 1/6. Answer: P(X < 5) = 0.9845.

b) P(X = 3) = P(X ≤ 3) – P(X ≤ 2)
Look these up in the tables.
[P (X = 0, 1, 2 or 3) – P (X = 0, 1 or 2)]. This gives P(X = 3) = 0.9303 – 0.7752 = 0.1551

c) P(X > 5) = 1 – P(X ≤ 5).
i.e. P (X = 6, 7, 8, 9 or 10) = 1 – P (X = 0, 1, 2, 3, 4 or 5) = 1 – 0.9976 = 0.0024

Find the **Number of 'Failures'** if you have to

Binomial tables usually just go up to p = 0.5. If p > 0.5, you have to look up q in the tables instead.

Example: If X ~ B(8, 0.6), find P(X = 3).

The question's asking for the probability of 3 successes in 8 trials, where P(success) = 0.6.
But the tables don't help, since p is too big.
Instead, you'll have to find the probability of 5 failures, where the probability of failure q = 0.4.

If Y is the number of failures, then Y ~ B(8, 0.4). You need P(Y = 5).
P(Y = 5) = P(Y ≤ 5) – P(Y ≤ 4) = 0.9502 – 0.8263 = 0.1239

Practice Questions

1) ***The random variable X has a binomial distribution with n = 5 and p = 0.3.***
Calculate (a) P(X = 2) (b) P(X ≤ 3) (c) P(X < 2) (d) E(X) (e) Var(X)

2) ***A fair dice is rolled 20 times. Find the probability that the number of sixes is (a) exactly 10, (b) at least 10.***

3) ***30% of the meals served in a canteen contain a salad. If 10 customers in the canteen are selected at random, what is the probability that fewer than 3 of them have a salad?***

Sample exam question:

4) In a room of 30 people, the probability that each person was born on a Saturday is 1/7.
(a) State the distribution of the number of people born on a Saturday and the conditions needed for this distribution. [5 marks]
(b) Calculate the probability that exactly 5 of the people in the room were born on a Saturday. [2 marks]

Binomial distributions are much like exams — all about success and failure...

Binomial distributions come up loads, so learning to love them is probably the best advice I could give you. There are 3 formulas you need to use — the one for a probability, plus those for the mean and variance. Then you need to get used to using the cumulative binomial tables — they tell you the probability that X is less than any value. The tables are slightly different depending on which Exam board you're doing, so have a practice with the actual tables you'll be using come the real Exam. Oh, and one more thing... Exam boards like you to quote your answers from tables in full — so don't round.

The Geometric Distribution

This page is for OCR S1

Imagine playing a board game where you need to throw a six to start. If you're lucky, you'll get a six first throw. If you're unlucky, you might have to wait a while. But if you're having one of those days, it could take forever...

The **Geometric Distribution** is about **Waiting for Success**

Now then... suppose X is the number of throws it takes you to get a six, then X can take the values 1, 2, 3, 4, 5, 6, 7, 8...

The probability of getting a six with the first throw, $P(X = 1) = \frac{1}{6}$

The probability of getting a six with the second throw, $P(X = 2) = \frac{5}{6} \times \frac{1}{6}$

Because you get 'not a six' first, and then you get a six.

The probability of getting a six with the third throw, $P(X = 3) = \frac{5}{6} \times \frac{5}{6} \times \frac{1}{6} = \left(\frac{5}{6}\right)^2 \times \frac{1}{6}$

More generally, the probability of needing r throws to get a six, $P(X = r) = \left(\frac{5}{6}\right)^{r-1} \times \frac{1}{6}$

This is what geometric distributions are all about — how many attempts it takes to achieve 'success'.

Here $X \sim \text{Geo}(\frac{1}{6})$ — meaning that X follows a geometric distribution, and the probability of success each time is $\frac{1}{6}$.

Learn the Geometric Distribution **Formula**

If $X \sim \text{Geo}(p)$ then the probability that it takes r goes to gain a success is:

$$P(X = r) = (1 - p)^{r-1} \times p$$

Just like with the binomial distribution, the events need to be independent and have a fixed probability of success (p) and failure ($1 - p$). A big difference though is that the number of trials can go on for ever and ever and ever...

Example: $X \sim \text{Geo}(0.2)$.
Find: a) $P(X = 3)$, b) $P(X \leq 3)$, c) $P(X > 3)$

a) Use the formula — that's all you need to do...

$P(X = 3) = (1 - 0.2)^{3-1} \times (0.2) = 0.8^2 \times 0.2 = 0.128$

b) This bit's not too bad either — you need to add some probabilities together:

$P(X \leq 3) = P(X = 1) + P(X = 2) + P(X = 3) = 0.2 + (0.8 \times 0.2) + (0.8^2 \times 0.2) = 0.488$

c) If X is greater than 3, then it's not less than or equal to 3 — so subtract the previous probability from 1:

$P(X > 3) = 1 - 0.488 = 0.512$

Example: During May the probability that it will rain on a given day is 0.15.
Find the probability that the first rainy day in May is May 5th.
(Assume the probability of rain each day is independent of what happens on other days.)

You're waiting for something to happen — must be a geometric distribution.
'Success' here means a rainy day, so $X \sim \text{Geo}(0.15)$.
And you need the probability that X is 5. Use the formula:

$P(X = 5) = (1 - 0.15)^{5-1} \times (0.15) = 0.85^4 \times 0.15 = 0.0783$ to 3 sig. fig.

The Geometric Distribution

This page is for OCR S1

Déjà vu. If you can do it with the binomial distribution then this geometric stuff will be a breeze.

The **Mean** is the **Expected Number of Goes** it takes to get a **Success**

$$\text{If } X \sim \text{Geo}(p) \text{ then } E(X) = \frac{1}{p}$$

So if you're playing Ludo and you need a six to start, on average it'll take $\frac{1}{\left(\frac{1}{6}\right)} = 6$ tries to get it.

Example: If you play once a week, how long would you expect it to take to win the Lotto jackpot?

First you need the probability of success:

$$P(\text{win the jackpot}) = \frac{6\times5\times4\times3\times2\times1}{49\times48\times47\times46\times45\times44} = \frac{1}{13\,983\,816}$$ — so $X \sim \text{Geo}\left(\frac{1}{13\,983\,816}\right)$

So you'd expect to have to wait 13 983 816 weeks — or about <u>269 000 years</u>.

Example: A student is waiting for a bus. Past experience shows that 2% of vehicles in the town are buses.

a) Write down the mean number of vehicles up to and including the first bus.
b) Calculate (to 3 sig. fig.) the probability that the first bus is the 8th vehicle to arrive.
c) Calculate the probability that there is at least one bus among the first 8 vehicles.

The student's waiting for something to happen, so it's a <u>geometric distribution</u>. In fact, $X \sim \text{Geo}(0.02)$.

a) The mean is $\frac{1}{0.02} = 50$

b) Now you need the probability that the first success is on the 8th trial — use the standard formula:

$$P(X = 8) = (1 - 0.02)^{8-1} \times 0.02 = 0.0174 \text{ (to 3 sig. fig.)}$$

c) Careful — you've got a <u>fixed</u> number of events now, so this needs a <u>binomial distribution</u>.
Let the random variable Y be 'the number of buses in the first 8 vehicles'. Then $Y \sim B(8, 0.02)$.
P(there is at least one bus in the first 8 vehicles) = 1 – P(no buses in the first 8 vehicles).
This is $1 - (0.98)^8 = 1 - 0.851 = 0.149$ (to 3 sig. fig.)

Practice Questions

1) $X \sim \text{Geo}(0.1)$. Find the probability that it takes 15 attempts to record a 'success'. What is the expected value of X?

<u>Sample exam question</u>:

2) A game involves throwing 2 dice and gaining a double to start.

(a) Find the mean number of throws needed to start the game. [2 marks]

(b) Calculate (to 3 significant figures) the probability that it takes:

(i) 4 throws, [2 marks]

(ii) at least 5 throws. [3 marks]

(c) Two players are about to play the game. Find the probability that after they have each had 4 throws of the dice at least one of them has not started the game. [2 marks]

So I should buy an extra Lottery ticket then, you reckon...

Geometric distributions have a few things in common with binomial distributions — but they're definitely not the same. Geometric distributions are about the question, "*How long do I have to wait to succeed once?*" But binomial distributions are about "*How often will I succeed if I have n goes?*" Geometric distributions are easier, but binomials are more common.

The Normal Distribution

This page is for AQA S1, Edexcel S1

The normal distribution is everywhere in statistics. Everywhere, I tell you. So learn this well...

For **Continuous** Distributions, **Area = Probability**

1) With discrete random variables, there are 'gaps' between the possible values (see page 94).
2) Continuous random variables are different — there are no gaps.
3) So for a continuous random variable, you can draw the probability density function (pdf) $f(x)$ as a line.
4) The probability of the random variable taking a value between two limits is the area under the graph between those limits.

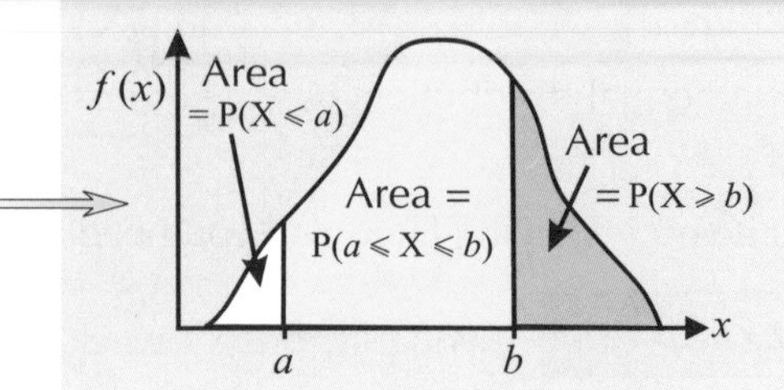

5) This means that for any single value b, P(X = b) = 0. (Since the area under a graph at a single point is zero).

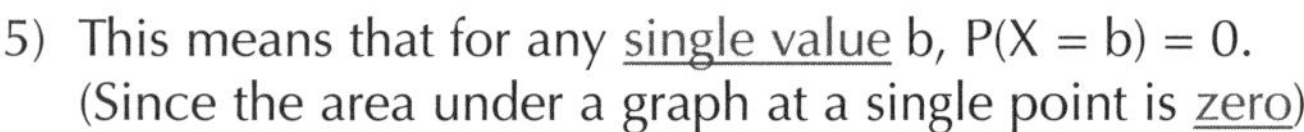

6) This also means that $P(X \le a)$ (or $P(X < a)$) is the area under the graph to the left of a. And $P(X \ge b)$ is the area to the right of b.
7) Since the total probability is 1, the total area under a pdf must also be 1.

The Normal Distribution has a **Peak** in the **Middle**

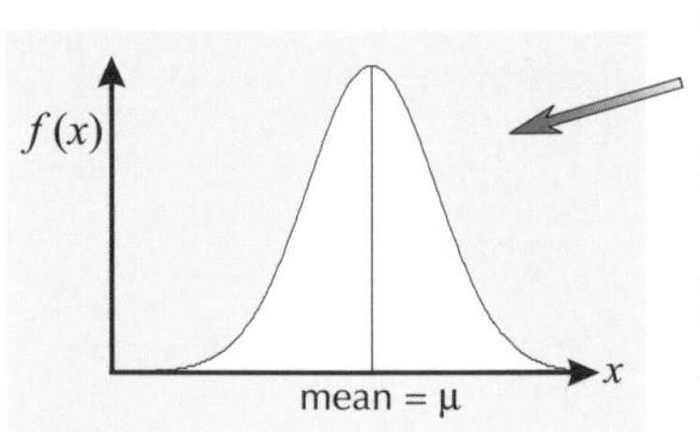

1) A random variable with a normal distribution has a pdf like this — a symmetrical bell-shaped curve.
2) The peak in the centre is at the mean (or expected value). The peak in the pdf tells you that values near the mean are most likely.
3) Further from the mean, the pdf falls — so values far from the mean are less likely.
4) The graph is symmetrical — so values the same distance above and below the mean are equally likely.
5) If X is normally distributed with **mean** μ and **variance** σ^2, it's written $X \sim N(\mu, \sigma^2)$.

Use **Normal Distribution Tables** and a **Sketch** to find probabilities

This is the practical bit.

1) Working out the area under a normal distribution curve is usually hard. But for the normally-distributed random variable Z, where **$Z \sim N(0, 1)$** (Z has mean 0 and variance 1), there are tables you can use.
2) You look up a value of z and these tables (usually labelled $\Phi(z)$) tell you the probability that $Z \le z$ (which is the area under the curve to the left of z).
3) You can convert any normally-distributed variable to Z by subtracting the mean and dividing by the standard deviation — this is called normalising. This means that if you normalise a variable, you can use the Z-tables.

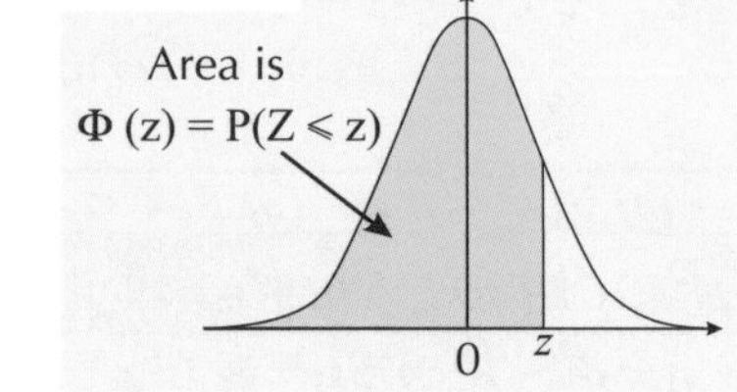

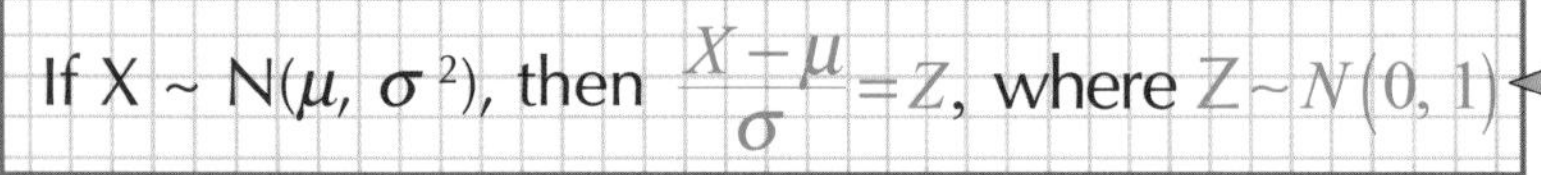

If $X \sim N(\mu, \sigma^2)$, then $\frac{X-\mu}{\sigma} = Z$, where $Z \sim N(0, 1)$

This means that if you subtract μ from any numbers in the question and then divide by σ — you can use your tables for Z.

Example: If $X \sim N(5, 16)$ find a) $P(X < 7)$, b) $P(X > 9)$, c) $P(5 < X < 11)$

Subtract μ (= 5) from any numbers and divide by σ (= $\sqrt{16} = 4$) — then you'll have a probability for $Z \sim N(0, 1)$.

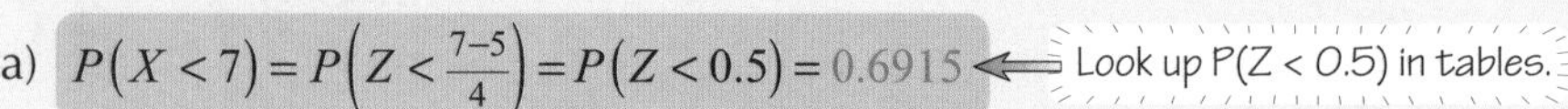

a) $P(X<7) = P\left(Z < \frac{7-5}{4}\right) = P(Z < 0.5) = 0.6915$ ← Look up P(Z < 0.5) in tables.

N(5, 16) means the variance is 16 — take the square root to find the standard deviation.

b) $P(X>9) = P\left(Z > \frac{9-5}{4}\right) = P(Z>1) = 1 - P(Z<1) = 1 - 0.8413 = 0.1587$

c) $P(5<X<11) = P\left(\frac{5-5}{4} < Z < \frac{11-5}{4}\right) = P(0<Z<1.5) = P(Z<1.5) - P(Z<0)$

$= 0.9332 - 0.5 = 0.4332$

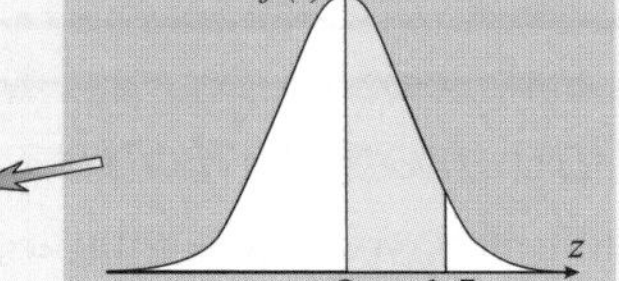

Find the area to the left of 1.5 and subtract the area to the left of 0.

The Normal Distribution

This page is for AQA S1, Edexcel S1

Draw a Sketch with Normal Probability Questions

There's usually a 'percentage points' table to help you work backwards — use it to answer questions like: How big is k if $P(Z < k) = 0.3$?

Example: $X \sim N(53, \sigma^2)$ and $P(X < 50) = 0.2$. Find σ.

It's a normal distribution — so your first thought should be to try and normalise it.

1. Subtract the mean and divide by the standard deviation:

$$P(X < 50) = P\left(Z < \frac{50-53}{\sigma}\right) = P\left(Z < -\frac{3}{\sigma}\right) = 0.2$$

Ideally, you'd look up 0.2 in the percentage points table to find $-\frac{3}{\sigma}$.
Unfortunately, in some tables it just ain't there, so you have to think a bit...

This area is 0.2... ...so this area must be 0.2 as well.

$-\frac{3}{\sigma}$ $\frac{3}{\sigma}$ Z

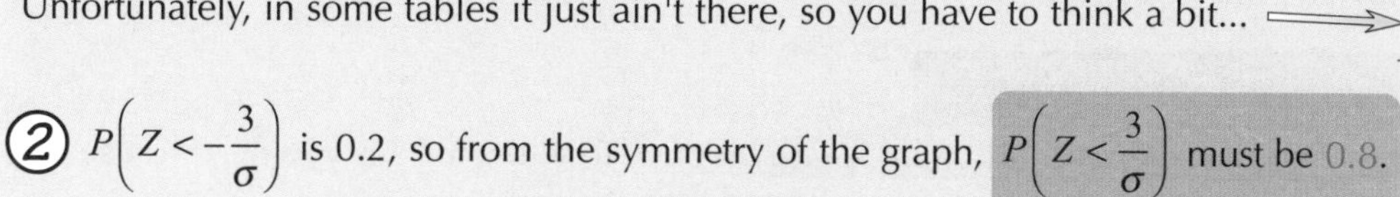

2. $P\left(Z < -\frac{3}{\sigma}\right)$ is 0.2, so from the symmetry of the graph, $P\left(Z < \frac{3}{\sigma}\right)$ must be 0.8.

So look up 0.8 in the 'percentage points' table to find that an area of 0.8 is to the left of $z = 0.8416$.

This tells you that $\frac{3}{\sigma} = 0.8416$, or $\sigma = 3.56$ (to 3 sig. fig.)

The 68–95–99.7% Rules

The following rules are true for all normal distribution curves:

"Approx. **68**% of observations lie within $\mu \pm \sigma$"
"Approx. **95**% of observations lie within $\mu \pm 2\sigma$"
"Approx. **99.8**% of observations lie within $\mu \pm 3\sigma$"

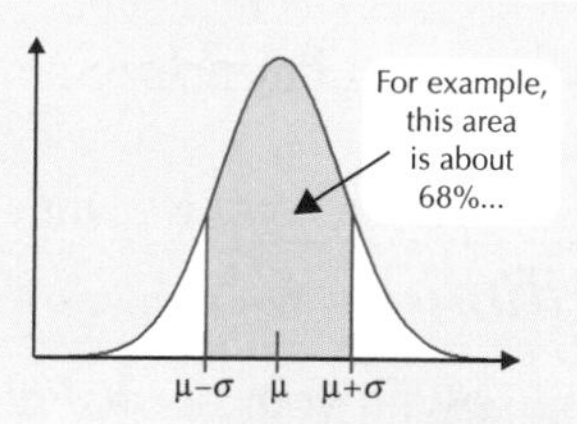
For example, this area is about 68%...
μ−σ
μ
μ+σ

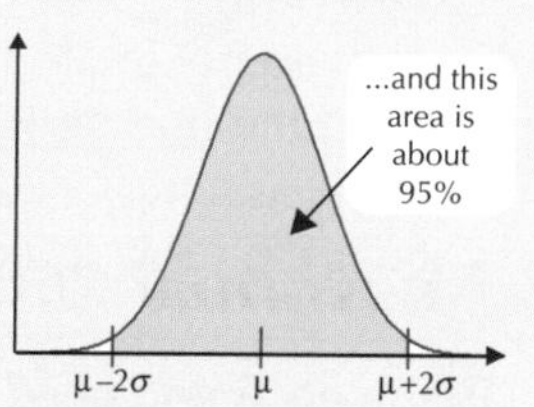
...and this area is about 95%
μ−2σ
μ
μ+2σ

Practice Questions

Practise these questions with the kind of tables you'll be using in the Exam, since the tables from the various Exam boards all look a bit different.

1) If $X \sim N(50, 16)$, find:
(a) $P(X < 55)$, (b) $P(X < 42)$, (c) $P(X > 56)$ and (d) $P(47 < X < 57)$.

2) $X \sim N(600, 202)$
(a) If $P(X < a) = 0.95$, find a.
(b) If $P(|X - 600| < b) = 0.8$, find b.

Sample exam questions:

3) The exam marks of 1000 candidates are normally distributed with mean 50 marks and standard deviation 30 marks.
(a) The pass mark is 41. Estimate the number of candidates who passed the exam. [3 marks]
(b) Find the mark required for an A-grade if 10% of the candidates achieved a grade A. [3 marks]

4) The lifetimes of a particular type of battery are normally distributed with mean μ and standard deviation σ.
A student using these batteries finds that 40% last less than 20 hours and 80% last less than 30 hours.
Find μ and σ. [7 marks]

The medium of a random variable follows a paranormal distribution...

Remember... it's definitely worth drawing a quick sketch when you're finding probabilities using a normal distribution — you're much less likely to make a daft mistake. Also, remember that it's $N(\mu, \sigma^2)$ — with the variance in the brackets and not the standard deviation. This topic isn't too bad once you're happy using the tables. So get hold of some and practise.

The Normal Distribution

This page is for AQA S1

The normal distribution is very useful. And trust me — there will be a question on it in the Exam.

The Normal Distribution is used to *Approximate* the *Binomial*

It can be a right pain to work out numbers to do with a binomial distribution. Luckily, as long as a few conditions are met, you can use a normal distribution instead — since it'll give you pretty much the same answers.

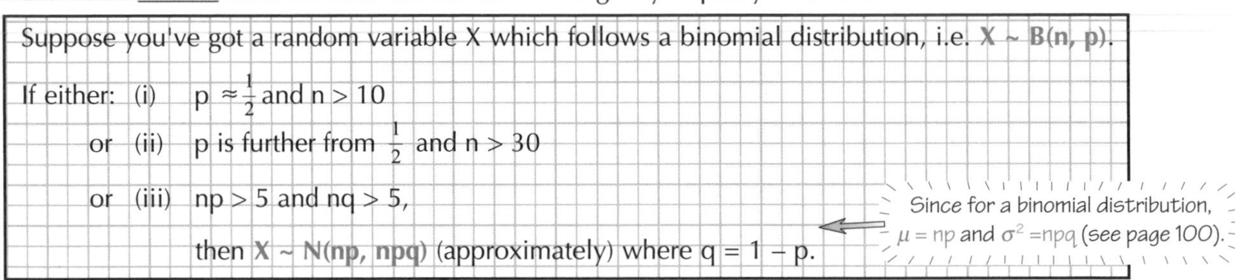

Suppose you've got a random variable X which follows a binomial distribution, i.e. **X ~ B(n, p)**.

If either: (i) $p \approx \frac{1}{2}$ and $n > 10$

or (ii) p is further from $\frac{1}{2}$ and $n > 30$

or (iii) $np > 5$ and $nq > 5$,

then **X ~ N(np, npq)** (approximately) where $q = 1 - p$.

Since for a binomial distribution, $\mu = np$ and $\sigma^2 = npq$ (see page 100).

You need to use a *Continuity Correction*

The binomial distribution is discrete but the normal distribution is continuous. To allow for this you need to use a continuity correction. Like a lot of this stuff, it sounds more complicated than it is.

A binomially-distributed variable X is discrete, so you can work out P(X = 0), P(X = 1), etc.

A normally-distributed variable is continuous, and so P(X = 0) = P(X = 1) = 0, etc. (see page 104).

So what you do is assume that the 'binomial 1' is spread out over the interval 0.5 - 1.5.

0 ⟵1⟶⟵2⟶⟵3⟶⟵4⟶

0.5 1.5 2.5 3.5 4.5

Then to approximate the binomial P(X = 1), you find the normal P(0.5 < X < 1.5).

Similarly, the 'binomial 2' is spread out over the interval 1.5 - 2.5 and so on.

The interval you need to use with your normal distribution depends on the binomial probability you're trying to find out.

The general principle is the same, though — each binomial value *b* covers the interval from $b-\frac{1}{2}$ up to $b+\frac{1}{2}$.

Binomial	Normal	
$P(X = b)$	$P(b - \frac{1}{2} < X < b + \frac{1}{2})$	
$P(X \leq b)$	$P(X < b + \frac{1}{2})$	...to include b
$P(X < b)$	$P(X < b - \frac{1}{2})$	...to exclude b
$P(X \geq b)$	$P(X > b - \frac{1}{2})$	...to include b
$P(X > b)$	$P(X > b + \frac{1}{2})$	...to exclude b

Example: If X ~ B(80, 0.4) find: (i) P(X < 45) and (ii) P(X ≥ 40)

1. You need to make sure first that the normal approximation is okay to use...
   n > 30 and p isn't too far from $\frac{1}{2}$ so the normal approximation is valid.

   Or you could say that both np = 32 and nq = 48 are greater than 5.

2. Next, work out np and npq: $np = 80 \times 0.4 = 32$ and $npq = 80 \times 0.4 \times (1 - 0.4) = 19.2$ ⟸ q = 1 − p
   So the approximation you need is: X ~ N(32, 19.2)

3. Now you can work out your probabilities by normalising everything and using your tables for Z $(= \frac{X-32}{\sqrt{19.2}})$

   (i) You need P(X < 45) — so with the continuity correction this is P(X < 44.5).

   $$P(X < 44.5) = P\left(Z < \frac{44.5-32}{\sqrt{19.2}}\right) = P(Z < 2.853) = 0.9978$$

   (ii) Now you need P(X ≥ 40) — with the continuity correction this is P(X > 39.5).

   $$P(X > 39.5) = P\left(Z > \frac{39.5-32}{\sqrt{19.2}}\right) = P(Z > 1.712) = 1 - P(Z < 1.712) = 1 - 0.9566 = 0.0434$$

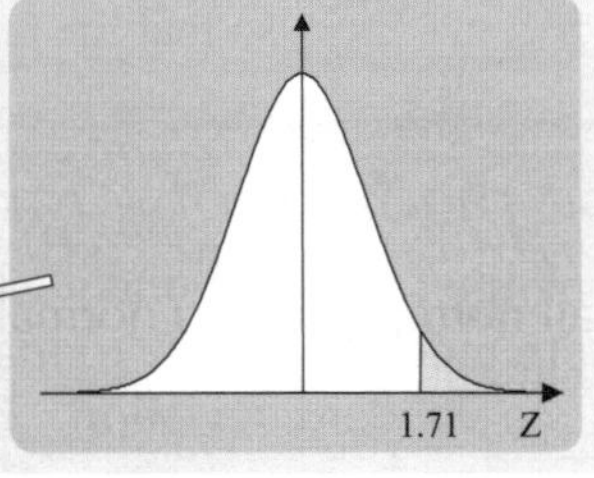

The Normal Distribution

The Sampling Distribution of $\bar{X}$

This page is for AQA S1

1) Suppose you've got a random variable $X \sim N(\mu, \sigma^2)$.
2) When you take a sample of n readings from the distribution of X, you can work out the sample mean, $\bar{X}$.
3) If you now keep taking samples of size n from that distribution and working out the sample means, then you get a collection of sample means — these sample means will also be normally distributed with mean μ and variance $\frac{\sigma^2}{n}$.

> The sampling distribution of $\bar{X}$ is $\bar{X} \sim N\left(\mu, \frac{\sigma^2}{n}\right)$

4) So the standard deviation of the sampling distribution of $\bar{X}$ is $\frac{\sigma}{\sqrt{n}}$ — this is the standard error of $\bar{X}$.
5) To estimate the standard error of $\bar{X}$ when the population variance σ^2 is unknown, you use the estimator:

> $\frac{S}{\sqrt{n}}$, where $S = \sqrt{\frac{\sum(X - \bar{X})^2}{n-1}}$

The Central Limit Theorem

This bit's useful.

Now suppose you take a sample of n readings from ANY distribution with mean μ and variance σ^2.

> For large n, the distribution of the sample mean, $\bar{X}$, is approximately normal with $\bar{X} \sim N\left(\mu, \frac{\sigma^2}{n}\right)$

The bigger n is, the better the approximation will be. (For $n > 30$ it's pretty good.)

Example: A sample of size 50 is taken from a population with mean 20 and variance 10. Find the probability that the sample mean is less than 19.

① Since n (= 50) is quite large, you can use the Central Limit Theorem. Here $\bar{X} \sim N\left(20, \frac{10}{50}\right)$ i.e. $\bar{X} \sim N(20, 0.2)$

② You need $P(\bar{X} < 19)$. Since $\bar{X}$ has an (approximately) normal distribution, you can normalise it and use tables for Z. So subtract the mean from all the numbers and divide by the standard deviation ($= \sqrt{0.2} = 0.4472$):

$$P(\bar{X} < 19) = P\left(Z < \frac{19-20}{0.4472}\right) = P(Z < -2.236)$$

③ Now it's best to draw a sketch:

$$P(Z < -2.236) = P(Z > 2.236) = 1 - P(Z < 2.236) = 1 - 0.9873 = 0.0127$$

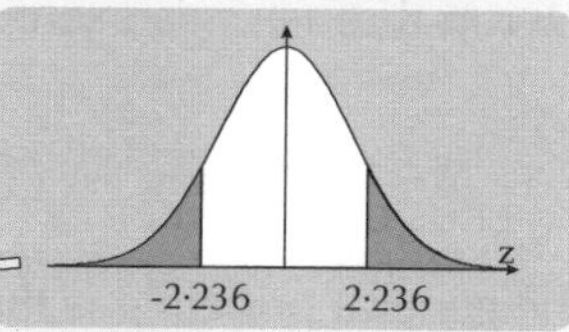

Practice Questions

1) A ~ B(200, 0.05), where A is a random variable. Use a suitable approximation to find the probability that A is less than 11.

2) 30% of the plants in a nursery are short of water. What is the probability that in a sample of 500 plants, the number of plants which are short of water is less than 149?

Sample exam questions:

3) A game involves two throws of a fair six-sided dice. If 200 people play, use a suitable approximation to find the probability that at least 10 people obtain two sixes. [6 marks]

4) A sample of 50 bags of sweets is taken from a supply which has known mean 30 g and standard deviation 5 g.
   (a) State the distribution of the sample mean. [1 mark]
   (b) Find the probability that the mean of the sample is less than 29 g. [2 marks]
   (c) If a sample of size n bags were to be taken from the same supply, find the minimum value of n so that the probability of getting the mean of the sample to less than 29 g is less than 5%. [5 marks]

Normal, normal everywhere — not a nice thought to think...

It's all normal this and normal that at the moment. If you haven't already, then now's the time to accept that you won't get far without knowing how to answer a normal question. Personally, I can't help but think the Central Limit Theorem's pretty amazing — you can take samples from any distribution you like and say something meaningful (excuse pun) about the mean.

Confidence Intervals

This page is for AQA S1

Confidence intervals are used in making predictions about populations based on sample data. Easier done than said.

A *Confidence Interval* is a Type of *Interval Estimation...*

First things first — what is a confidence interval? Look at this example:

"(11.2, 14.4) is a 95% confidence interval for μ"

This means that there's a 95% chance that the interval (11.2, 14.4) includes the true value of μ.

(It's not the same as saying that there's a 95% chance that the true value of μ lies between 11.2 and 14.4. If you take 10 samples from the same distribution and calculate a 95% confidence interval for μ from each sample, you'll probably get 10 different intervals. So it's the interval which varies, not the value of μ, which is fixed.)

Estimating the *Mean* of a *Normal Distribution* when You *Know* the Variance

1) Suppose you've got a normally distributed variable X with mean μ (which you don't know) and variance σ^2 / standard deviation σ (which you do know).
2) You take a sample of n readings from the distribution of X and work out the sample mean, $\bar{X}$.
3) You can now work out a confidence interval for μ:

The confidence interval for μ is $\left(\bar{X}-z\left(\frac{\sigma}{\sqrt{n}}\right), \bar{X}+z\left(\frac{\sigma}{\sqrt{n}}\right)\right)$

z = the z-value for the % confidence you want.

Look Your *z-value* Up in the *z-tables*

I'll run through this example for 95% confidence. For other % values, engage brain and follow the same method.

1) You're looking for $P(-z < Z < z) = 0.95$.
2) From the diagram, you can see that $P(Z < z) = 0.95 + 0.025 = 0.975$.
3) Looking up 0.975 in the z-tables gives you z = 1.9600.

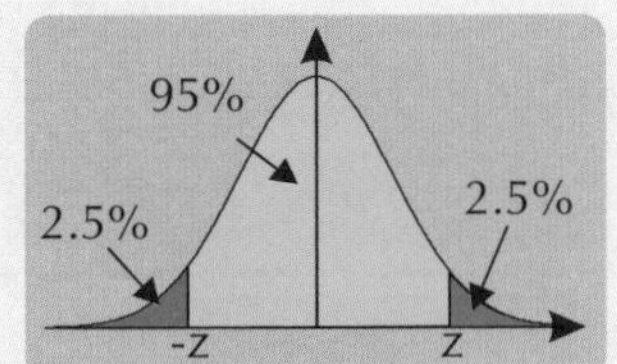

Example: A machine makes chocolate bars with weights that are normally distributed with a standard deviation of 8.2 grams. A random sample of 10 bars is chosen, with weights (in grams) as follows: 251.3 257.2 249.5 254.3 252.6 256.1 254.7 255.3 254.8 252.6.
Calculate a 95% confidence interval for the mean of all the bars made by the machine.

① Start by working out $\bar{X}$:

(251.3 + 257.2 + 249.5 + 254.3 + 252.6 + 256.1 + 254.7 + 255.3 + 254.8 + 252.6) ÷ 10 = 253.84

② Next, find the value you need for z in your z-tables. In this case, z = 1.96.

③ Now just plug all your values into the confidence interval expression:

$$\left(253.84-1.96\left(\frac{8.2}{\sqrt{10}}\right), 253.84+1.96\left(\frac{8.2}{\sqrt{10}}\right)\right) = (248.76, 258.92)$$

...And of course, you can use exactly this same method if the variable X isn't normally distributed — as long as n is over about 30 (see p.107).

Confidence Intervals

This page is for AQA S1

If You Don't Know the Variance of the Population, You Have to Estimate It

1) Suppose you've got a random variable X of any distribution and you don't know the mean μ or the variance σ^2.
2) If your sample size, n, is over about 30, you can work out confidence intervals in the same way as before, except now you have to estimate the population variance σ^2.
3) Since n is large, you can estimate σ^2 as being equal to S^2, the variance of your sample. The confidence interval for μ becomes:

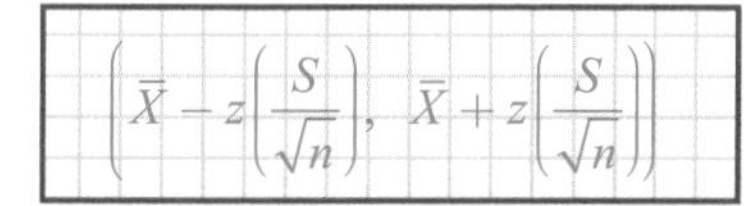

$$\left(\bar{X} - z\left(\frac{S}{\sqrt{n}}\right),\ \bar{X} + z\left(\frac{S}{\sqrt{n}}\right)\right)$$

Confidence Intervals can be Used to Check a Claim of the Population Mean

1) If you've found a confidence interval for a sample mean, you can use it to check a value that someone claims is the population mean.
2) If the claimed population mean lies outside the confidence interval, then that's evidence the claim might not be true.
3) If the claimed population mean lies within the confidence interval, there is no evidence the claim isn't true.
4) For instance, in the example on p.108, your 95% confidence interval for the mean mass of the bars of chocolate is from 248.76 grams to 258.92 grams.
5) If a manager claims the mean weight of bars produced is 250 grams, this is within the confidence interval — you have no evidence to doubt the claim.
6) If a customer says they think the bars in fact have a mean of 245 grams, this is outside the confidence interval — your sample does not give evidence to support this claim.

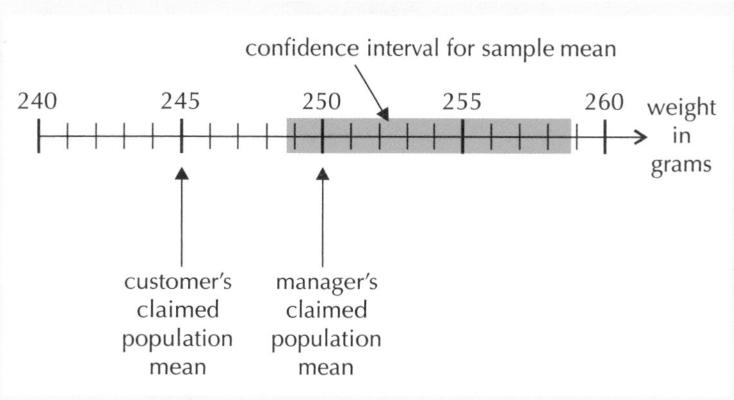

Practice Questions

1) ***Based on past experience, the contents of packets of a brand of crisps have weights that are normally distributed, with a standard deviation of 3.4 grams. The weights of the contents of a random sample of eight packets (in grams) are as follows:***

35.7 38.5 40.3 39.4 37.2 32.1 30.4 36.3

(a) Find a 95% confidence interval for the mean weight of the contents of the packets of crisps, based on this sample.

(b) A customer claims that the mean weight of the contents of the packets of crisps is under 35 grams. Investigate this claim.

Sample exam question:

2) A random sample of 50 people chosen from a population were asked how far their workplace was from their home. The mean and standard deviation of their replies were 4.5 km and 3.2 km respectively.

(a) Find a 90% confidence interval for the mean distance travelled to work by people in this population, based on this sample. [4 marks]

(b) Why isn't it necessary to assume that the distribution of the distances travelled is normal for this confidence interval to be valid? [2 marks]

Mmm, chocolate...

Wow, fantastically useful. Actually, it is. Using just 30 samples from a population, you can say quite a lot about the mean and variance of the population as a whole. Of course, bear in mind that your 95% confidence interval does have a 5% chance of being off the mark. The only thing you can say with 100% confidence is "$-\infty < \mu < \infty$" (which isn't much help).

Correlation

This page is for AQA S1, OCR S1, Edexcel S1

Correlation is all about how closely two quantities are linked. And it can involve a fairly hefty formula.

Draw a **Scatter Diagram** to see **Patterns** in Data

Sometimes variables are measured in pairs — maybe because you want to find out how closely they're linked.
These pairs of variables might be things like: — 'my age' and 'length of my feet', or
— 'temperature' and 'number of accidents on a stretch of road'.

You can plot readings from a pair of variables on a scatter diagram — this'll tell you something about the data.

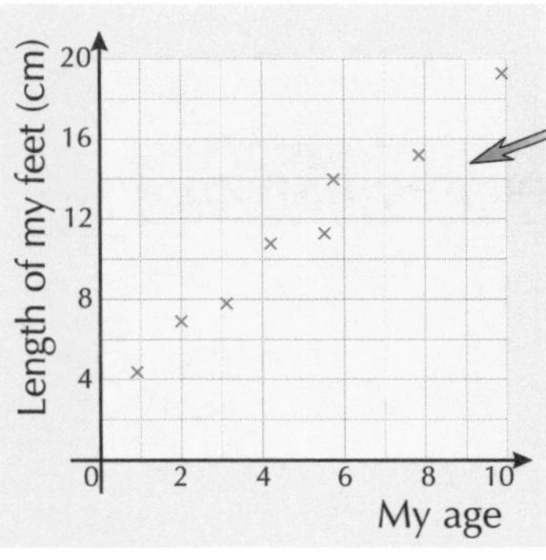

The variables 'my age' and 'length of my feet' seem linked — all the points lie close to a line. As I got older, my feet got bigger and bigger (though I stopped measuring when I was 10).

It's a lot harder to see any connection between the variables 'temperature' and 'number of accidents' — the data seems scattered pretty much everywhere.

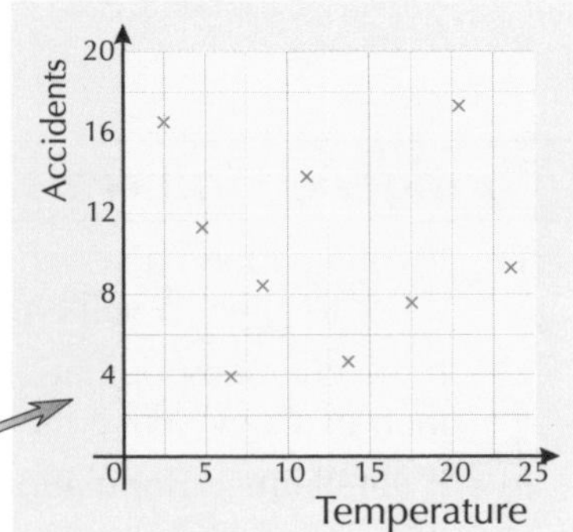

Correlation is a measure of **How Closely** variables are **Linked**

1) Sometimes, as one variable gets bigger, the other one also gets bigger — then the scatter diagram might look like the one on the right. Here, a line of best fit would have a positive gradient. The two variables are positively correlated (or there's a positive correlation between them).

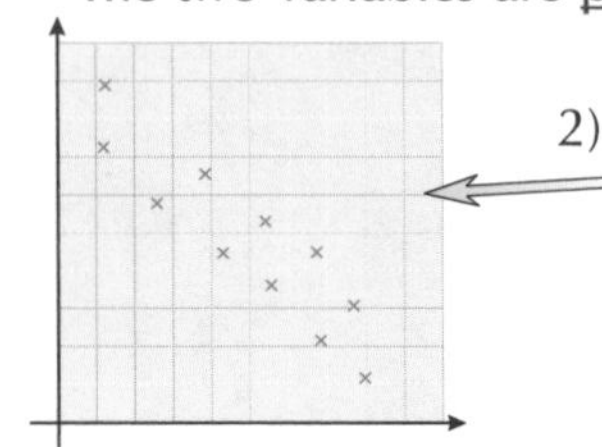

2) But if one variable gets smaller as the other one gets bigger, then the scatter diagram might look like this one — and the line of best fit would have a negative gradient. The two variables are negatively correlated (or there's a negative correlation between them).

3) And if the two variables aren't linked at all, you'd expect a random scattering of points — it's hard to say where the line of best fit would be. The variables aren't correlated (or there's no correlation).

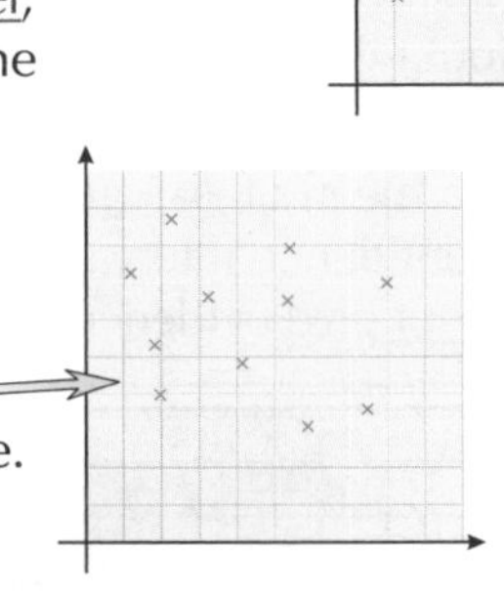

The **Product-Moment Correlation Coefficient** (*r*) measures Correlation

1) The Product-Moment Correlation Coefficient (PMCC, or r, for short) measures how close to a straight line the points on a scatter graph lie.

2) The PMCC is always between +1 and –1.
If all your points lie exactly on a straight line with a positive gradient (perfect positive correlation), $r = +1$.
If all your points lie exactly on a straight line with a negative gradient (perfect negative correlation), $r = -1$.
(In reality, you'd never expect to get a PMCC of +1 or –1 — your scatter graph points might lie pretty close to a straight line, but it's unlikely they'd all be on it.)

3) If $r = 0$ (or more likely, pretty close to 0), that would mean the variables aren't correlated.

4) The formula for the PMCC is a real stinker. The best thing is to type the pairs of readings into your calculator, and let it work the PMCC out for you. Otherwise, just take it nice and slow and use the formula.

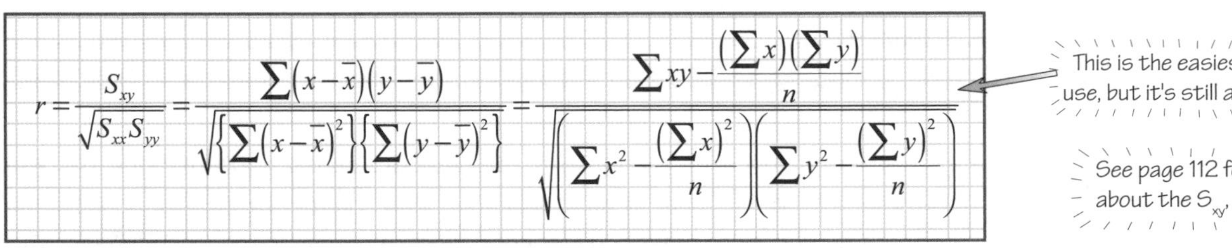

$$r = \frac{S_{xy}}{\sqrt{S_{xx}S_{yy}}} = \frac{\sum(x-\bar{x})(y-\bar{y})}{\sqrt{\left\{\sum(x-\bar{x})^2\right\}\left\{\sum(y-\bar{y})^2\right\}}} = \frac{\sum xy - \frac{(\sum x)(\sum y)}{n}}{\sqrt{\left(\sum x^2 - \frac{(\sum x)^2}{n}\right)\left(\sum y^2 - \frac{(\sum y)^2}{n}\right)}}$$

This is the easiest one to use, but it's still a bit hefty.

See page 112 for more about the S_{xy}, S_{xx} one.

Correlation

This page is for AQA S1, OCR S1, Edexcel S1

The correlation formula's a little on the large side. So don't rush this kind of question...

It's best to make a **Table**

Example: Illustrate the following data with a scatter diagram, and find the product-moment correlation coefficient (r) between the variables x and y.

x	1.6	2.0	2.1	2.1	2.5	2.8	2.9	3.3	3.4	3.8	4.1	4.4
y	11.4	11.8	11.5	12.2	12.5	12.0	12.9	13.4	12.8	13.4	14.2	14.3

1) The scatter diagram's the easy bit — just plot the points.

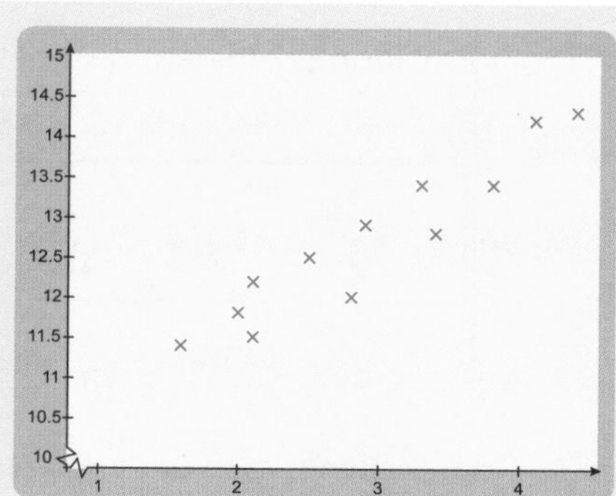

Now for the correlation coefficient. From the scatter diagram, the points lie pretty close to a straight line with a positive gradient — so if the correlation coefficient doesn't come out pretty close to +1, we'd need to worry...

2) There are 12 pairs of readings, so n = 12. That bit's easy — now you have to work out a load of sums. It's best to add a few extra rows to your table...

x	1.6	2	2.1	2.1	2.5	2.8	2.9	3.3	3.4	3.8	4.1	4.4	$35 = \Sigma x$
y	11.4	11.8	11.5	12.2	12.5	12	12.9	13.4	12.8	13.4	14.2	14.3	$152.4 = \Sigma y$
x^2	2.56	4	4.41	4.41	6.25	7.84	8.41	10.89	11.56	14.44	16.81	19.36	$110.94 = \Sigma x^2$
y^2	129.96	139.24	132.25	148.84	156.25	144	166.41	179.56	163.84	179.56	201.64	204.49	$1946.04 = \Sigma y^2$
xy	18.24	23.6	24.15	25.62	31.25	33.6	37.41	44.22	43.52	50.92	58.22	62.92	$453.67 = \Sigma xy$

Stick all these in the formula to get: $r = \dfrac{\left(453.67 - \frac{35\times152.4}{12}\right)}{\sqrt{\left(110.94 - \frac{(35)^2}{12}\right)\times\left(1946.04 - \frac{(152.4)^2}{12}\right)}} = \dfrac{9.17}{\sqrt{8.857\times10.56}} = 0.948$ (to 3 s.f.)

This is pretty close to 1, so there's a high positive correlation between x and y.

Practice Questions

1. Plot a scatter diagram and calculate the product-moment coefficient of correlation for the data below.

Height (cm)	***165***	***176***	***159***	***167***	***174***	***171***	***169***	***168***	***169***	***172***
Weight (kg)	***72***	***90***	***70***	***75***	***86***	***84***	***80***	***81***	***82***	***83***

What does the value of the product-moment coefficient of correlation tell you about the data?

Sample Exam Question:

2. Values of two variables x and y obtained from a survey are recorded in the table below.

x	1	2	3	4	5	6	7	8
y	0.50	0.70	0.10	0.82	0.64	0.36	0.16	0.80

Represent these data on a scatter diagram, and obtain the product-moment correlation coefficient (PMCC) between the two variables.

What does this tell you about the variables? [9 marks]

What's a statistician's favourite soap — Correlation Street... *(Boom boom)*

It's worth remembering that the PMCC assumes that both variables are normally distributed — chances are you won't get asked a question about that, but there's always the possibility that you might, so learn it. If you know the original readings, then you can use a calculator to get the PMCC directly, but that's not an excuse for not knowing how to use the formula.

Correlation

This page is for AQA S1, OCR S1, Edexcel S1

Sometimes they don't give you all the data — just summations (things with Σ in). Just bung the numbers in the formula.

Sometimes they give you **Summary Data**

Example: I spread different amounts of manure (x) around my 10 tomato plants last summer. When I picked my toms, I measured the weight of tomatoes I got from each plant (y). Work out the correlation coefficient between x and y using the figures below.

$$\sum x = 16.5,\ \sum y = 22.1,\ \sum x^2 = 40.69,\ \sum y^2 = 59.81,\ \sum xy = 48.29$$

The correlation coefficient is sometimes written as: $r = \dfrac{S_{xy}}{\sqrt{S_{xx}S_{yy}}}$,

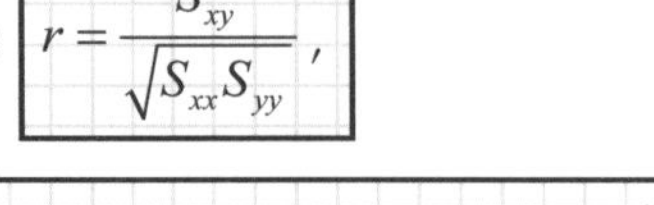

where $S_{xy} = \sum(x-\bar{x})(y-\bar{y}) = \sum xy - \dfrac{\sum x \sum y}{n}$ and $S_{xx} = \sum(x-\bar{x})^2 = \sum x^2 - \dfrac{\left(\sum x\right)^2}{n}$

S_{yy} is the same as this, but with the x's replaced with y's.

The versions in red are easier to use.

These same formulas come up on pages 114-115 as well — in regression.

Same as ever — just stick the numbers in the formula to get the correlation coefficient:

$$r = \frac{S_{xy}}{\sqrt{S_{xx}S_{yy}}} = \frac{48.29 - \frac{16.5\times 22.1}{10}}{\sqrt{\left(40.69 - \frac{16.5^2}{10}\right)\left(59.81 - \frac{22.1^2}{10}\right)}} = \frac{11.825}{\sqrt{13.465\times 10.969}} = 0.973 \text{ (to 3 sig. fig.)}$$

The **PMCC** isn't affected by a **Linear Transformation**

Correlation coefficients aren't affected by linear transformations. In normal language, this means you can multiply variables by a number, and add a number to them — and you won't change the PMCC between them.

Example: The variables p and q are given by: p = 4x – 3 and q = 9y + 17, where x and y are as in the example above. Work out the PMCC between p and q.

Easy — p and q are linear transformations of x and y (i.e. there are no x^2 or xy bits flying around). So the PMCC between p and q is the same as the PMCC between x and y.

The PMCC between p and q is given by r = 0.973 (to 3 sig. fig.)

Don't make **Sweeping Statements** using Statistics

This mini-section's all about not making 'sweeping statements' (always a temptation where statistics are concerned).

1) A high correlation coefficient doesn't necessarily mean that one quantity causes the other.

Example: The number of televisions sold in Japan and the number of cars sold in America may well be correlated, but that doesn't mean that high TV sales in Japan cause high car sales in the US.

2) The PMCC is only a measure of a linear relationship between two variables (i.e. how close they'd be to a line if you plotted a scatter diagram).

Example: In the diagram on the right, the PMCC would probably be pretty low, but the two variables definitely look linked. It looks like the points lie on a parabola (the shape of an x^2 curve) — not a line.

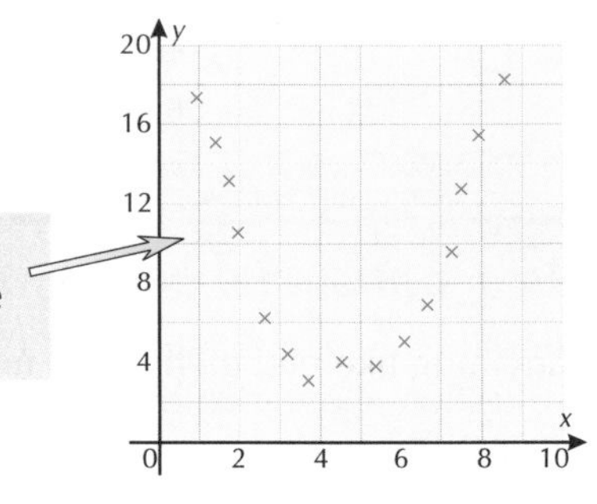

Correlation

This page is for OCR S1

This is another correlation coefficient, but you can use it where you couldn't use the PMCC.

Spearman's Rank Correlation Coefficient (**SRCC** or r_s) works with **Ranks**

You can use the SRCC (or r_s, for short) when your data is a set of ranks. (Ranks are the positions of the values when you put them in order — e.g. from biggest to smallest, or from best to worst, etc.)

Example: At a dog show, two judges put 8 labradors (A-H) in the following orders, from best to worst. Calculate the SRCC between the sets of ranks.

Position	1st	2nd	3rd	4th	5th	6th	7th	8th
Judge 1:	B	C	E	A	D	F	G	H
Judge 2:	C	B	E	D	F	A	G	H

First, make a table of the ranks of the 8 labradors — i.e. for each dog, write down where it came in the show.

Dog	A	B	C	D	E	F	G	H
Rank from Judge 1:	4	1	2	5	3	6	7	8
Rank from Judge 2:	6	2	1	4	3	5	7	8

Now for each dog, work out the difference (d) between the ranks from the two judges — you can ignore minus signs.

Dog	A	B	C	D	E	F	G	H
d	2	1	1	1	0	1	0	0

Take a deep breath, and add another line to your table — this time for d^2:

Dog	A	B	C	D	E	F	G	H	Total = Σd^2
d^2	4	1	1	1	0	1	0	0	8

Then the SRCC is:

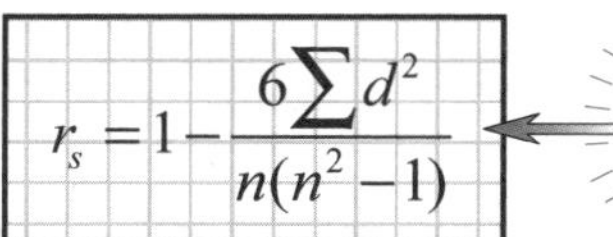

$$r_s = 1 - \frac{6\sum d^2}{n(n^2-1)}$$

You can ignore minus signs when you work out d, since only d^2 is used to work out the SRCC.

So here, $r_s = 1 - \frac{6\times 8}{8\times(8^2-1)} = 1 - \frac{48}{504} = 0.905$ (to 3 sig. fig.)

— this is close to +1, so the judges ranked the dogs in a pretty similar way.

If there'd been 2 or more equal values, you'd need to find the average rank. E.g. if Judge 1 had awarded joint 7th place to G and H, then they'd each have rank $\frac{7+8}{2} = 7.5$ (i.e. they would have been 7th and 8th)

Practice Questions

1. These are the marks obtained by 10 pupils in their Physics and English exams. Calculate Spearman's rank correlation coefficient.

Physics	54	34	23	58	52	58	13	65	69	52
English	16	73	89	81	23	81	56	62	61	37

Sample Exam question:

2. A chocolate manufacturer sells eight products. The sales of each product in a certain period (y) and the amount of money spent advertising each product (x) were recorded.

(a) Using the information below, calculate the product-moment correlation coefficient between the variables 'sales' and 'advertising costs'.

$\sum x = 386$, $\sum y = 460$, $\sum x^2 = 25426$, $\sum y^2 = 28867$, $\sum xy = 26161$ [5 marks]

(b) What does this result tell you about the quantities x and y? [2 marks]

I don't like this page — it's a bit rank...

The good thing about the SRCC is that you can use it even when your data isn't normally distributed. More generally, if you're anything like me, then you'll find it easy to scribble a load of important-looking numbers down on a piece of paper, and then forget what any of them actually are. So do yourself a favour — take tricky questions slowly, be nice and methodical, and write down every step of your working clearly. It makes it easier to find any mistakes later.

Linear Regression

This page is for AQA S1, OCR S1, Edexcel S1

Linear regression is just fancy stats-speak for 'finding lines of best fit'. Not so scary now, eh...

Decide which is the *Independent Variable* and which is the *Dependent*

Example: The data below show the load on a lorry, x (in tonnes), and the fuel consumption, y (in km per litre).

x	5.1	5.6	5.9	6.3	6.8	7.4	7.8	8.5	9.1	9.8
y	9.6	9.5	8.6	8.0	7.8	6.8	6.7	6	5.4	5.4

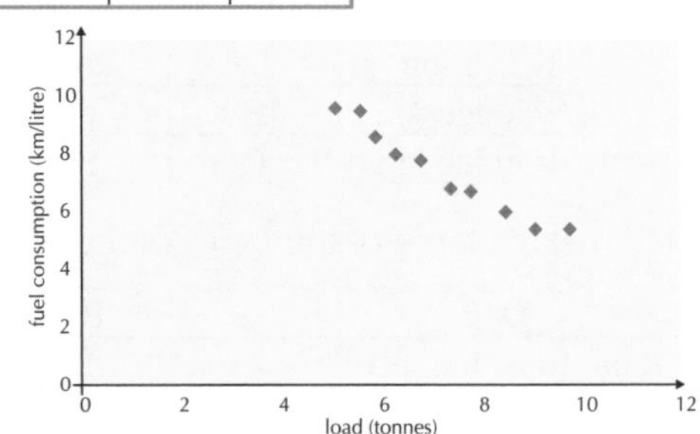

1) The variable along the x-axis is the explanatory or independent variable — it's the variable you can control, or the one that you think is affecting the other.
   The variable 'load' goes along the x-axis here.
2) The variable up the y-axis is the response or dependent variable — it's the variable you think is being affected.
   In this example, this is the fuel consumption.

The *Regression Line* (Line of Best Fit) is in the form *y = a + bx*

To find the line of best fit for the above data you need to work out some sums.
Then it's quite easy to work out the equation of the line. If your line of best fit is $y = a + bx$, this is what you do...

(1) First work out these four sums — a table is probably the best way: $\sum x$, $\sum y$, $\sum x^2$, $\sum xy$.

x	5.1	5.6	5.9	6.3	6.8	7.4	7.8	8.5	9.1	9.8	$72.3 = \Sigma x$
y	9.6	9.5	8.6	8	7.8	6.8	6.7	6	5.4	5.4	$73.8 = \Sigma y$
x^2	26.01	31.36	34.81	39.69	46.24	54.76	60.84	72.25	82.81	96.04	$544.81 = \Sigma x^2$
xy	48.96	53.2	50.74	50.4	53.04	50.32	52.26	51	49.14	52.92	$511.98 = \Sigma xy$

(2) Then work out S_{xy}, given by: $S_{xy} = \sum(x-\bar{x})(y-\bar{y}) = \sum xy - \frac{(\sum x)(\sum y)}{n}$

and S_{xx}, given by: $S_{xx} = \sum(x-\bar{x})^2 = \sum x^2 - \frac{(\sum x)^2}{n}$

These are the same as the terms used to work out the PMCC (see p. 112).

(3) The gradient (b) of your regression line is given by: $b = \frac{S_{xy}}{S_{xx}}$

Loads of calculators will work out regression lines for you — but you still need to know this method, since they might give you just the sums from Step 1.

(4) And the intercept (a) is given by: $a = \bar{y} - b\bar{x}$.

(5) Then the regression line is just: $y = a + bx$.

The '**regression line of y on x**' means x is the independent variable, and y is the dependent variable. The '**regression line of x on y**' would be where that y is independent, and x is dependent — it'd have a different gradient.

Example: Find the equation of the regression line of y on x for the data above.

1) Work out the sums: $\sum x = 72.3$, $\sum y = 73.8$, $\sum x^2 = 544.81$, $\sum xy = 511.98$.
2) Then work out S_{xy} and S_{xx}: $S_{xy} = 511.98 - \frac{72.3 \times 73.8}{10} = -21.594$ $\quad S_{xx} = 544.81 - \frac{72.3^2}{10} = 22.081$
3) So the gradient of the regression line is: $b = \frac{-21.594}{22.081} = -0.978$ (to 3 sig. fig.)
4) And the intercept is: $a = \frac{\sum y}{n} - b\frac{\sum x}{n} = \frac{73.8}{10} - (-0.978) \times \frac{72.3}{10} = 14.451$
5) This all means that your regression line is: $y = 14.451 - 0.978x$

Remember: $\bar{x} = \frac{\sum x}{n}$

The regression line always goes through the point $(\bar{x}, \bar{y})$.

Linear Regression

This page is for AQA S1, OCR S1, Edexcel S1

*Residuals — the difference between **Practice** and **Theory***

A residual is the difference between an observed y-value and the y-value predicted by the regression line.

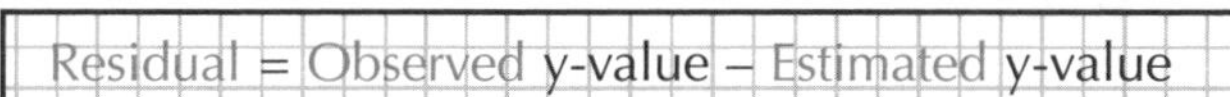
Residual = Observed y-value – Estimated y-value

1) Residuals show the experimental error between the y-value that's observed and the y-value your regression line says it should be.
2) Residuals are shown by a vertical line from the actual point to the regression line.

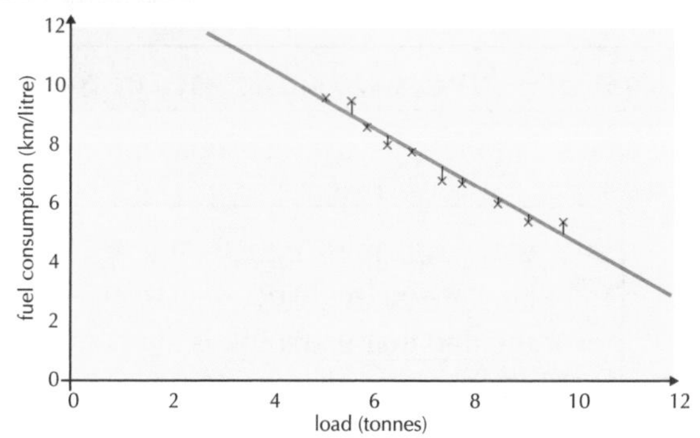

Example: For the fuel consumption example opposite, calculate the residuals for: (i) x = 5.6, (ii) x = 7.4.

(i) When x = 5.6, the residual = 9.5 – (-0.978 × 5.6 + 14.451) = 0.526 (to 3 sig. fig.)
(ii) When x = 7.4, the residual = 6.8 – (-0.978 × 7.4 + 14.451) = -0.414 (to 3 sig. fig.)

A positive residual means the regression line is too low for that value of x.
A negative residual means the regression line is too high.
If the residual is relatively large (positive or negative), this might indicate an outlier.
(NB: The regression line takes into account all values, so outliers might have a big effect on the line.)

This kind of regression line is called Least Squares Regression, because you're finding the equation of the line which minimises the sum of the squares of the residuals (i.e. $\sum e_k^2$ is as small as possible, where the e_k are the residuals).

*Use Regression Lines **With Care***

You can use your regression line to predict values of y. But it's best not to do this for x-values outside the range of your original table of values.

Example: Use your regression equation to estimate the value of y when: (i) x = 7.6, (ii) x = 12.6

(i) When x = 7.6, y = –0.978 × 7.6 + 14.451 = 7.02 (to 3 sig. fig.). This should be a pretty reliable guess, since x = 7.6 falls in the range of x we already have readings for — this is called interpolation.

(ii) When x = 12.6, y = –0.978 × 12.6 + 14.451 = 2.13 (to 3 sig. fig.). This may well be unreliable since x = 12.6 is bigger than the biggest x-value we already have — this is called extrapolation.

Practice Questions

Sample Exam question:

1. The following times (in seconds) were taken by eight different runners to complete distances of 20 metres and 60 metres.

Runner	A	B	C	D	E	F	G	H
20-metre time (x)	3.39	3.20	3.09	3.32	3.33	3.27	3.44	3.08
60-metre time (y)	8.78	7.73	8.28	8.25	8.91	8.59	8.90	8.05

(a) Plot a scatter diagram to represent the data. [3 marks]
(b) Find the equation of the regression line of y on x and plot it on your scatter diagram. [8 marks]
(c) Use the equation of the regression line to estimate the value of y when: (i) $x = 3.15$, (ii) $x = 3.88$ and comment on the reliability of your estimates. [6 marks]
(d) Find the residuals for: (i) $x = 3.32$, (ii) $x = 3.27$. Illustrate them on your scatter diagram. [4 marks]

99% of all statisticians make sweeping statements...

Be careful with that extrapolation business — it's like me saying that because I grew at an average rate of 10 cm a year for the first few years of my life, by the time I'm 50 I should be 5 metres tall. Residuals are always errors in the values of y — these equations for working out the regression line all assume that you can measure x perfectly all the time.

Constant Acceleration Equations

This page is for AQA M1, OCR M1, Edexcel M1

Welcome to the technicolour world of Mechanics 1. Fashions may change, but there will always be M1 questions that involve objects travelling in a straight line. It's just a case of picking the right equations to solve the problem.

There are *Five Constant Acceleration Equations*

Examiners call these "uvast" questions (pronounced ewe-vast, like a large sheep) because of the five variables involved:

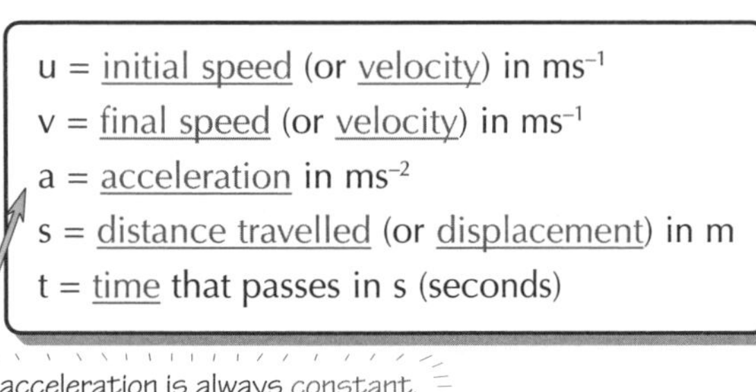

u = initial speed (or velocity) in ms^{-1}
v = final speed (or velocity) in ms^{-1}
a = acceleration in ms^{-2}
s = distance travelled (or displacement) in m
t = time that passes in s (seconds)

The acceleration is always constant.

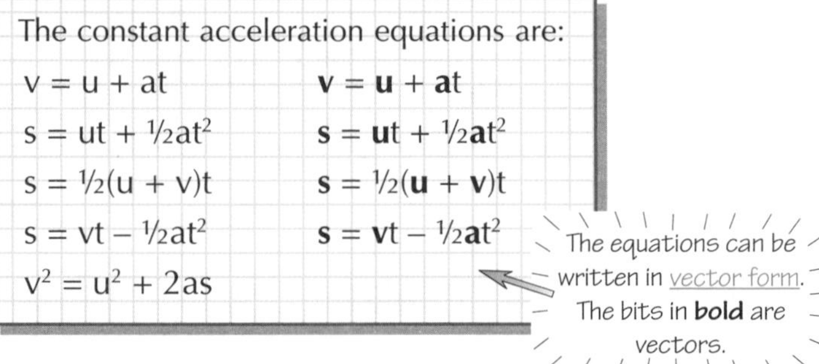

The constant acceleration equations are:

$v = u + at$	$\mathbf{v} = \mathbf{u} + \mathbf{a}t$
$s = ut + \frac{1}{2}at^2$	$\mathbf{s} = \mathbf{u}t + \frac{1}{2}\mathbf{a}t^2$
$s = \frac{1}{2}(u + v)t$	$\mathbf{s} = \frac{1}{2}(\mathbf{u} + \mathbf{v})t$
$s = vt - \frac{1}{2}at^2$	$\mathbf{s} = \mathbf{v}t - \frac{1}{2}\mathbf{a}t^2$
$v^2 = u^2 + 2as$	

The equations can be written in vector form. The bits in **bold** are vectors.

None of those equations are in the formula book, so you're going to have to learn them. Questions will usually give you three variables — your job is to choose the equation that will find you the missing fourth variable.

Example: A jet ski travels in a straight line along a river. It passes under two bridges 200 m apart and is observed to be travelling at 5 ms^{-1} under the first bridge and at 9 ms^{-1} under the second bridge. Calculate its acceleration (assuming it is constant).

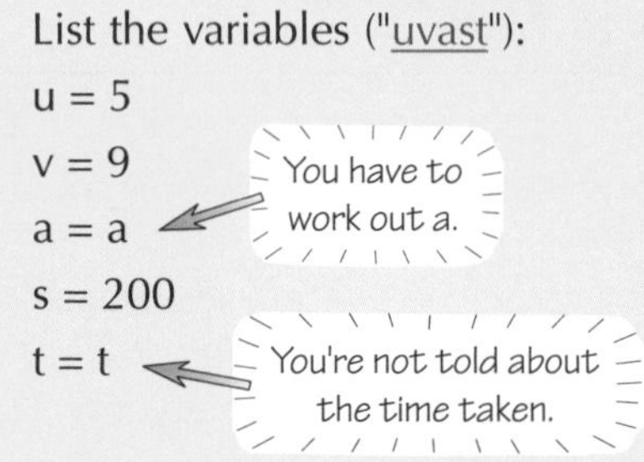

List the variables ("uvast"):
u = 5
v = 9
a = a
s = 200
t = t

You have to work out a.
You're not told about the time taken.

Choose the equation with u, v, s and a in it: $v^2 = u^2 + 2as$
Substitute values: $9^2 = 5^2 + (2 \times a \times 200)$
Simplify: $81 = 25 + 400a$
Rearrange: $400a = 81 - 25 = 56$
Then solve: $a = \frac{56}{400} = 0.14\ ms^{-2}$

Check you're using the right units — m, s, ms^{-1} and ms^{-2}.

Motion under Gravity just means taking a = g

Don't be put off by questions involving objects moving freely under gravity — they're just telling you the acceleration is g.

Use the value of g given on the front of the paper or in the question. If you don't, you risk losing a mark because your answer won't match the examiners' answer.

Example: A pebble is dropped into a well 18 m deep and moves freely under gravity until it hits the bottom. Calculate the time it takes to reach the bottom. (Take g = 9.8 ms^{-2}.)

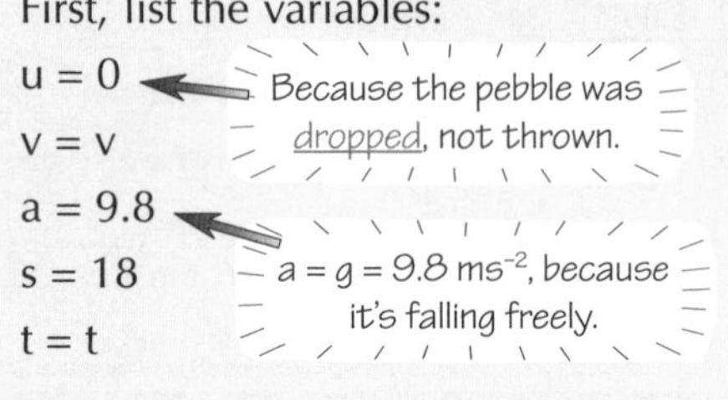

First, list the variables:
u = 0
v = v
a = 9.8
s = 18
t = t

Because the pebble was dropped, not thrown.
a = g = 9.8 ms^{-2}, because it's falling freely.

You need the equation with u, a, s and t in it: $s = ut + \frac{1}{2}at^2$
Substitute values: $18 = (0 \times t) + (\frac{1}{2} \times 9.8 \times t^2)$
Simplify: $18 = 4.9t^2$
Rearrange to give t^2: $t^2 = \frac{18}{4.9} = 3.67...$
Solve by square-rooting: $t = \sqrt{3.67...} = 1.92$ s

Watch out for tricky questions like this — at first it looks like they've only given you one variable.
You have to spot that the pebble was dropped (so it started with no velocity) and that it's moving freely under gravity.

...In truth, acceleration due to gravity does vary — but you can assume it's constant so long as you stay close to the Earth's surface.

Constant Acceleration Equations

This page is for AQA M1, OCR M1, Edexcel M1

Sometimes there's **More Than One Object Moving** at the **Same Time**

For these questions, t is often the same (or connected as in this example) because time ticks along for both objects at the same rate. The distance travelled might also be connected.

Example:

A car, A, travelling along a straight road at a steady 30 ms^{-1} passes point R at t = 0. Exactly 2 seconds later, a second car, B, travelling at 25 ms^{-1}, moves in the same direction from point R. Car B accelerates at a constant 2 ms^{-2}. Show that the two cars are level when $\mathbf{t^2 - 9t - 46 = 0}$ where t is the time taken by car A.

For each car, there are different "uvast" equations, so you write separate lists and separate equations.

CAR A	CAR B
$u_A = 30$	$u_B = 25$
$v_A = 30$	$v_B = v$
$a_A = 0$	$a_B = 2$
$s_A = s$	$s_B = s$
$t_A = t$	$t_B = (t - 2)$

Constant speed so $a_A = 0$

s is the same for both cars because they're level.

B starts moving 2 seconds after A passes point R.

The two cars are level, so choose an equation with s in it:

$s = ut + \frac{1}{2}at^2$

Substitute values: $s = 30t + (\frac{1}{2} \times 0 \times t^2)$

Simplify: $\mathbf{s = 30t}$

Use the same equation for Car B: $s = ut + \frac{1}{2}at^2$

Substitute values: $s = 25(t - 2) + (\frac{1}{2} \times 2 \times (t - 2)^2)$

Simplify: $s = 25t - 50 + (t - 2)(t - 2)$

$s = 25t - 50 + (t^2 - 4t + 4)$

$\mathbf{s = t^2 + 21t - 46}$

The distance travelled by both cars is equal, so put the equations for s equal to each other:

$30t = t^2 + 21t - 46$

$t^2 - 9t - 46 = 0$

That's the result you were asked to find.

Constant acceleration equation questions involve modelling assumptions (simplifications to real life so you can use the equations):

1) The object is a particle — this just means it's very small and so isn't affected by air resistance as cars or stones would be in real life.
2) Acceleration is constant — without it, the equations couldn't be used.

Practice Questions

1) A motorcyclist accelerates uniformly from 3 ms^{-1} to 9 ms^{-1} in 2 seconds. What is the distance travelled by the motorcyclist during this acceleration?

2) A ball is projected vertically upwards at 3 ms^{-1}. How long does it take to reach its maximum height?

3) Sample exam question:

The window cleaner of a high-rise block accidentally drops his sandwich, which then falls freely to the ground. The height between the consecutive floors of the building is h and the speed of the sandwich as it passes a high floor is u. The sandwich takes 1.2 seconds to fall a further 4 floors. Use g = 10ms^{-2}.

a) Show that $4h = 1.2u + 7.2$ [2 marks]

b) The sandwich takes another 0.6 seconds to fall a further 4 floors.

i) Obtain another equation in u and h. [2 marks]

ii) Hence calculate the values of u and h. [2 marks]

c) Comment on modelling assumptions you have made. [1 mark]

As Socrates once said, "The unexamined life is not worth living"... *but what did he know...*

Make sure you:

1) Make a list of the uvast variables EVERY time you get one of these questions.
2) Look out for "hidden" values — e.g. "particle initially at rest..." means u = 0.
3) Choose and solve the equation that goes with the variables you've got.

Motion Graphs

This page is for AQA M1, OCR M1, Edexcel M1

You can use displacement-time (X/T), velocity-time (V/T) and acceleration-time (A/T) graphs to represent all sorts of motion.

Displacement-time Graphs: Height = Distance and Gradient = Velocity

The steeper the line, the greater the velocity. A horizontal line has a zero gradient, so that means the object isn't moving.

Example: A cyclist's journey is shown on this X/T graph. Describe the motion.

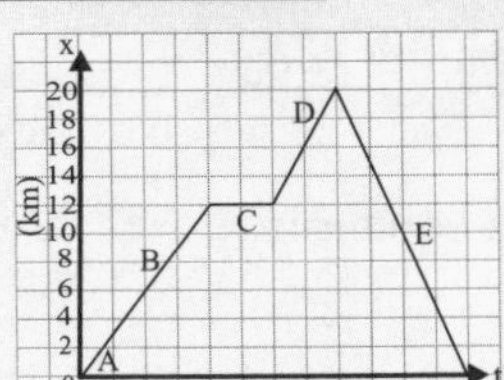

A: Starts from rest (when t = 0, x = 0)

B: Travels 12 km in 1 hour at a velocity of 12 kmh^{-1}

C: Rests for ½ hour (v = 0)

D: Cycles 8 km in ½ hour at a velocity of 16 kmh^{-1}

E: Returns to starting position, cycling 20 km in 1 hour at a velocity of –20 kmh^{-1} (i.e. 20 kmh^{-1} in the opposite direction)

Example: A girl jogs 2 km in 15 minutes and a boy runs 1.5 km in 6 min, rests for 1 min then walks the last 0.5 km in 8 min. Show the two journeys on an X/T graph.

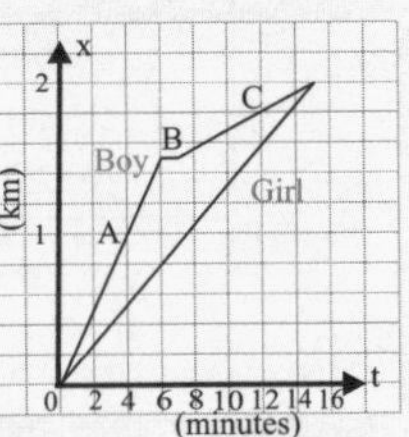

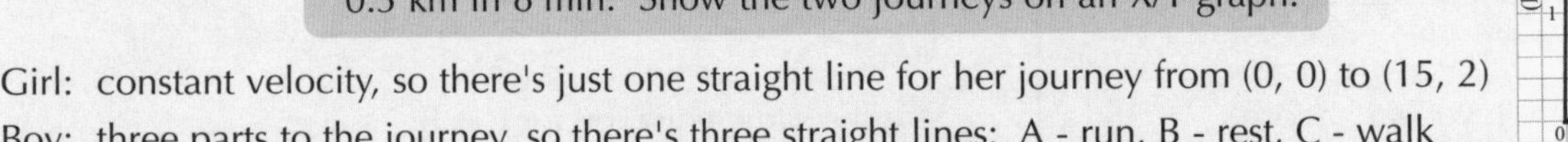

Girl: constant velocity, so there's just one straight line for her journey from (0, 0) to (15, 2)

Boy: three parts to the journey, so there's three straight lines: A - run, B - rest, C - walk

Velocity-time Graphs: Area = Distance and Gradient = Acceleration

The area under the graph can be calculated by splitting the area into rectangles, triangles or trapeziums. Work out the areas separately, then add them all up at the end.

Example: A train journey is shown on the V/T graph on the right. Find the distance travelled and the rate of deceleration as the train comes to a stop.

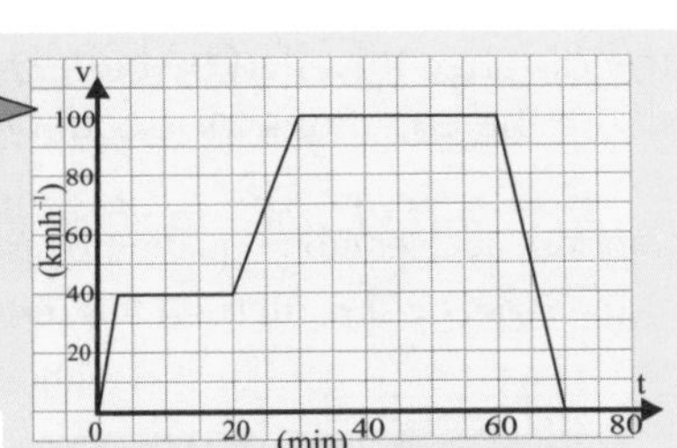

The time is given in minutes and the velocity as kilometres per hour, so divide the time in minutes by 60 to get the time in hours.

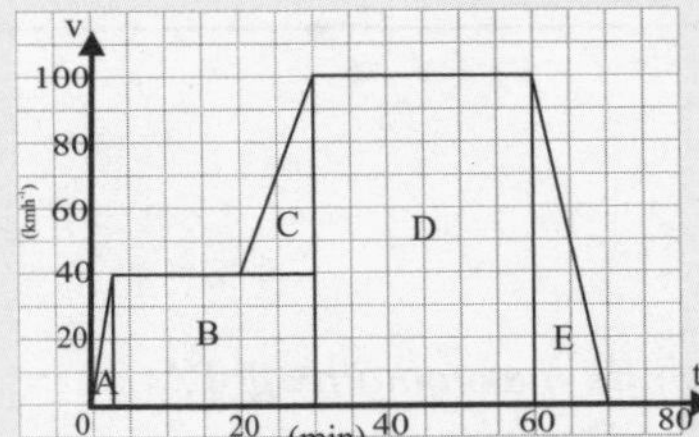

Area of A: $(2.5 \div 60 \times 40) \div 2 = 0.833...$

Area of B: $27.5 \div 60 \times 40 = 18.33...$

Area of C: $(10 \div 60 \times 60) \div 2 = 5$

Area of D: $30 \div 60 \times 100 = 50$

Area of E: $(10 \div 60 \times 100) \div 2 = 8.33...$

Total area = 82.5 so distance is 82.5 km

You might get a speed-time graph instead of a velocity-time graph — they're pretty much the same, except speeds are always positive, whereas you **can** have negative velocities.

The gradient of the graph at the end of the journey is $-100 \text{ kmh}^{-1} \div (10 \div 60)\text{h} = -600 \text{ kmh}^{-2}$

So the train decelerates at 600 kmh^{-2}.

Acceleration-time Graphs: Area = Velocity

Example: The acceleration of a parachutist who jumps from a plane is shown on the A/T graph on the right. Describe the motion of the parachutist and find the final velocity as the parachutist falls towards the ground.

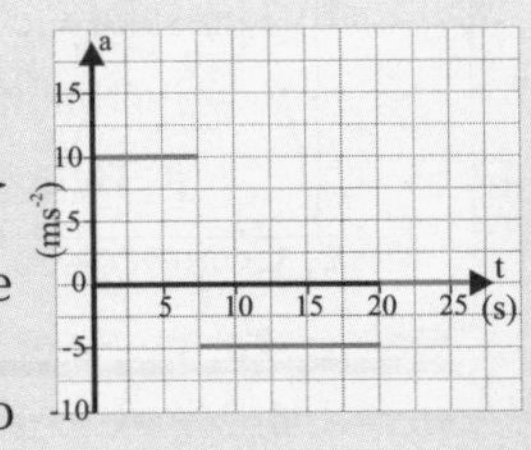

He falls with acceleration due to gravity of 10 ms^{-2} for 7.5 s. The parachute opens and the acceleration due to the air resistance of the parachute is 5 ms^{-2} acting upwards for 12.5 s. After 20 s, the acceleration is zero and so he falls with constant velocity. You need to find the area under the graph:

Area A: $10 \times 7.5 = 75 \text{ ms}^{-1}$ Area B: $5 \times 12.5 = 62.5 \text{ ms}^{-1}$

Area B is under the horizontal axis, so subtract area B from area A:

Velocity = $75 \text{ ms}^{-1} - 62.5 \text{ ms}^{-1} = 12.5 \text{ ms}^{-1}$

Motion Graphs

This page is for AQA M1, OCR M1, Edexcel M1

Graphs can be used to **Solve Complicated Problems**

As well as working out distance, velocity and acceleration from graphs, you can also solve more complicated problems. These might involve working out information not shown directly on the graph.

Example:

A jogger and a cyclist set off at the same time. The jogger runs with a constant velocity. The cyclist accelerates from rest, reaching a velocity of 5 ms^{-1} after 6 s and then continues at this velocity. The cyclist overtakes the jogger after 15 s.

Draw a graph of the motion and find the velocity of the jogger.

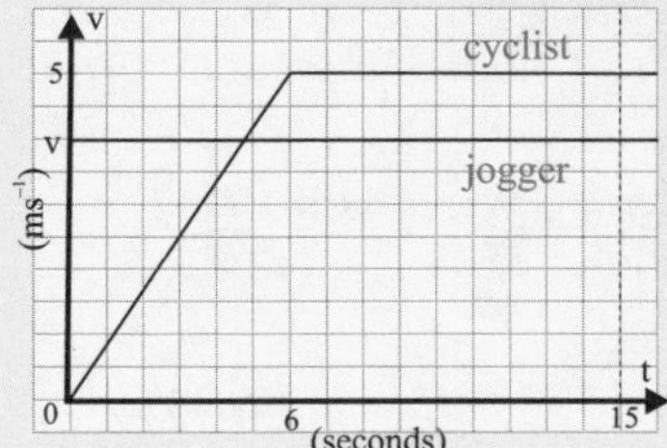

Call the velocity of the jogger v.

After 15 s the distance each has travelled is the same, so you can work out the area under the two graphs to get the distances:

Jogger: Area = distance = 15v

Cyclist: Area = distance = $(5 \times 6) \div 2 + (9 \times 5) = 60$ (area of triangle + area of rectangle)

So $15v = 60$

$v = 4\ ms^{-1}$

Practice Questions

1) **A runner starts from rest and accelerates at 0.5 ms^{-2} for 5 seconds. She maintains a constant velocity for 20 seconds then decelerates to a stop at 0.25 ms^{-2}.**

Draw a (t,v) graph to show the motion and find the distance the runner travelled.

2) **The start of a journey is shown on the (t,a) graph below.**

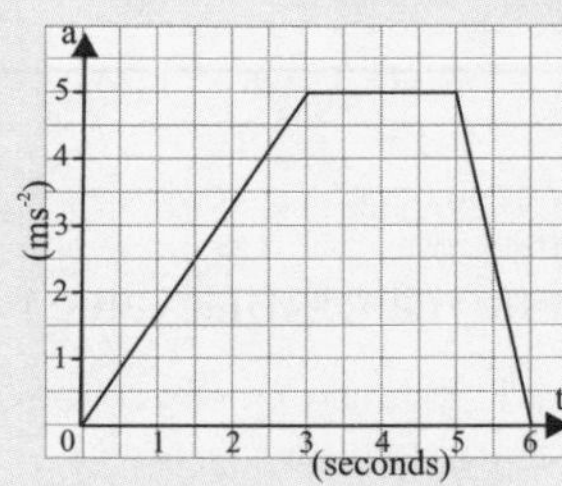

Find the velocity when: a) t = 3 b) t = 5 c) t = 6

3) **Sample exam question:**

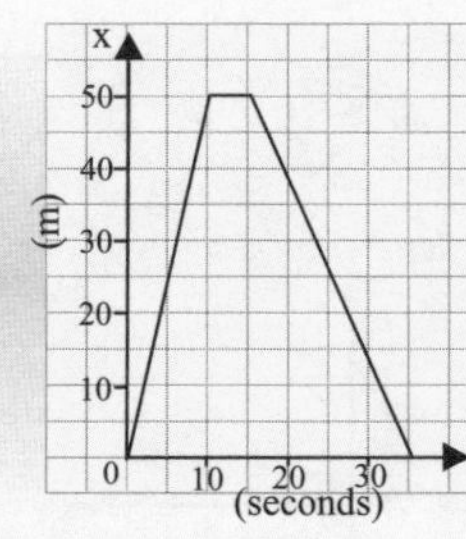

A sprinter runs 50 m, rests and then jogs back. His motion is shown on this (x,t) graph.

a) When is the runner at rest? Explain how the graph shows this. [1 mark]

b) Find the runner's velocity in the first part of the motion. [2 marks]

c) What is the runner's velocity as he returns home? [2 marks]

Random tongue-twister #1 — I wish to wash my Irish wristwatch...

If a picture can tell a thousand words then a graph can tell... um... a thousand and one. Make sure you know what type of graph you're using, and learn what the gradient and the area under each type of graph tell you.

Variable Acceleration

This page is for OCR M1

Find out all there is to know about how something is moving — use differentiation and integration to switch between position, velocity and acceleration.

Differentiating lets you change from Position to Velocity to Acceleration

The position, velocity and acceleration of a moving object can be written as expressions in terms of time, t.
Velocity is the rate of change of position, so you can differentiate position to find velocity.

In the same way, acceleration is the rate of change of velocity, so differentiating velocity gives you acceleration.

Example:

An object moves in a straight line. Its position at time t is given by $r = 3t^2 + 5t - 1$.
Find the velocity and acceleration of the object at time t.

$\frac{d}{dt}(3t^2+5t-1) = 6t+5$ so $v = 6t + 5$

$\frac{d}{dt}(6t+5) = 6$ so $a = 6$

Differentiate Both Components of a Vector Separately

See Section 14 for more about vectors.

Example:

The position vector of an object is $\boldsymbol{r} = \begin{bmatrix} 7t+3 \\ 5t^2-2t+1 \end{bmatrix}$.

Find the velocity and acceleration of the object when t = 0.

The top line is movement parallel to the x-axis, the bottom line parallel to the y-axis.

Velocity: $\frac{d}{dt}(7t+3) = 7$ and $\frac{d}{dt}(5t^2-2t+1) = 10t - 2$ so $\boldsymbol{v} = \begin{bmatrix} 7 \\ 10t-2 \end{bmatrix}$

when t = 0, $\boldsymbol{v} = \begin{bmatrix} 7 \\ 10\times0-2 \end{bmatrix} = \begin{bmatrix} 7 \\ -2 \end{bmatrix}$

Acceleration: $\frac{d}{dt}7 = 0$ and $\frac{d}{dt}(10t-2) = 10$ so $\boldsymbol{a} = \begin{bmatrix} 0 \\ 10 \end{bmatrix}$

When t = 0, $\boldsymbol{a} = \begin{bmatrix} 0 \\ 10 \end{bmatrix}$ i.e. the acceleration is constant (the same no matter what value t is).

Work Backwards using Integration

I'm sure this is more than a little obvious to you, but integration is the inverse of differentiation, so you can integrate acceleration to get velocity, and integrate velocity to get position.

Example:

An object moves in a straight line with velocity in ms^{-1} at time t given by $v = 3t^2 + 5$. Find the distance (r) it travels from t = 3 to t = 5.

$$r = \int_{t=3}^{t=5} v\,dt = \int_{t=3}^{t=5} (3t^2+5)\,dt$$

Use the values of t as limits.

$$= \left[t^3+5t\right]_{t=3}^{t=5}$$

$$= (125 + 25) - (27 + 15)$$

$$= 108 \text{ m}$$

Variable Acceleration

This page is for OCR M1

Use **Values** to work out **General Formulas** when **Integrating**

If you integrate to find a general formula (i.e. when there are no limits), you have to add on a constant. If you know extra information such as the initial velocity or the position at a certain time, then you can use it to work out what the constant is.

Example:

The acceleration of an object is given by $\mathbf{a} = \begin{bmatrix} 4 \\ -2 \end{bmatrix}$ and its initial velocity is $\begin{bmatrix} 1 \\ 3 \end{bmatrix}$.

Find the velocity at time t.

$$\mathbf{v} = \int \mathbf{a}\, dt = \begin{bmatrix} 4t + c_1 \\ -2t + c_2 \end{bmatrix}$$

c_1 and c_2 are the constants of integration

When t = 0, $\mathbf{v} = \begin{bmatrix} 4 \times 0 + c_1 \\ -2 \times 0 + c_2 \end{bmatrix} = \begin{bmatrix} 1 \\ 3 \end{bmatrix}$

You were told what the initial velocity was in the question.

So $c_1 = 1$ and $c_2 = 3$.

You already worked out that $\mathbf{v} = \begin{bmatrix} 4t + c_1 \\ -2t + c_2 \end{bmatrix}$

Now just plug the constants in: $\mathbf{v} = \begin{bmatrix} 4t + 1 \\ -2t + 3 \end{bmatrix}$

Sample exam questions:

1) A particle moves in a straight line. At time t after starting its motion, its velocity is given by $v = t^2(15 - 2t)$.
   a) Find the time t when the acceleration is zero. [3 marks]
   b) Find how far the particle has moved when t = 4. [4 marks]

2) A particle moves with velocity at time t given by $\mathbf{v} = \begin{bmatrix} 6t - 2 \\ 5 \end{bmatrix}$.
   a) Given that the initial position vector of the particle is $\begin{bmatrix} 4 \\ 3 \end{bmatrix}$, find the position vector for the particle at time t. [5 marks]
   b) Show that the acceleration of the particle is constant. [2 marks]

I've got variable acceleration if I tap my feet to music when I'm driving...

Examiners are usually pretty predictable (no offence). Questions on this topic usually follow the same pattern: differentiate or integrate, work out any constants and then maybe work out a few values for particular times. Look out for the initial conditions (initial velocity or position) and remember the order of differentiation: position; velocity; acceleration.

Vectors

This page is for AQA M1, OCR M1, Edexcel M1

'Vector' might sound like a really dull Bond villain, but... well, it's not. Vectors have both size (or magnitude) and direction. If a measurement just has size but not direction, it's called a scalar.

- So, Vector, you expect me to talk?
- No, Mr Bond, I expect you to die! Ak ak ak!

Vectors have a Magnitude — Scalars Don't

Examples of vectors: velocity, displacement, acceleration, force

Examples of scalars: speed, distance

A really important thing to remember is that an object's speed and velocity aren't always the same:

Example: A runner sprints 100 m along a track at a speed of 8 ms^{-1} and then she jogs back 50 m at 4 ms^{-1}.

Average Speed

Speed = Distance ÷ Time

The runner takes (100 ÷ 8) + (50 ÷ 4) = 25 s to travel **150 m**.

So the average speed is 150 ÷ 25 = 6 ms^{-1}

Average Velocity

Velocity = Change in displacement ÷ Time

In total, the runner ends up **50 m** away from her start point and it takes 25 s.

So the average velocity is 50 ÷ 25 = 2 ms^{-1} in the direction of the sprint.

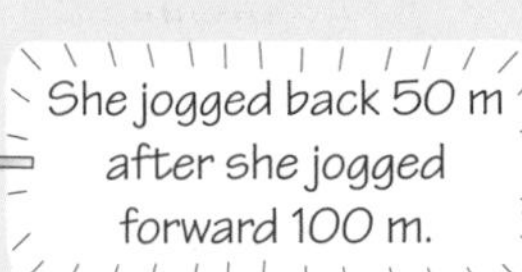

The Length of the Arrow shows the Magnitude of a Vector

You can add vectors together by drawing the arrows nose to tail.
The single vector that goes from the start to the end of the vectors is called the resultant vector.

It doesn't matter which order you draw the vectors.

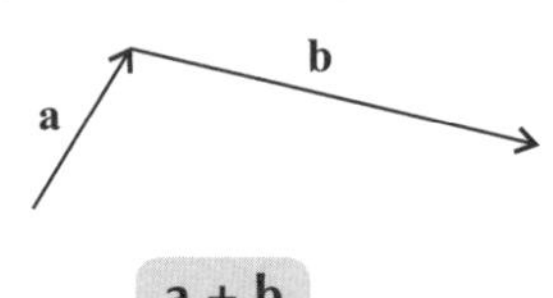

a + b

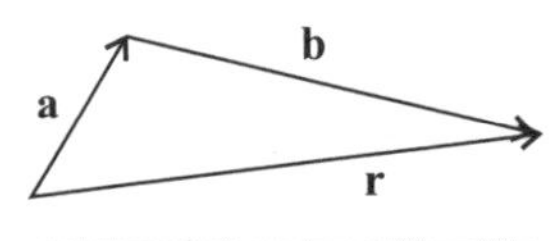

Resultant: **r = a + b**

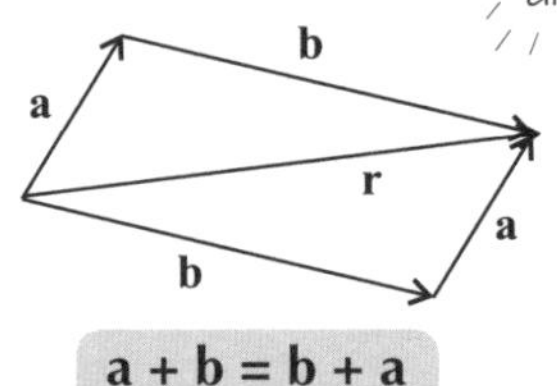

a + b = b + a

You can also multiply a vector by a scalar (just a number): the length changes but the direction stays the same.

a   3**a**

Resolving means writing a vector as Component Vectors

Splitting a vector up like this means you can work things out with one component at a time.
The unit vectors **i** and **j** are often used for resolving. They're called unit vectors because they each have a magnitude of 1. **i** is in the direction of the x-axis and **j** is in the direction of the y-axis.

Example: $\overrightarrow{AB} = 3\mathbf{i} + 2\mathbf{j}$ and $\overrightarrow{BC} = 5\mathbf{i} - 3\mathbf{j}$. Work out $\overrightarrow{AC}$.

Add the horizontal and vertical components separately.

$$\overrightarrow{AC} = \overrightarrow{AB} + \overrightarrow{BC} = (3\mathbf{i} + 2\mathbf{j}) + (5\mathbf{i} - 3\mathbf{j}) = 8\mathbf{i} - \mathbf{j}$$

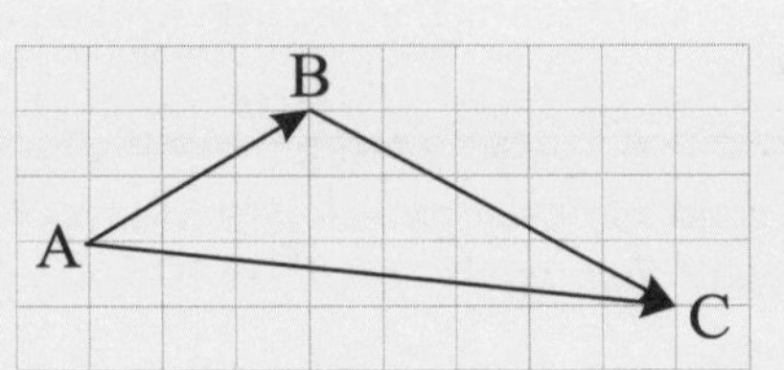

AQA M1 and Edexcel M1 only

Vectors can also be written as two components in a column vector:

$$\overrightarrow{AC} = \overrightarrow{AB} + \overrightarrow{BC} = \begin{bmatrix} 3 \\ 2 \end{bmatrix} + \begin{bmatrix} 5 \\ -3 \end{bmatrix} = \begin{bmatrix} 8 \\ -1 \end{bmatrix}$$

Vectors

This page is for AQA M1, OCR M1, Edexcel M1

Use **Trig and Pythagoras** to **Change** a vector into **Component Form**

Example: A ball travels with speed 5 ms^{-1} at an angle of 30° to the horizontal. Find the horizontal and vertical components of the ball's velocity, **v**.

First, draw a diagram and make a right-angle triangle:

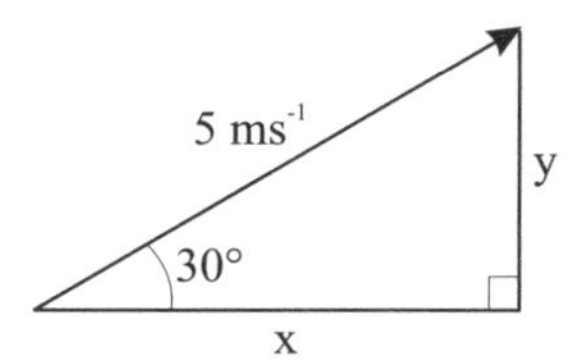

Using trigonometry, we can find x and y:

$\cos 30° = \frac{x}{5}$ so $x = 5\cos 30°$

$\sin 30° = \frac{y}{5}$ so $y = 5\sin 30°$

So $\mathbf{v} = (5\cos 30°\ \mathbf{i} + 5\sin 30°\ \mathbf{j})$ ms^{-1}

Example: The acceleration of a body is given by the vector $\mathbf{a} = 6\mathbf{i} - 2\mathbf{j}$. Find the magnitude and direction of the acceleration.

Start with a diagram again. Remember, the y-component "-2" means "down 2".

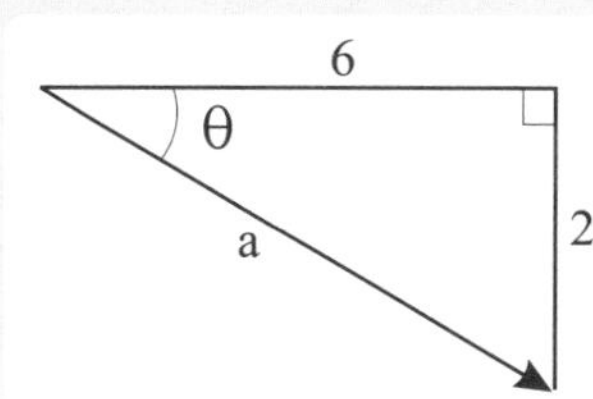

Using Pythagoras' Theorem, you can work out the magnitude of **a**:

$a^2 = 6^2 + (-2)^2 = 40$

so $a = \sqrt{40} = 6.3$ (2 sf)

Use trigonometry to work out the angle:

$\tan\theta = \frac{2}{6}$ so $\theta = \tan^{-1}\frac{2}{6} = 18.4°$

So vector **a** has magnitude 6.3 and direction 18.4° below the horizontal.

> In general, a vector with magnitude r and direction θ can be written as $r\cos\theta\mathbf{i} + r\sin\theta\mathbf{j}$
>
> The vector $x\mathbf{i} + y\mathbf{j}$ has magnitude $r = \sqrt{x^2 + y^2}$ and direction $\theta = \tan^{-1}\left(\frac{y}{x}\right)$

You can **Resolve** in any two **Perpendicular Directions** — not just x and y

Example: Find the resultant of the forces shown in the diagram.

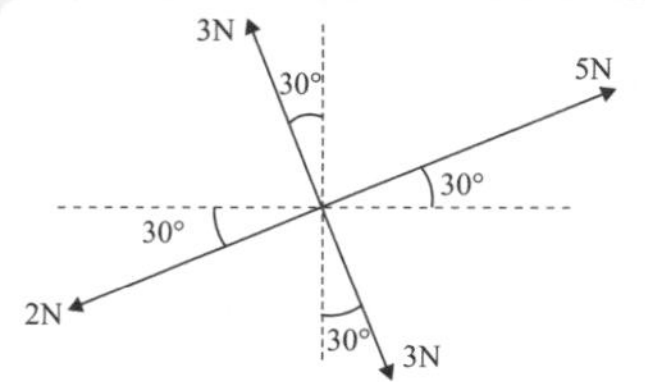

Resolving in ↖ direction: 3 N – 3 N = 0

Resolving in ↗ direction: 5 N – 2 N = 3 N

The two forces of 3 N balance each other.

So the resultant force is 3 N in the direction of the force of 5 N.

Practice Questions

1) ***Find the average velocity of a cyclist who cycles at 15 kmh^{-1} north for 15 minutes and then cycles south at 10 km/h for 45 minutes.***

2) ***Find a + 2b – 3c where a = 3i + 7j; b = -2i + 2j; c = i – 3j.***

3) ***Sample exam question:***

4N
30°
5N

The diagram shows two forces acting on a particle. Find the magnitude and direction of the resultant force. [5 marks]

Bet you can't say 'perpendicular' 10 times fast...

Next time you're baffled by vectors, just start resolving and Bob's your mother's brother.

Vector Motion

This page is for AQA M1, OCR M1, Edexcel M1

Vectors are much more than just a pretty face (or arrow). Once you deal with all the waffle in the question, all sorts of problems involving displacement, velocity, acceleration and forces can be solved using vectors.

Draw a Diagram if there's Lots of Vectors floating around

In fact, draw a diagram even when there's only a couple of vectors. But it's vital when there are lots of the little beggars.

Example:

A ship travels 100 km at a bearing of 025°, then 75 km at 140° before going 125 km at 215°. What is the displacement of the ship from its starting point?

Remember that bearings are always measured starting from north.

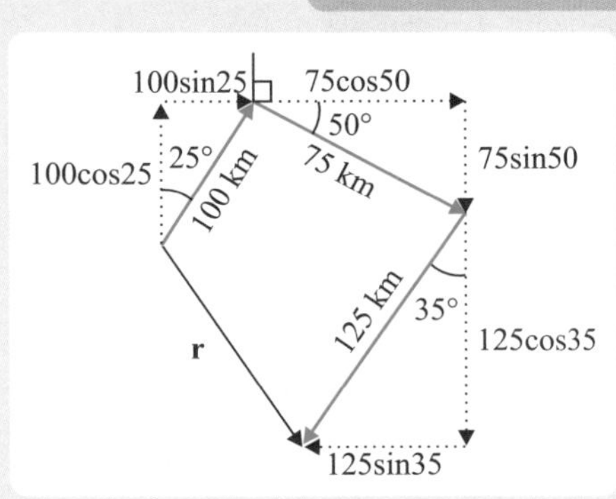

Resolve East: 100sin25 + 75cos50 – 125sin35 = 18.8 km

Resolve North: 100cos25 – 75sin50 – 125cos35 = -69.2 km

Magnitude of **r** = $\sqrt{18.8^2 + (-69.2)^2}$ = 71.7 km

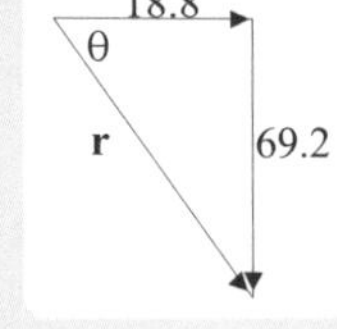

direction $\theta = \tan^{-1}\left(\frac{69.2}{18.8}\right) = 74.8°$

Bearing is 90° + 74.8° = 164.8°

So the displacement is 71.7 km on a bearing of 164.8°

The *Direction* part of a vector is *Really Important*

...and that means that you've got to make sure your diagram is spot on.

These two problems look similar, and the final answers are pretty similar too.
But look closely at the diagrams and you will see they are a bit different.

Example:

A canoe is paddled at 4 ms^{-1} in a direction perpendicular to the seashore. The sea current has a velocity of 1 ms^{-1} parallel to the shore. Find the resultant velocity **r** of the canoe.

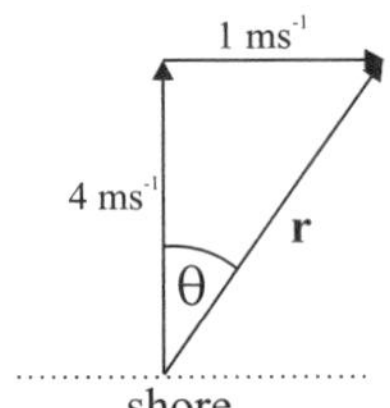

The resultant velocity **r** is the hypotenuse of the right-angle triangle.

Magnitude of **r** = $\sqrt{4^2 + 1^2}$ = 4.1 ms^{-1}

Direction: $\theta = \tan^{-1}\left(\frac{1}{4}\right) = 14.0°$

Example:

A canoe can be paddled at 4 ms^{-1} in still water. The sea current has a velocity of 1 ms^{-1} parallel to the shore. Find the angle at which the canoe must be paddled in order to travel in a direction perpendicular to the shore and the magnitude of the resultant velocity.

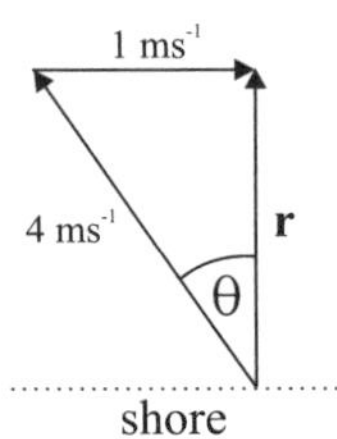

The resultant velocity **r** isn't the hypotenuse this time.

Magnitude of **r** = $\sqrt{4^2 - 1^2}$ = 3.9 ms^{-1}

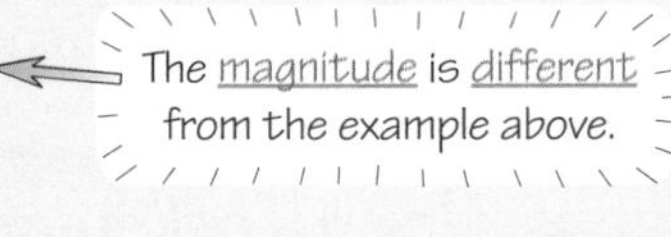

The magnitude is different from the example above.

Direction: $\theta = \sin^{-1}\left(\frac{1}{4}\right) = 14.5°$

Vector Motion

This page is for AQA M1, OCR M1, Edexcel M1

First **Decide** which vectors you need to **Work Out**

It's no good ploughing into a question if you're working out the wrong value. Once again, the diagram's the key.

Example:

A sledge of weight 1000 N is being held on a rough slope at an angle of 35° by a force parallel to the slope of 700 N. Find the normal contact force N and the frictional force F acting on the sledge.

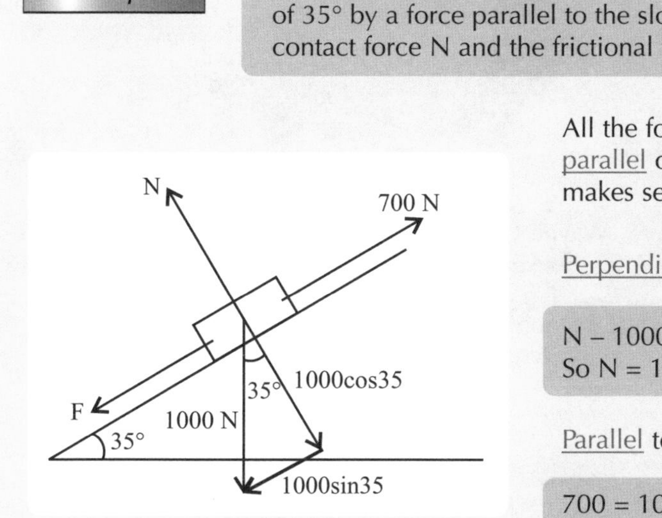

All the forces involved here act either parallel or perpendicular to the slope so it makes sense to resolve in these directions.

Perpendicular to the slope:

$N - 1000\cos35 = 0$
So $N = 1000\cos35 \quad = 819.2$ N

Parallel to the slope:

$700 = 1000\sin35 + F$
So $F = 700 - 1000\sin35° \quad = 126.4$ N

Practice Questions

1) ***A plane flies 40 miles due south, then 60 miles southeast before going 70 miles on a bearing of 020°. Find the distance and bearing on which the plane must fly to return to its starting point.***

2) ***A toy train of weight 25 N is pulled up a slope of 20°. The tension in the string is 25 N and a frictional force of 5 N acts on the train. Find the normal contact force and the resultant force acting on the train.***

3) ***Sample exam question:***

> A girl can swim at 2 ms^{-1} in still water. She is swimming across a river in a direction perpendicular to the riverbank. The river is flowing at 3 ms^{-1} and so it carries the girl downstream.
> Find the magnitude of the resultant velocity of the girl and the angle it makes with the riverbank. [6 marks]

'Vector motion' is an anagram of 'croon, vote tim'...

Diagrams are really important and really useful for solving problems using vectors. Sketch out a nice clear diagram before you decide how to tackle the problem. There's usually loads of different things you could work out, so always have a check back at the end to make sure you have worked out what the question was actually asking you for.

Mathematical Modelling

This page is for AQA M1, OCR M1, Edexcel M1

You'll have to make lots of assumptions in M1. Doing this is called 'modelling', and you do it to make a sticky real-life situation simpler.

Hint: 'modelling' in maths doesn't have anything to do with plastic aeroplane kits... or catwalks.

Example: **The ice hockey player**

You might have to assume:

- no friction between the skates and the ice
- no drag (air resistance)
- the skater generates a constant forward force S
- the skater is very small (a point mass)
- there is only one point of contact with the ice
- the weight acts downwards

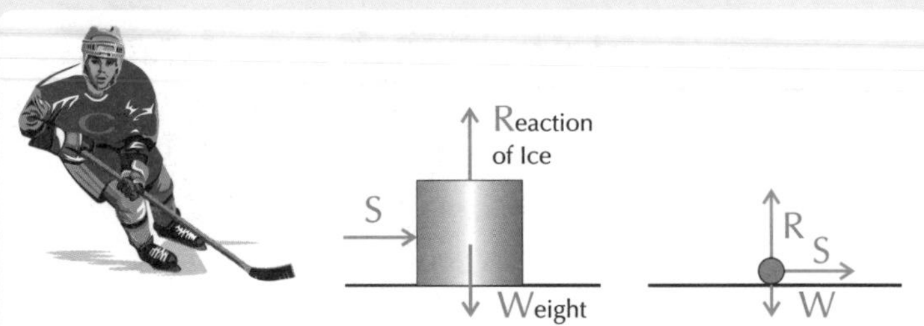

Wayne Grotski — ice hockey superstar

Wayne as a block, with forces shown

Wayne as a point mass, with forces

The complex hockey player on the left has become a simple mathematical model on the right with only three forces. Easy.

Modelling is a Cycle

Having created a model you can later improve it by making more (or less) assumptions.

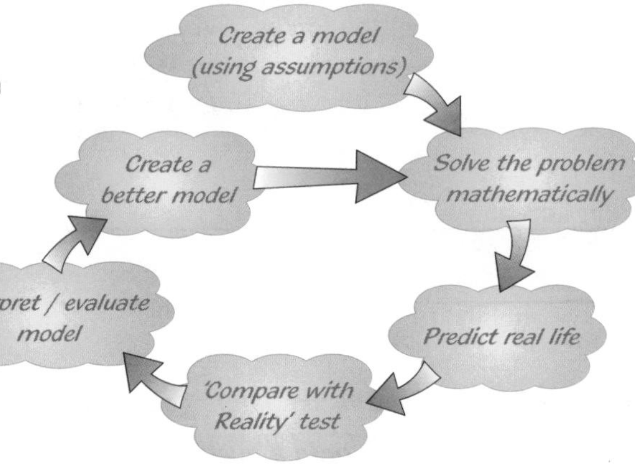

Mmm... green and red. I think this graphic wins the coveted Award for Most Tasteless Colour Scheme.

Talk the Talk

Maths questions in M1 use a lot of words that you already know, but here they're used to mean something very precise. Learn these definitions so you don't get caught out:

Particle — the 'mass' or 'body' drawn as a rectangle or dot (e.g. Wayne the hockey player)
Light — the body has no mass (e.g. a feather)
Smooth — the surface doesn't have friction / drag opposing motion (e.g. an ice rink)
Rough — the surface will oppose motion with friction / drag (e.g. a table top)
Beam or Rod — a long particle (e.g. a carpenter's plank)
Uniform — the mass is evenly spread out throughout the body (e.g. a school ruler)
Non-uniform — the mass is unevenly spread out (e.g. along a tennis racket)
Rigid — the body does not bend (e.g. a metal ruler)
Thin — the body has no thickness
Lamina — a surface that is thin (e.g. a sheet of A4 paper)
Equilibrium — nothing's moving
Plane — a flat surface (e.g. a table top)
Tension — the force in a taut wire, rope or string
Inextensible — the body can't be stretched (e.g. a metal rod)
Static — not moving

Mathematical Modelling

This page is for AQA M1, OCR M1, Edexcel M1

*Always start by drawing a **Simple Diagram** of the **Model***

Here's a couple of old chestnuts that often turn up in M1 exams in one form or another.

Example: **The book on a table**

A book is put flat on a table. One end of the table is slowly lifted and the angle to the horizontal is measured when the book starts to slide. What assumptions might you make?

Assumptions:
The book is rigid, so it doesn't bend or open.
The book is a particle, so its dimensions don't matter.
There's no wind or other external forces involved.

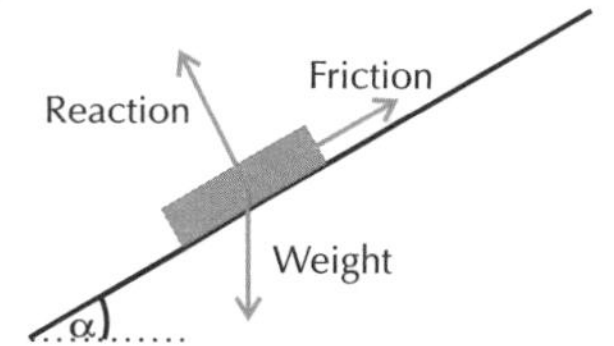

Example: **The balance**

A pencil is placed on a table and a 12″ ruler is put across it so that it balances. A 1p coin is placed on one side and a 10p coin on the other so that the ruler still balances. Draw a model of the forces. What assumptions have you made?

Assumptions:
The coins are point masses.
The ruler is rigid.
The support acts at a single point.

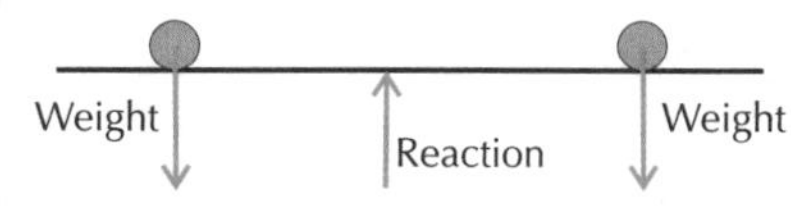

Practice Questions

1) The following items are dropped from a height of 2 m onto a cushion:

a) a full 330ml drinks can b) an empty drinks can c) a table tennis ball

The time each takes to fall is measured.

Draw a model of each situation and list any assumptions which you've made.

2) A car is travelling at 25 mph along a level road.

Draw a model of the situation and list any assumptions which you've made.

3) A skydiver is falling to earth before his parachute opens.

Draw a model of the situation and list any assumptions which you've made.

Phobia #1 — pteronophobia: fear of being tickled with feathers...

I've said it before and I'll say it again: if it makes their lives easier, examiners always stick the same kinds of questions into M1 Exams year after year. If you practise enough 'books on tables' and 'balancing pencil' questions, you'll have no nasty surprises in the Exam. And if you feel you've just learned more about ice hockey then you've missed the point slightly.

Forces are Vectors

This page is for AQA M1, OCR M1, Edexcel M1

Forces have magnitude and direction. Only force arrows should be attached to a particle. Geddit? Gottit. Good.

Forces have Components

You've done a fair amount of trigonometry already, so this should be as straightforward as watching dry paint.

Example: A particle is acted on by a force of 15 N at 30° above the horizontal. Find the horizontal and vertical components of the force.

A bit of trigonometry is all that's required:

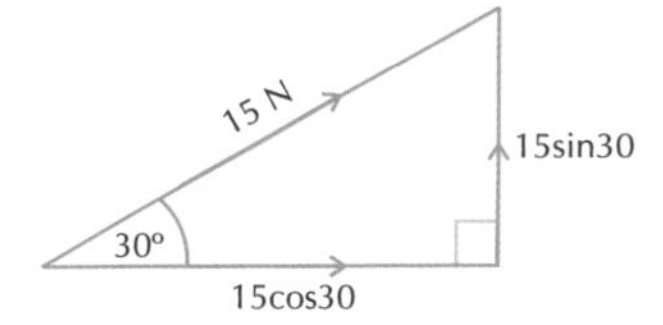

$$\text{Force} = \begin{pmatrix} 15\cos 30° \\ 15\sin 30° \end{pmatrix}$$
$$= (13.0\,\mathbf{i} + 7.5\,\mathbf{j})\text{ N}$$

(i.e. 13 N to the right and 7.5 N upwards)

Add Forces Top to Tail to get the Resultant

The important bit when you're drawing a diagram to find the resultant is to make sure the arrows are the right way round. Repeat after me: top to tail, top to tail, top to tail.

Example: A second horizontal force of 20 N is also applied to the particle in the example above. Find the resultant of these forces.

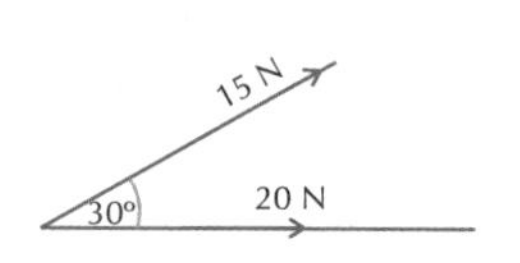

You need to put the arrows top to tail:

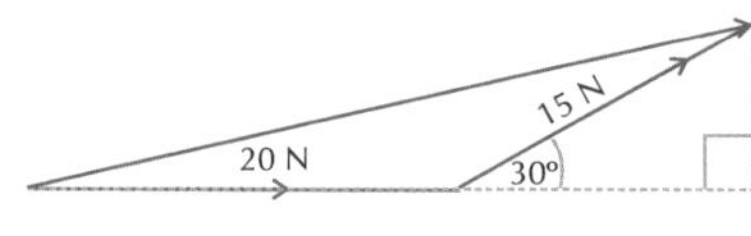

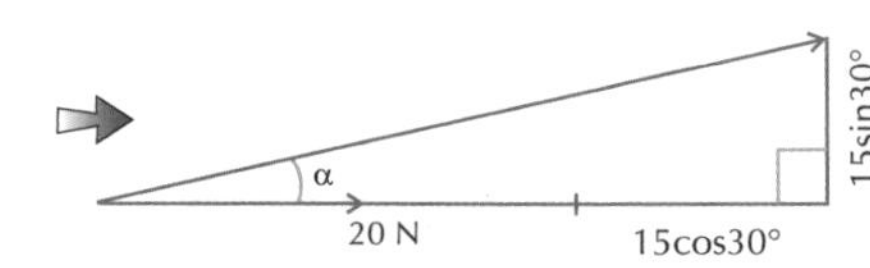

Using Pythagoras and trigonometry:

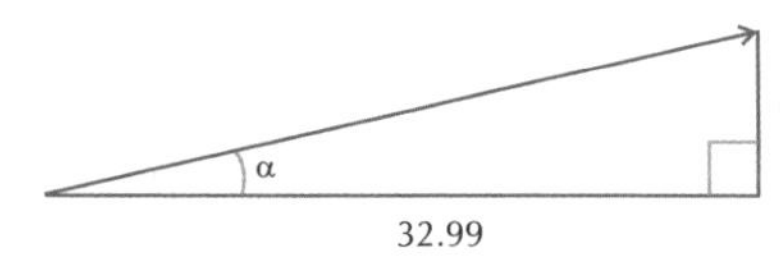

$$R = \sqrt{32.99^2 + 7.5^2} = 33.8\text{ N}$$
$$\alpha = \tan^{-1}\left(\frac{7.5}{32.99}\right) = 12.8° \text{ above the horizontal}$$

Example: Find the magnitude and direction of the resultant of the forces shown acting on the particle.

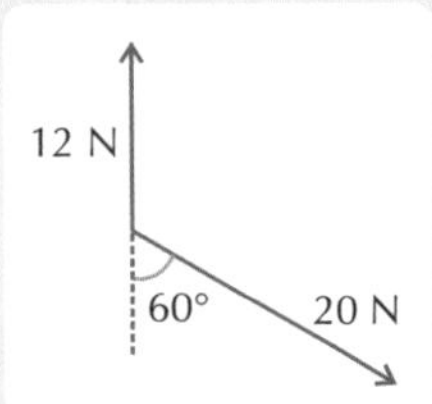

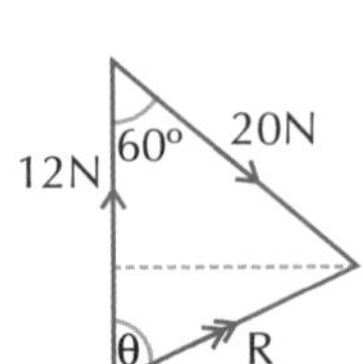

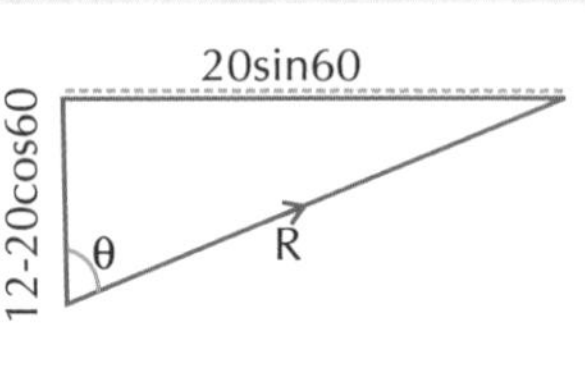

Hint: you could use the cosine rule here to get R.

$$R = \sqrt{(12 - 20\cos 60°)^2 + (20\sin 60°)^2} = 17.4\text{ N}$$
$$\theta = \tan^{-1}\left(\frac{20\sin 60°}{12 - 20\cos 60°}\right) = 83.4° \text{ to vertical}$$

If a particle is released it will move in the direction of the resultant.

Forces are Vectors

This page is for AQA M1, OCR M1, Edexcel M1

Particles in **Equilibrium Don't Move**

Forces acting on a particle in equilibrium add up to zero force. That means when you draw all the arrows top to tail, you finish up where you started. That's why diagrams showing equilibrium are called 'polygons of forces'.

Example: Find the magnitude of force P for equilibrium.

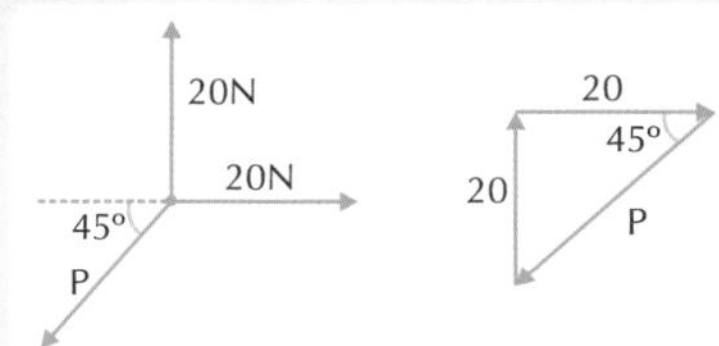

$P\cos 45 = 20$

$P = 28.3$ N

Example: Find S and T for equilibrium.

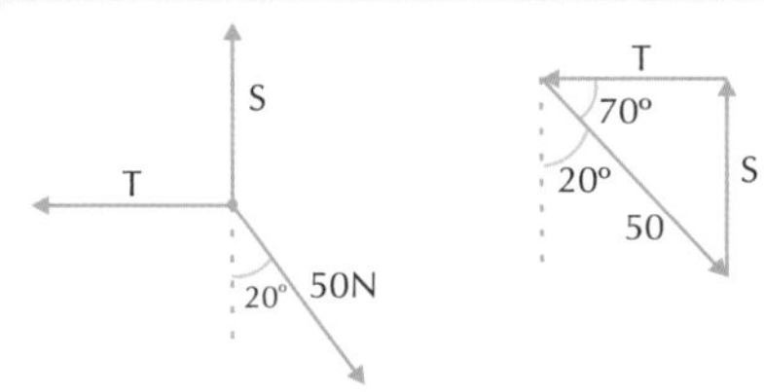

$S = 50\sin 70° = 47.0$ N

$T = 50\cos 70° = 17.1$ N

Example: Find P and Q for equilibrium.

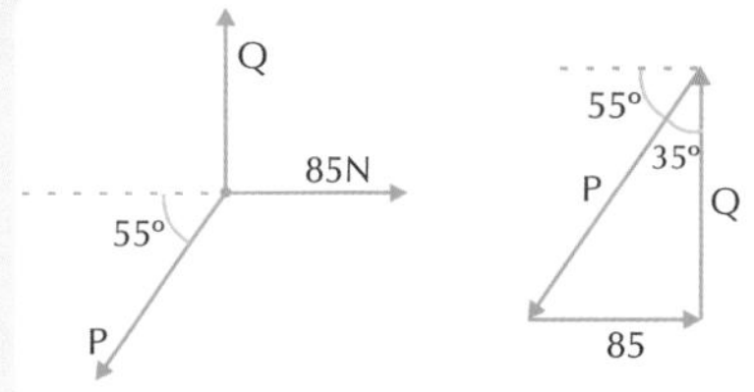

$\sin 35° = \frac{85}{P}$ so $P = 148$ N

$\tan 35° = \frac{85}{Q}$ so $Q = 121$ N

Practice Questions

1) Find the magnitudes and directions to the horizontal of the resultant force in each situation.

a) 3N, 4N

b)

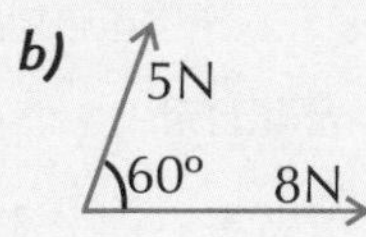

c)

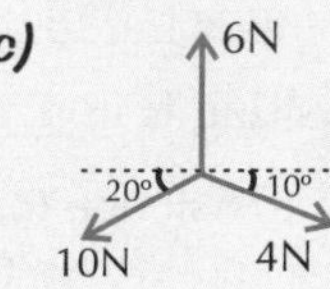

Sample exam questions:

2) A force of magnitude 7 N acts horizontally on a particle. Another force, of magnitude 4 N, acts on the particle at an angle of 30° to the horizontal. The resultant of the two forces has a magnitude R at an angle α to the horizontal.

Find a) The force R [3 marks]

b) The angle α [3 marks]

3) Three forces of magnitudes 15 N, 12 N and W act on a particle as shown.
Given that the particle is in equilibrium, find:

a) The value of θ [2 marks]

b) The force W [2 marks]

The force W is now removed.
State the magnitude and direction of the resultant of the two remaining forces. [2 marks]

15N θ 12N W

Phobia #2 — coulrophobia: fear of clowns...

...the red noses, the baggy pants, the pratfalls... uh. Better have a lie down and forget all about it.

Types of Forces

This page is for AQA M1, OCR M1, Edexcel M1

Different types of forces act on a body for different reasons...

Weight (W)

Due to the particle's mass, m and the force of gravity, g: **W = mg** — weight always acts downwards.

*The **Normal Reaction** (R or N)*

The reaction from a surface. Reaction is always at 90° to the surface.

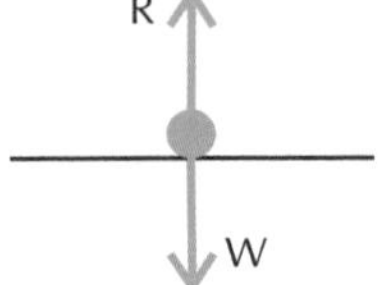

Tension (T)

Force in a taut rope, wire or string.

Friction (F)

Due to the roughness between a body and a surface. If the body isn't moving, it's called static friction. When the body is moving relative to the surface, it's called dynamic friction. Friction always acts against motion, or likely motion.

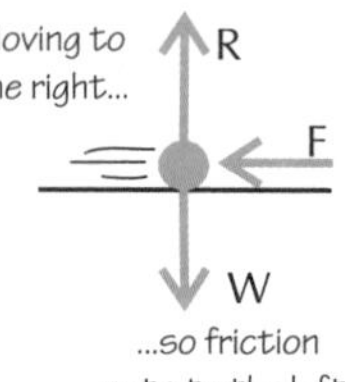

Moving to the right...
R
F
W
...so friction acts to the left.

Thrust

Force in a rod (e.g. the pole of an open umbrella).

Example: A sledge is being steadily pulled by a small child on horizontal snow. Draw a force diagram for a model of the sledge. List your assumptions.

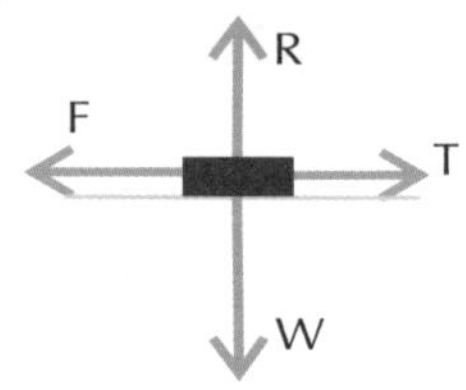

Assumptions:

1) Friction is too big to be ignored (i.e. it's not ice).
2) The string is horizontal (it's a small child).
3) Take the sledge to be a small particle (so its size doesn't matter).

Example: A mass of 12 kg is held by two light strings, P and Q, acting at 40° and 20° to the vertical as shown. Find the tension in each string. Take $g = 9.8\text{ ms}^{-2}$.

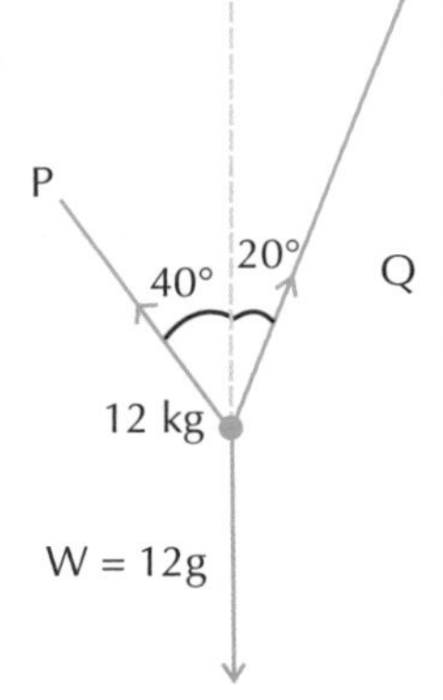

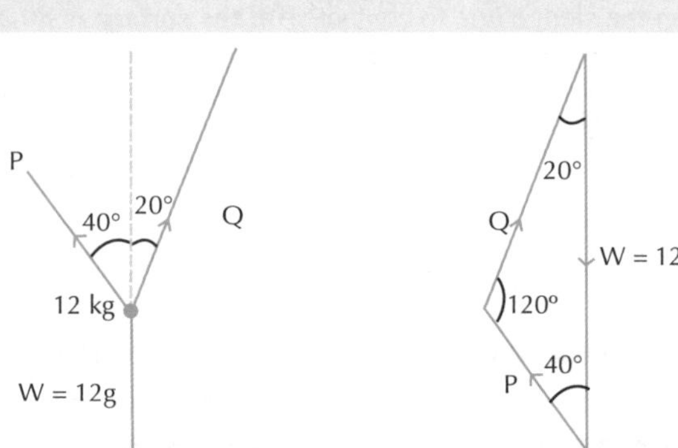

Sine rule:

$$\frac{P}{\sin 20} = \frac{12g}{\sin 120}$$

So P = 46.4 N

$$\frac{Q}{\sin 40} = \frac{12g}{\sin 120}$$

So Q = 87.3 N

Always look out for sine rule triangles in your polygons of forces.

Types of Forces

This page is for AQA M1, OCR M1, Edexcel M1

An **Inclined Plane** is a **Sloping Surface**

Example:

A particle of mass 0.1 kg is held at rest on a rough plane inclined at 20° to the horizontal by a friction force acting up the plane. Find the magnitude of this friction force and the normal reaction. (Take g = 9.8 ms⁻².)

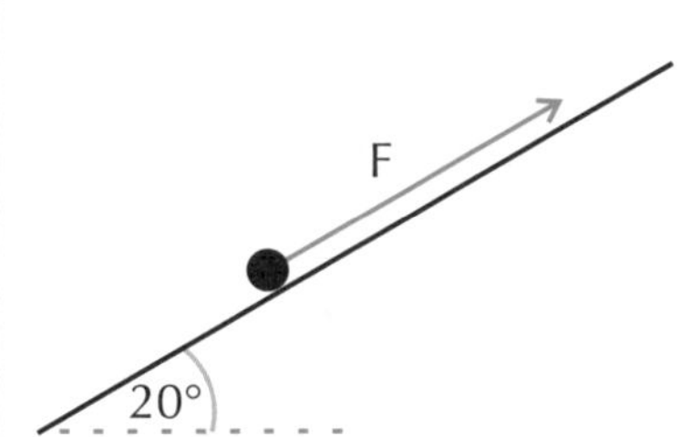

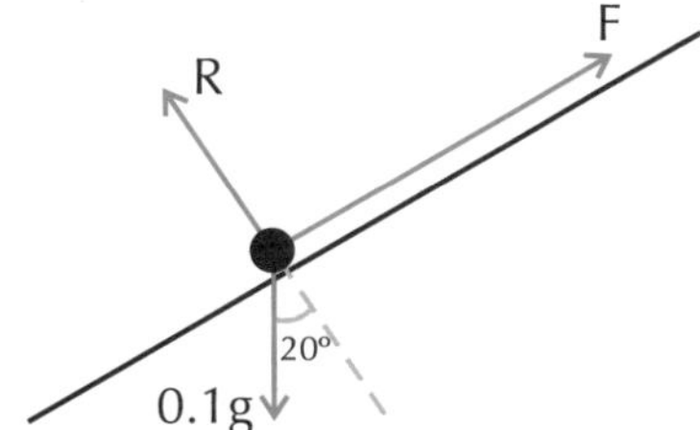

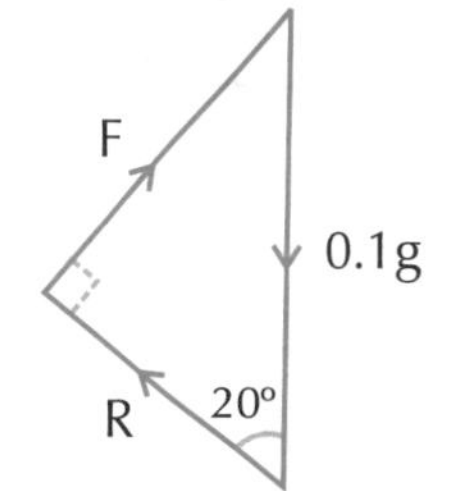

$F = 0.1g\sin 20°$
$= 0.335$ N
$R = 0.1g\cos 20°$
$= 0.921$ N

Practice Questions

Take g = 9.8 ms⁻² in each of these questions.

1) A mass of M kg is suspended by two light wires A and B, with angles 60° and 30° to the vertical respectively, as shown. The tension in A is 20 N. Find:

   a) the tension in wire B

   b) the mass M

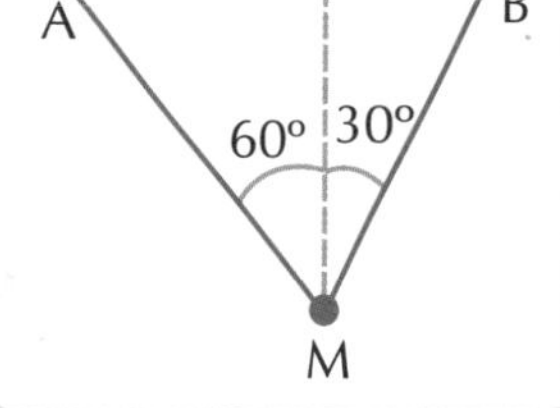

2) A particle, Q, of mass m kg, is in equilibrium on a smooth plane which makes an angle of 60° to the vertical. This is achieved by an attached string S, with tension 70 N, angled at 10° to the plane as shown. Draw a force diagram and find both the mass of Q and the reaction on it from the surface.

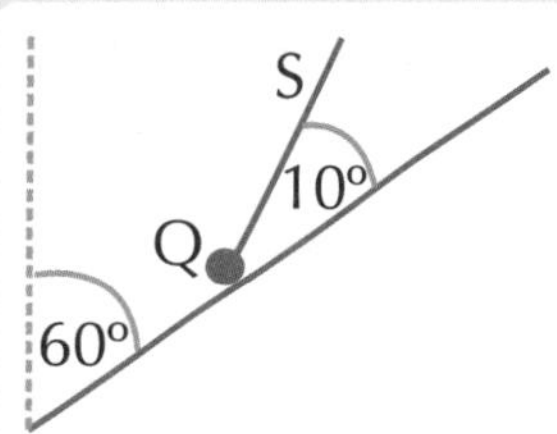

3) Sample exam question:

A sledge is held at rest on a smooth slope which makes an angle of 25° with the horizontal. The rope is at an angle of 20° to the slope. It is given that the normal reaction acting on the sledge due to contact with the surface is 80 N.

Find:

a) The tension, T, in the rope. [3 marks]

b) The weight of the sledge. [3 marks]

Phobia #3 — pogonophobia: fear of beards...

Those 5 types of forces on the opposite page crop up loads in M1 Exams (except thrust, which isn't so common). You need to be completely familiar with all the jargon that gets bandied around in this subject. And it's absolutely vital that you can do inclined planes questions really well, because there's no way you'll get out of the Exam hall without getting one of those...

Friction

This page is for AQA M1, OCR M1, Edexcel M1

Friction Tries to **Prevent Motion**

Push hard enough and a particle will move, even though there's friction opposing the motion — so a friction force, F, has a maximum value. This depends on the roughness of the surface and the value of the normal reaction from the surface.

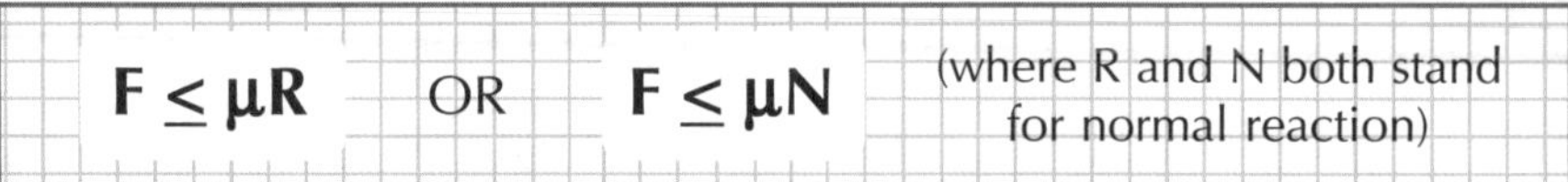

μ has no units.
μ is pronounced as 'mu'.

μ is called the "coefficient of friction". The rougher the surface, the bigger μ gets. The values of μ for different surfaces are found experimentally — they can't be found through calculations alone.

Example:

What range of values can a friction force take in resisting a horizontal force P acting on a particle Q, of mass 12 kg, resting on a rough horizontal plane which has a coefficient of friction of 0.4? (Take g = 9.9 ms^{-2}.)

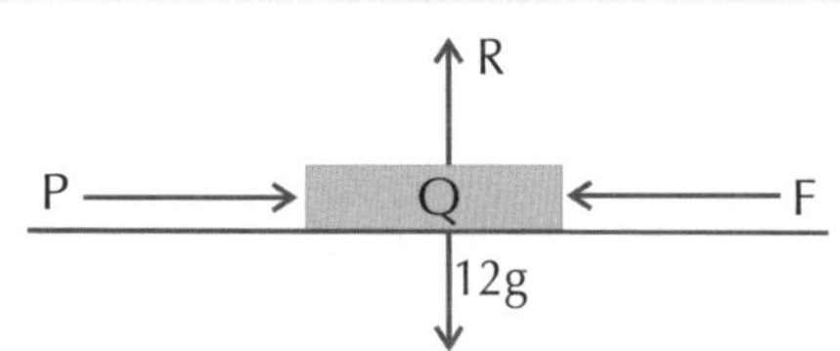

Resolving vertically: $R = 12g$

Use formula from above: $F \leq \mu R$

$F \leq 0.4(12g)$

$F \leq 47.04$ N

So friction can take any value between 0 and 47.04 N, depending on how large P is.
If $P \leq 47.04$ N then Q remains in equilibrium. If P = 47.04 N then Q is on the point of sliding — i.e. friction is at its limit. If P > 47.04 then Q will start to move.

Limiting Friction *is When Friction is* ***Maximum (F = μR)***

When the particle is about to start moving, the frictional force F reaches its maximum value of μR.

Example:

A particle of mass 6 kg is placed on a rough horizontal plane which has a coefficient of friction 0.3. A horizontal force Q is applied to the mass. Describe what happens if Q is:

a) 16 N
b) 20 N

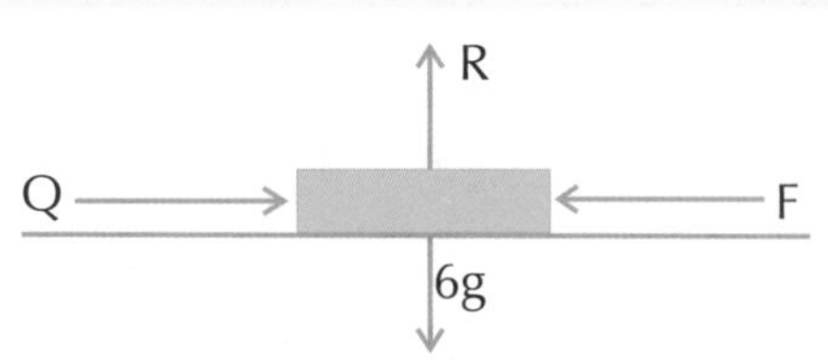

Resolving vertically: $R = 6g$

Using formula above: $F \leq \mu R$

$F \leq 0.3(6g)$

$F \leq 17.64$ N

a) Since Q < 17.64 it won't move.

b) Since Q > 17.64 it'll start moving. No probs.

Example:

A particle of mass 4 kg at rest on a rough horizontal plane is being pushed by a horizontal force of 30 N. Given that the particle is on the point of moving, find the coefficient of friction.

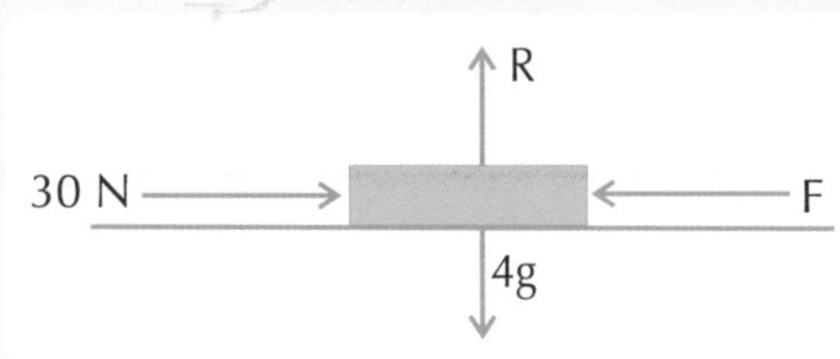

Resolving horizontally: $F = 30$

Resolving vertically: $R = 4g$

The particle's about to move, so friction is at its limit:

$F = \mu R$

$30 = \mu(4g)$

$$\mu = \frac{30}{4g} = 0.77$$

Friction

This page is for AQA M1, OCR M1, Edexcel M1

Example:

Look back to the example on p 131. Given that the mass is only just held in equilibrium (i.e. it's about to slide down the plane), find μ.

Since it's limiting friction: $F = \mu R$

Therefore: $0.1g\sin 20° = \mu(0.1g\cos 20°)$

So: $\mu = \frac{0.1g\sin 20°}{0.1g\cos 20°} = \tan 20° = 0.36$

Both the 0.1 mass and g cancel — to find μ you didn't need to know the mass.

In fact, when a particle is about to slide on a rough plane inclined at α to the horizontal, and there are no other forces involved other than W (the weight), F and R, then $\mu = \tan \alpha$.

Practice Questions

1) ***Describe the motion of a mass of 12 kg pushed by a force of 50 N parallel to the rough horizontal plane on which the mass is placed. The plane has coefficient of friction $\mu = \frac{1}{2}$. (Take $g = 9.8$ ms^{-2}.)***

2) ***What minimum force would be needed to move the mass in Q1?***

Sample exam questions:

3) A particle is placed on a rough inclined plane with coefficient of friction $\mu = 0.2$. At what angle to the horizontal is the plane, if the particle is about to slide? [2 marks]

4)

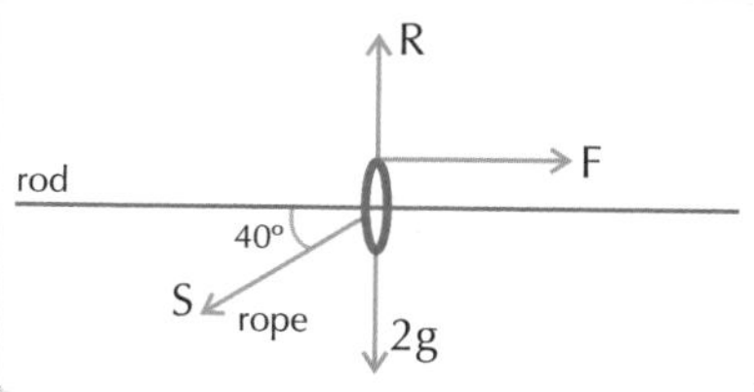

A 2 kg ring threaded on a rough horizontal rod is pulled sideways by a rope having tension S, as shown. The coefficient of friction between the rod and the ring is $\frac{3}{10}$. Given that the ring is about to slide, find the magnitude of S. [5 marks]

Phobia #4 — xenoglossophobia: fear of foreign languages...

Friction is a pain in the posterior for two reasons:

1) it stops you being able to slide all the way home with your eyes shut after a long night out;
2) it makes M1 questions just that little bit more complicated.

Moments

This page is for Edexcel M1

Moments are **Clockwise** or **Anti-Clockwise**

A '<u>moment</u>' is the <u>turning effect</u> a force has <u>around a point</u>.
The <u>larger the force</u>, and the <u>greater the distance</u> from a point, then the <u>larger the moment</u>.

Moment = Force × Perpendicular Distance

The units for moments are just newtons × metres = Nm.
Couldn't they have thought of a cleverer name?

In these examples you need to take moments about point O each time:

Example:

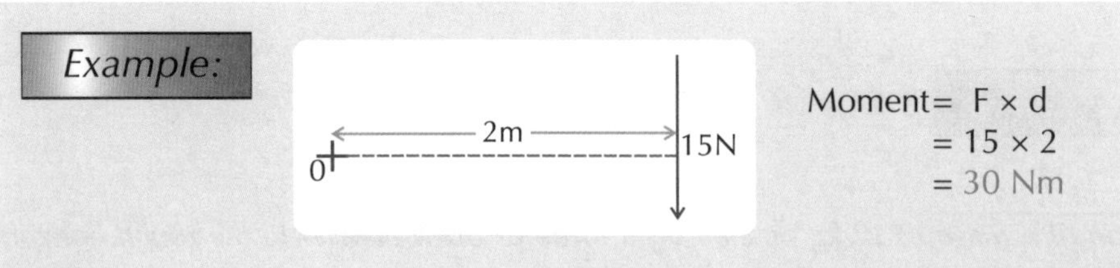

Moment = F × d
= 15 × 2
= 30 Nm

Example:

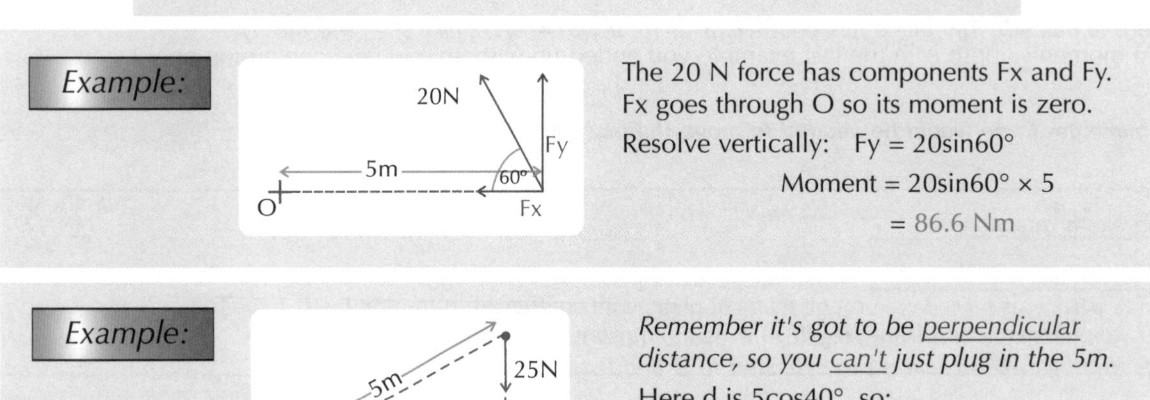

The 20 N force has components Fx and Fy.
Fx goes through O so its moment is zero.
Resolve vertically: Fy = 20sin60°
Moment = 20sin60° × 5
= 86.6 Nm

Example:

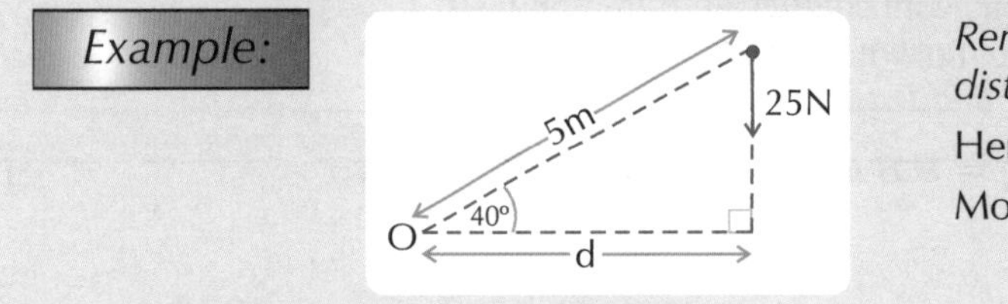

Remember it's got to be <u>perpendicular</u> distance, so you <u>can't</u> just plug in the 5m.
Here d is 5cos40°, so:
Moment = 25 × 5cos40°
= 95.8 Nm

In **Equilibrium** Moments Total **Zero**

...and that means that for a body in <u>equilibrium</u> the total moments either way must be <u>equal</u>:

Total Clockwise Moment = Total Anticlockwise Moment

Example: Two weights of 30 N and 45 N are placed on a light 8 m beam. The 30 N weight is at one end of the beam as shown whilst the other weight is a distance d from the midpoint M. The beam is in equilibrium held by a single wire with tension T attached at M. Find T and the distance d.

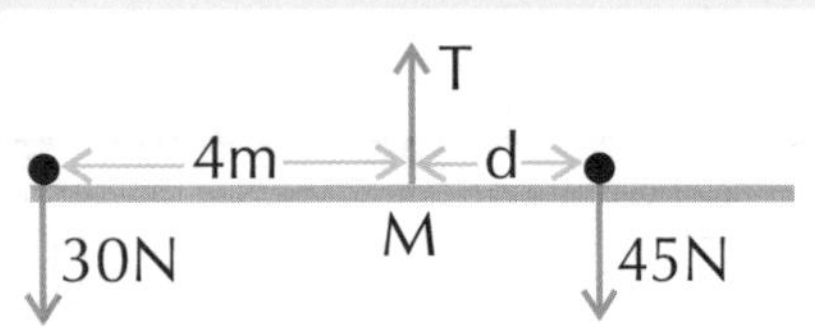

Resolving vertically: 30 + 45 = T = 75N
Take moments about M:
Clockwise Moment = Anticlockwise Moment
$45 \times d = 30 \times 4$

$$d = \frac{120}{45} = 2\frac{2}{3} \text{ m}$$

Moments

This page is for Edexcel M1

The Weight of a Uniform Beam Acts at its Middle

Example: A 6 m long uniform beam AB of weight 40 N is supported at A by a vertical reaction R. AB is held horizontal by a vertical wire attached 1 m from the other end. A weight of 30 N is placed 2 m from the support R. Find the tension T in the wire and the force R.

Even if you're not told, the weight of the beam acts at its centre.

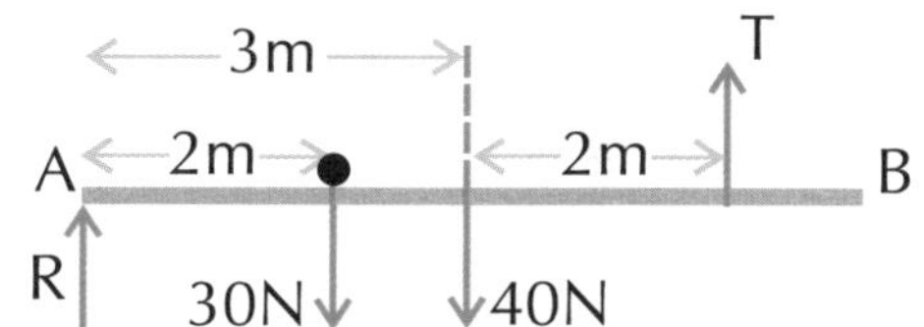

Take moments about A.

Clockwise Moment = Anticlockwise Moment

$(30 \times 2) + (40 \times 3) = T \times 5$

$T = 36$ N

Resolve vertically: $T + R = 30 + 40$

So: $R = 34$ N

Take Moments Wisely

By taking moments about A in the last example you ended up with an equation containing only T. That's because R goes through A, so has no moment about it.

It's always easier to do questions if you take moments about a point that has an unknown force going through it.

Practice Questions

1) ***A 60 kg uniform beam AE of length 14 m is in equilibrium, supported by two vertical ropes attached to B and D as shown.***

Find the tensions in the ropes to 1 d.p. Take $g = 9.8$ ms^{-2}.

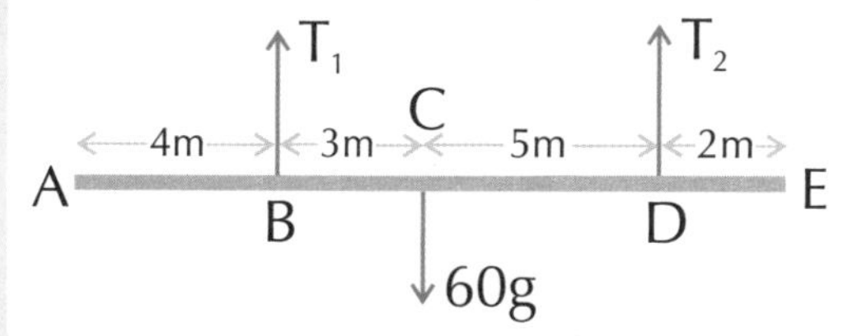

2) ***Sample exam question:***

A 6 m long uniform beam of mass 20 kg is in equilibrium. One end is resting on a vertical pole, and the other end is held up by a vertical wire attached to that end so that the beam rests horizontally. There are two 10 kg weights attached to the beam, situated 2 m from either end.

Draw a diagram of the beam including all the forces acting on it. [2 marks]

Find, in terms of g:

a) T, the tension in the wire; [3 marks]

b) N, the normal reaction at the pole. [2 marks]

Phobia #5 — alliumphobia: fear of garlic...

Learn those formulas in the checked boxes — without knowing them you won't get very far with questions about moments. Whenever I go to an Italian restaurant I always order a Quattro Stagioni pizza, and M1 examiners are just the same. They'll almost always stick with what they know.

Newton's Laws

This page is for AQA M1, OCR M1, Edexcel M1

That clever chap Isaac Newton established 3 laws involving motion. You need to know all of them.

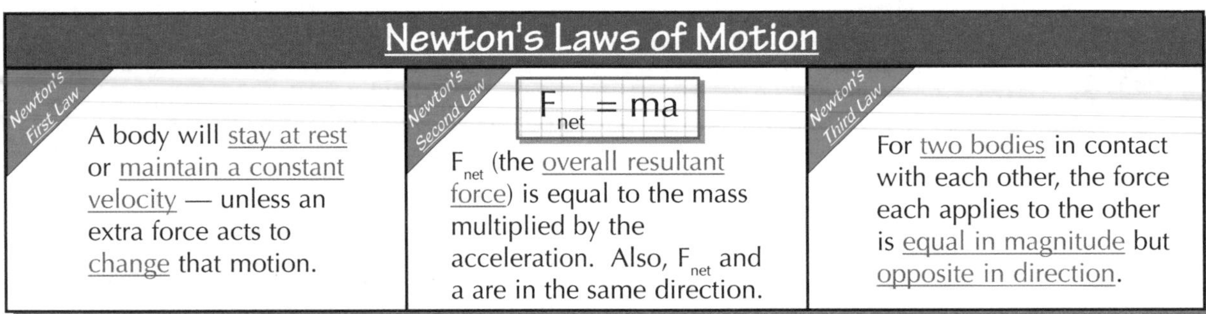

Newton's Laws of Motion
Newton's First Law
A body will stay at rest or maintain a constant velocity — unless an extra force acts to change that motion.
Newton's Second Law
F net = ma
F net (the overall resultant force) is equal to the mass multiplied by the acceleration. Also, F net and a are in the same direction.
Newton's Third Law
For two bodies in contact with each other, the force each applies to the other is equal in magnitude but opposite in direction.

Hint: $\mathbf{F_{net} = ma}$ is sometimes just written as **F = ma**, but it means the same thing.

Resolve Forces in Perpendicular Directions

Example: A mass of 4 kg at rest on a smooth horizontal plane is acted on by a horizontal force of 5 N. Find the acceleration of the particle and the normal reaction from the plane. Take $g = 9.8$ ms^{-2}.

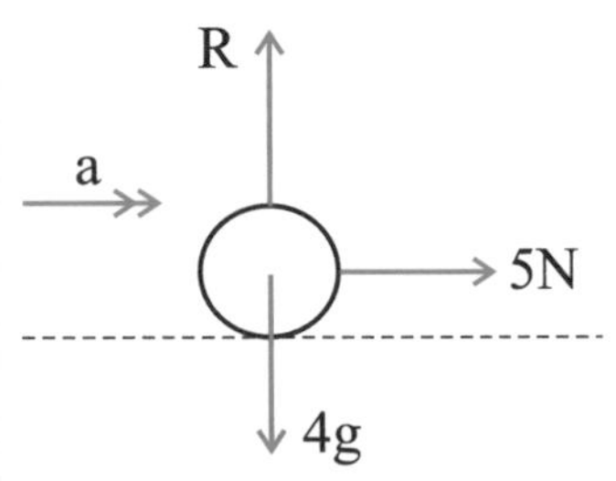

Resolve horizontally:
$F_{net} = ma$ ← Always write $F_{net} = ma$ first.
$5 = 4a$
$a = 1.25$ ms^{-2} to the right

Resolve vertically:
$F_{net} = ma$
$R - 4g = 4 \times 0$
$R = 4g = 39.2$ N

Example: A particle of weight 30 N is being accelerated across a smooth plane by a force of 6 N acting at an angle of 25° to the horizontal. Given that the particle starts from rest, find:
a) its speed after 4 seconds,
b) the magnitude of the normal reaction with the plane. What assumptions have you made?

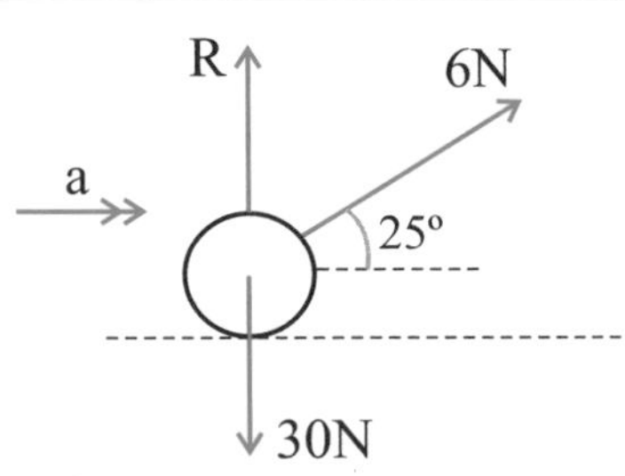

a) Resolve horizontally:
$F_{net} = ma$

$6\cos 25° = \frac{30}{g}a$ so $a = 1.78$ ms^{-2}

$v = u + at$
$v = 0 + 1.78 \times 4 = 7.11$ ms^{-1}

b) Resolve vertically:
$F_{net} = ma$

$R + 6\sin 25° - 30 = \frac{30}{g} \times 0$

So $R = 30 - 6\sin 25° = 27.5$ N (to 3 s.f.)

Assumptions:
- particle is considered as a point mass,
- there's no air resistance,
- it's a constant acceleration.

Newton's Laws

This page is for AQA M1, OCR M1, Edexcel M1

You can resolve forces **Parallel** and **Perpendicular** to **Planes**

Example:

A mass of 600 g is propelled up the line of greatest slope of a smooth plane inclined at 30° to the horizontal. If its initial velocity is 3 ms^{-1} find the distance it travels before coming to rest and the magnitude of the normal reaction.

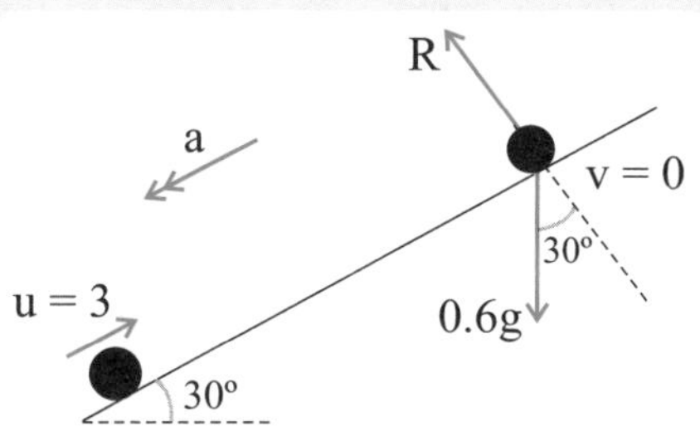

Resolve in ↗ direction:

$F_{net} = ma$

$-0.6g\sin 30 = 0.6a$

$\mathbf{a = -4.9\ ms^{-2}}$

$v^2 = u^2 + 2as$

$0 = 3^2 + 2(-4.9)s$

So $s = 0.92$ m

Taking up plane as + ve.

Resolve in ↖ direction:

$F_{net} = ma$

$R - 0.6g\cos 30 = 0.6 \times 0$

So $R = 5.09$ N

Practice Questions

1) ***A horizontal force of 2 N acts on a 1.5 kg particle initially at rest on a smooth horizontal plane. Find the speed of the particle 3 seconds later.***

2) ***Two forces act on a particle of mass 8 kg which is initially at rest on a smooth horizontal plane. The two forces are (24i+18j) N and (6i+22j) N (with i and j being perpendicular unit vectors in the plane). Find the magnitude and direction of the resulting acceleration of the particle and its displacement after 3 seconds.***

3) ***A horizontal force P acting on a 2 kg mass generates an acceleration of 0.3 ms^{-2}. Given that the mass is in contact with a rough horizontal plane which resists motion with a force of 1 N, find P. Then find the coefficient of friction, μ, to 2 d.p.***

4) ***Sample exam question:***

A crane moves a mass of 300 kg, which is modelled as a particle A suspended by two cables AB and AC attached to a movable beam BC. The mass is moved in the direction of the line of the supporting beam BC during which time the cables maintain a constant angle of 40° to the horizontal as shown.

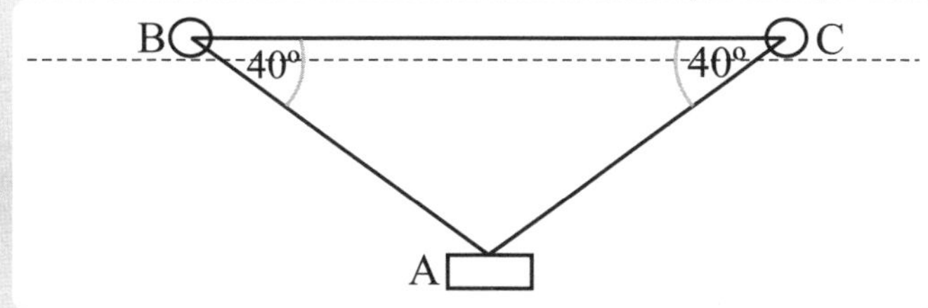

a) The mass is initially moving with constant speed. Find the tension in each cable. [4 marks]

b) The crane then moves the mass with a constant acceleration of 0.4 ms^{-2}. Find the tension in each cable. [6 marks]

c) What modelling assumptions have you made in part b? [2 marks]

Interesting Newton fact: Isaac Newton had a dog called Diamond...

Did you know that Isaac Newton held the same position at Cambridge that Stephen Hawking holds today? And the dog fact about Newton is true — don't ask me how I know such things, just bask in my amazing knowledge of all things trivial.

Friction and Inclined Planes

This page is for AQA M1, OCR M1, Edexcel M1

Solving these problems involves careful use of $F_{net} = ma$, $F \leq \mu R$ and the equations of motion.

*Use F = ma in **Two Directions** for **Inclined Plane** questions*

For inclined slope questions, it's much easier to resolve forces parallel and perpendicular to the plane's surface.

Example:

A mass of 3 kg is being pulled up a plane inclined at 20° to the horizontal by a rope parallel to the surface. Given that the mass is accelerating at 0.6 ms⁻² and that the coefficient of friction is 0.4, find the tension in the rope. Take g = 9.8 ms⁻².

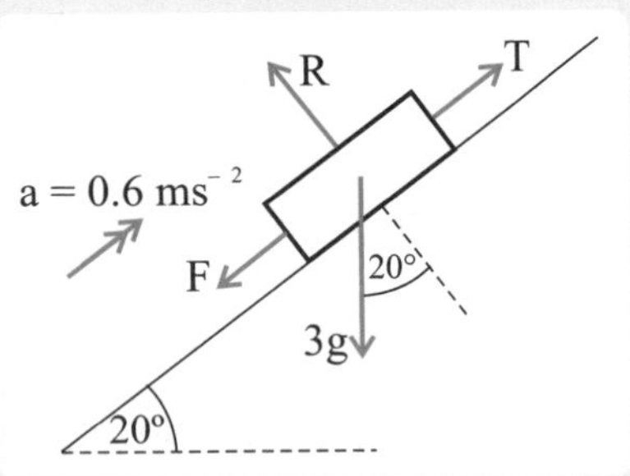

Resolving in ↖ direction: $F_{net} = ma$

$R - 3g\cos 20° = 3 \times 0$

so: $R = 3g\cos 20° = 27.63 \text{ N}$

The mass is sliding, so $F = \mu R$

$= 0.4 \times 27.62 = 11.05 \text{ N}$

Resolving in ↗ direction: $F_{net} = ma$

$T - F - 3g\sin 20° = 3 \times 0.6$

$T = 1.8 + 11.05 + 3g\sin 20° = 22.9 \text{ N}$

Remember that friction always acts in the opposite direction to the motion.

Example:

A small body of weight 20 N accelerates from rest and moves a distance of 5 m down a rough plane angled at 15° to the horizontal. Draw a force diagram and find the coefficient of friction between the body and the plane given that the motion takes 6 seconds. Take g = 9.8 ms⁻².

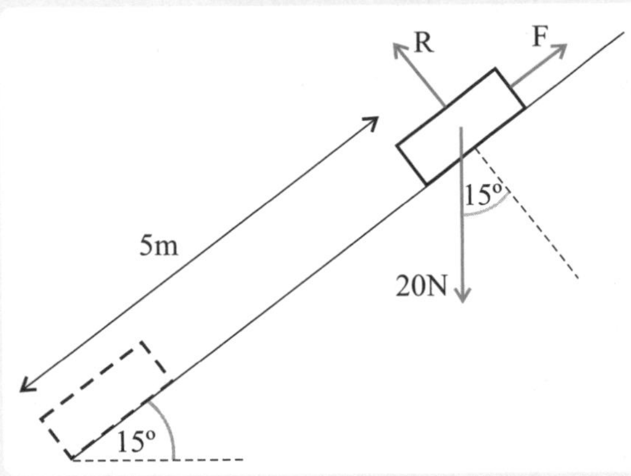

$u = 0, \quad s = 5, \quad t = 6, \quad a = ?$

Use one of the equations of motion: $s = ut + \frac{1}{2}at^2$

$5 = (0 \times 6) + (\frac{1}{2}a \times 6^2)$ so $\mathbf{a = 0.2778 \text{ ms}^{-2}}$

Resolving in ↖ direction: $F_{net} = ma$

$R - 20\cos 15° = \frac{20}{g} \times 0$

So: $R = 20\cos 15° = \mathbf{19.32 \text{ N}}$

Resolving in ↙ direction: $F_{net} = ma$

$20\sin 15° - F = \frac{20}{g} \times 0.28$

$\mathbf{F = 4.609 \text{ N}}$

It's sliding, so $F = \mu R$

$4.609 = \mu \times 19.32$

$\mu = 0.24$ (to 2 d.p.)

Friction and Inclined Planes

This page is for AQA M1, OCR M1, Edexcel M1

Friction Opposes Limiting Motion

For a body at rest but on the point of moving down a plane, the friction force is up the plane.
A body about to move up a plane is opposed by friction down the plane.

Example: A 4 kg box is placed on a 30° plane where $\mu = 0.4$. A force Q maintains equilibrium by acting up the plane parallel to the line of greatest slope.
Find Q if the box is on the point of sliding a) up the plane, b) down the plane.

a)

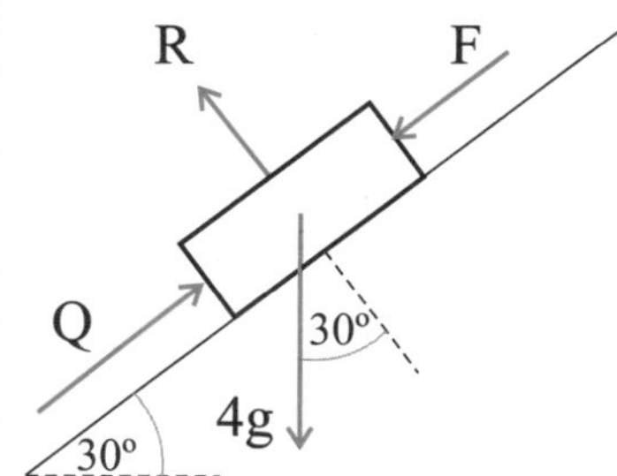

$F_{net} = ma$

Resolving in ↖ direction:

$R - 4g\cos 30° = 0$

$R = 4g\cos 30°$

$F = \mu R$

$= 0.4 \times 4g\cos 30$

$= 1.6g\cos 30$

Resolving in ↗ direction:

$Q - 4g\sin 30 - F = 4 \times 0$

$Q = 4g\sin 30 + 1.6g\cos 30$

$= 33.2$ N

b)

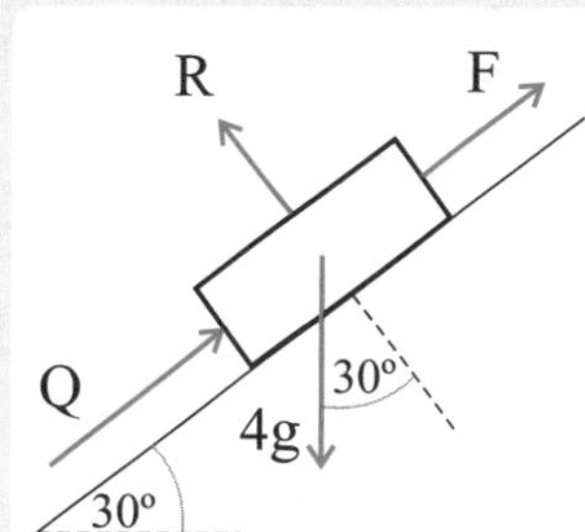

Resolving in ↖ direction:

$R = 4g\cos 30$

$F = 1.6g\cos 30$

Resolving in ↗ direction:

$Q - 4g\sin 30 + F = 4 \times 0$

$Q = 4g\sin 30 - 1.6g\cos 30$

$Q = 6.02$ N

So for equilibrium $6.02 \text{ N} \leq Q \leq 33.2 \text{ N}$

Practice Questions

1) ***A brick of mass 1.2 kg is sliding down a rough plane which is inclined at 25° to the horizontal. Given that its acceleration is 0.3 ms^{-2}, find the coefficient of friction between the brick and the plane. What assumptions have you made?***

2) ***An army recruit of weight 600 N steps off a tower and accelerates down a "death slide" wire as shown. The recruit hangs from a light rope held between her hands and looped over the wire. The coefficient of friction between the rope and wire is 0.5. Given that the wire is 20 m long and makes an angle of 30° to the horizontal throughout its length, find how fast the recruit is travelling when she reaches the end of the wire.***

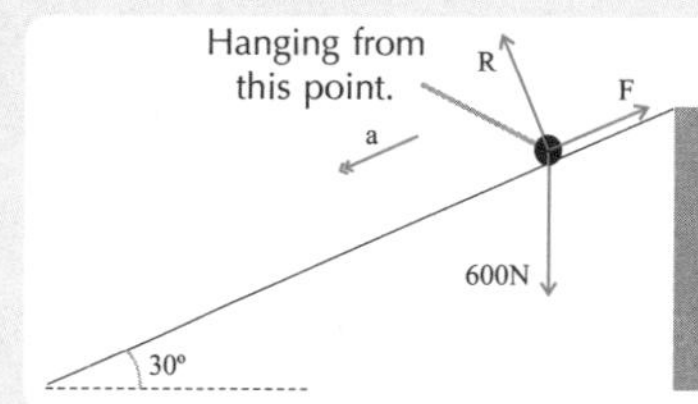

3) ***Sample exam question:***

A horizontal force of 8 N just stops a mass of 7 kg from sliding down a plane inclined at 15° to the horizontal as shown.

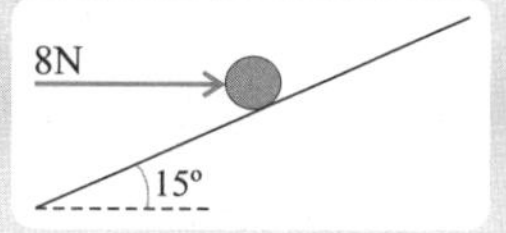

a) Calculate the coefficient of friction between the mass and the plane to 2 d.p. [6 marks]

b) The 8 N force is now removed.
Find how long the mass takes to slide a distance of 3 m down the line of greatest slope. [7 marks]

Inclined planes — nothing to do with suggestible Boeing 737s...

The main thing to remember is that you can choose to resolve in any two directions as long as they're perpendicular.
It makes sense to choose the directions that involve doing as little work as possible. Obviously.

Constant Acceleration and Path Equations

This page is for AQA M1, Edexcel M1

You can use column vectors and **i,j** notation when working with equations of motion. Thought you'd be pleased.

*All vectors have **Magnitude** and **Direction***

Example:

A particle P moves on a smooth horizontal plane with constant acceleration. Initially the particle has a velocity of 2**i** ms^{-1} and starts from an origin O. 5 seconds later its velocity is 10**j** ms^{-1}, where **i** and **j** are perpendicular unit vectors in the plane.

Find the magnitude and direction of the acceleration, plus the position of P, 5 seconds after motion begins.

First list the variables for equations of motion:

In this question, letters in bold (e.g. **a**) refer to vectors. Letters not in bold (e.g. a) mean the magnitude of that vector.

$$\mathbf{u} = \begin{pmatrix} 2 \\ 0 \end{pmatrix}; \quad \mathbf{v} = \begin{pmatrix} 0 \\ 10 \end{pmatrix}; \quad t = 5; \quad \mathbf{a} = ?$$

You need to use an equation containing **u**, **v**, t and **a**:

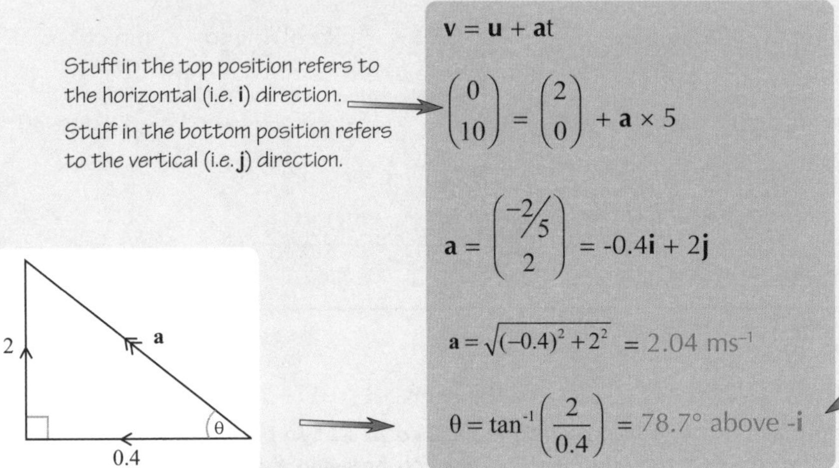

Stuff in the top position refers to the horizontal (i.e. **i**) direction.
Stuff in the bottom position refers to the vertical (i.e. **j**) direction.
2
a
θ
0.4

$$\mathbf{v} = \mathbf{u} + \mathbf{a}t$$

$$\begin{pmatrix} 0 \\ 10 \end{pmatrix} = \begin{pmatrix} 2 \\ 0 \end{pmatrix} + \mathbf{a} \times 5$$

$$\mathbf{a} = \begin{pmatrix} -2/5 \\ 2 \end{pmatrix} = -0.4\mathbf{i} + 2\mathbf{j}$$

$$a = \sqrt{(-0.4)^2 + 2^2} = 2.04 \text{ ms}^{-1}$$

$$\theta = \tan^{-1}\left(\frac{2}{0.4}\right) = 78.7^\circ \text{ above } -\mathbf{i}$$

Don't just write the angle — you need to say where it is in relation to **i** or **j**.

So you've got the direction and magnitude of the acceleration sorted out — now you need the position of P after 5 seconds of motion.

Play it safe and make another list of all the variables you know, including the one you've just worked out:

$$\mathbf{u} = \begin{pmatrix} 2 \\ 0 \end{pmatrix}; \quad \mathbf{v} = \begin{pmatrix} 0 \\ 10 \end{pmatrix}; \quad t = 5; \quad \mathbf{a} = \begin{pmatrix} -2/5 \\ 2 \end{pmatrix}$$

You need to know **s**, so choose an equation containing **s**, t, **a** and either **u** or **v**:

$$\mathbf{s} = \mathbf{u}t + \frac{1}{2}\mathbf{a}t^2$$

$$\begin{pmatrix} x \\ y \end{pmatrix} = \begin{pmatrix} 2 \\ 0 \end{pmatrix} \times 5 + \frac{1}{2}\begin{pmatrix} -0.4 \\ 2 \end{pmatrix} \times 5^2 = \begin{pmatrix} 5 \\ 25 \end{pmatrix}$$

i.e. when t = 5, $\overrightarrow{OP} = 5\mathbf{i} + 25\mathbf{j}$

Constant Acceleration and Path Equations

This page is for AQA M1, Edexcel M1

Path Equations *give the* ***Position of a Particle*** *in terms of* ***Time***

Path equations always contain t, so that you can work out the x and y coordinates of the particle at a particular time. If you eliminate t, you can get values for the coordinates of the particle.

Example:

The position vector, **r**, of a particle is given by **r** = 5t**i** – 5t²**j** relative to perpendicular unit vectors **i** and **j**, based on an origin 0.
Assuming motion begins at 0, sketch the path of the particle during the first 2 seconds of motion and find its path equation for the general position (x,y).

First things first... Start by sketching a quick table showing the x and y values at t = 0, 1 and 2.

	t	0	1	2
x	5t	0	5	10
y	$-5t^2$	0	-5	-20

Then you can plot the points as a graph:

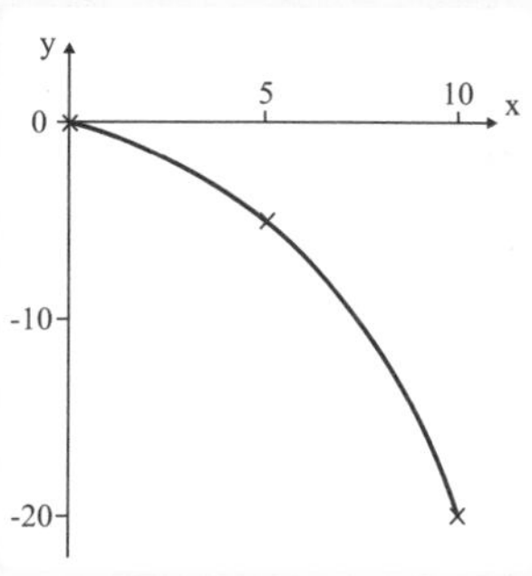

You know for any time t that $x = 5t$ and $y = -5t^2$. Just rearrange to eliminate t, to give you an equation for the general position (x,y).

$x = 5t$ so $t = \frac{x}{5}$

Plug this into y: $y = -5t^2 = -5\left(\frac{x}{5}\right)^2 = -\frac{1}{5}x^2$

So $y = -\frac{1}{5}x^2$

Now you don't even need to know what t is at any point.

Practice Questions

1) ***A particle moves with constant acceleration away from an origin O. Initially it is moving due south with a speed of 6 ms⁻¹. 20 seconds later it is moving due east with a speed of 8 ms⁻¹. Unit vectors i and j are in directions east and north respectively.***

a) Find the acceleration of the particle in i, j form and find its magnitude and direction.
b) Find the position of the particle after 5 seconds.

2) ***Sample exam question:***

A small aircraft flies low over a field. As it pulls sharply up, its path relative to an origin 0 at the end of the field is modelled by the path equation $\mathbf{r} = (15t + 10)\mathbf{i} + (15\sqrt{3}\,t - 5t^2)\mathbf{j}$ for $t > 1$ where **i** and **j** are unit vectors relative to 0 in horizontal and vertical directions respectively. Distances are in metres.

a) How far is the aircraft from the origin when t = 3? [2 marks]
b) Sketch the path of the aircraft's flight for $1 \leq t \leq 3$. [4 marks]
c) Show that the path equation for the aircraft's motion can be approximated to $y = 2.2x - 20 - 0.02x^2$. [4 marks]
d) Find the expressions for the aircraft's velocity and acceleration at time, t. [3 marks]
e) Suggest why the model is not appropriate for all values of t for $t > 1$. [1 marks]

I promise I'm not leading you up the garden path equation...

Even if you're working with column vectors, these questions are all OK — just plug the values into the equation of motion that gives you the values you want. And path equations are good for working out exactly where that particle's got to.

Connected Particles

This page is for AQA M1, OCR M1, Edexcel M1

Like Laurel goes with Hardy and Posh goes with Becks, some particles are destined to be together...

Connected Particles* act like *One Mass

Particles connected together have the same speeds and accelerations as each other, unless the connection fails. Train carriages moving together have the same acceleration.

Example: A 30 tonne locomotive engine is pulling a single 10 tonne carriage as shown. They are accelerating at 0.3 ms^{-2} due to the force P generated by the engine. It's assumed that there are no forces resistant to motion. Find P and the tension in the coupling.

Here's the pretty picture:

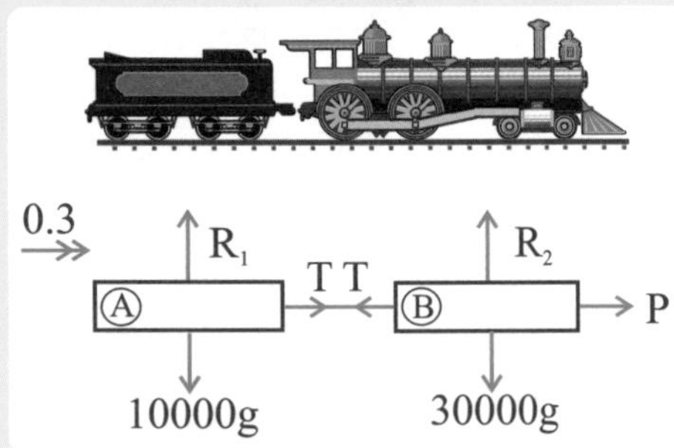

And here's the ugly maths:

For A: $F_{net} = ma$
$T = 10\,000 \times 0.3$
$T = 3000$ N

For B: $F_{net} = ma$
$P - T = 30\,000 \times 0.3$
$P = 12000$ N

Pulleys* (and *'Pegs'*) are always *Smooth

In M1 questions, you can always assume that the tension in a string will be the same either side of a smooth pulley.

Example: Masses of 3 kg and 5 kg are connected by an inextensible string and hang vertically either side of a smooth pulley. They are released from rest. Find their acceleration and the time it takes for each to move 40 cm. State any assumptions made in your model.

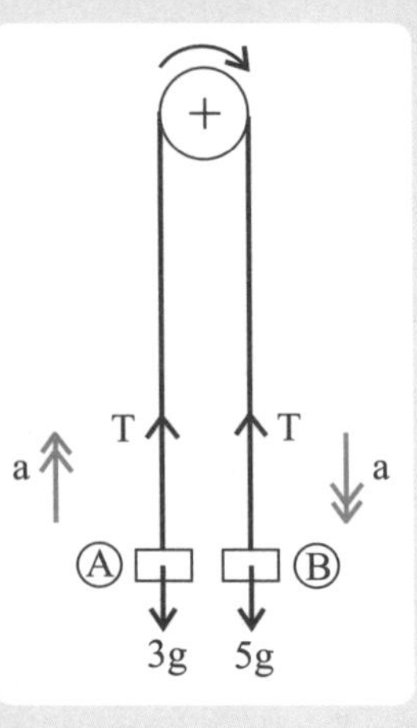

For A: $F_{net} = ma$
Resolving upwards: $T - 3g = 3a$ ①

For B: $F_{net} = ma$
Resolving downwards: $5g - T = 5a$
$T = 5g - 5a$ ②

Sub ② into ①: $(5g - 5a) - 3g = 3a$
$a = 2.45$ ms^{-2}

List variables: $u = 0$; $a = 2.45$; $s = 0.4$
Use an equation with u, a, s and t in it:
$s = ut + \frac{1}{2}at^2$

$0.4 = (0 \times t) + (\frac{1}{2} \times 2.45 \times t^2)$ So $t = \sqrt{\frac{0.8}{2.45}} = 0.57$ s

Assumptions: The 3 kg mass does not hit the pulley; there's no air resistance; the string is 'light' and doesn't break.

Connected Particles

This page is for AQA M1, OCR M1, Edexcel M1

Use F = ma in the **Direction Each Particle Moves**

Example:

A mass of 3 kg is placed on a smooth horizontal table. A light inextensible string connects it over a smooth peg to a 5 kg mass which hangs vertically as shown. Find the tension in the string if the system is released from rest. Take $g = 9.8\ ms^{-2}$.

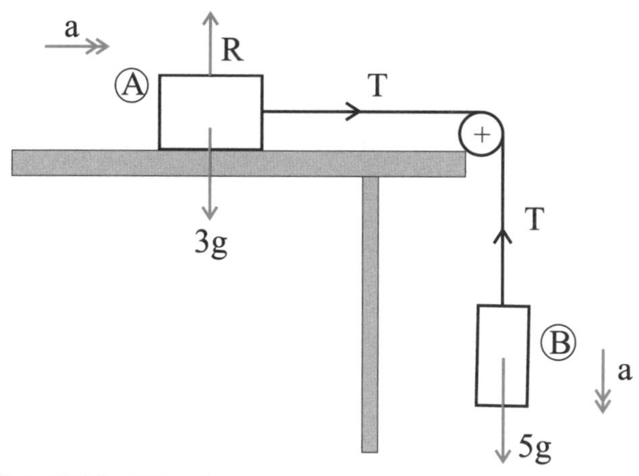

For A:
Resolve horizontally: $F_{net} = ma$

$T = 3a$

$a = \frac{T}{3}$ ①

For B:
Resolve vertically: $F_{net} = ma$

$5g - T = 5a$ ②

Sub ① into ②: $5g - T = 5 \times \frac{T}{3}$

So $\frac{8}{3}T = 5g$

$T = 18.4\ N$ (to 3 s.f.)

Practice Questions

1) ***A 2 tonne tractor experiences a resistance force of 1000 N whilst pulling a 1 tonne trailer along a horizontal road. If the tractor engine provides a forward force of 1500 N find the resistance force acting on the trailer, and the tension in the coupling between tractor and trailer, if they are moving with constant speed.***

2) ***Two particles are connected by a light inextensible string, and hang in a vertical plane either side of a smooth pulley. When released from rest the particles accelerate at 1.2 ms^{-2}. If the heavier mass is 4 kg, find the weight of the other.***

3) ***Sample exam question:***

A car of mass 1500 kg is pulling a caravan of mass 500 kg.
They experience resistance forces totalling 1000 N and 200 N respectively.
The forward force generated by the car's engine is 2500 N. The coupling between the two does not break.

a) Find the acceleration of the car and caravan. [3 marks]

b) Find the tension in the coupling. [3 marks]

Useful if you're hanging over a Batman-style killer crocodile pit...

It makes things a lot easier when you know that connected particles act like one mass, and that in M1 pulleys can always be treated as smooth. Those examiners occasionally do try and make your life easier, honestly.

Connected Particles

This page is for AQA M1, OCR M1, Edexcel M1

More complicated pulley and peg questions have friction and inclined planes for you to enjoy too.

*Remember to use $F \le \mu R$ on **Rough Planes***

For a particle on a plane, don't forget to resolve the forces in two directions.

Example:

The peg system of the example on p143 is set up again. However, this time a friction force, F, acts on the 3 kg mass due to the table top now being rough, with coefficient of friction $\mu = 0.5$. Find the new tension in the string when the particles are released from rest.

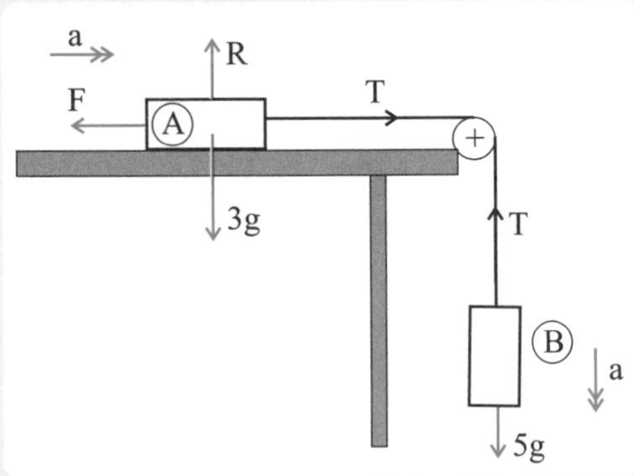

For B: Resolving vertically: $5g - T = 5a$ ①

For A: Resolving horizontally: $F_{net} = ma$

$T - F = 3a$ ②

Resolving vertically: $R - 3g = 0$

$\mathbf{R = 3g}$

The particles are moving, so $F = \mu R = 0.5 \times 3g$

$= \mathbf{14.7\ N}$

Sub this into ②: $T - 14.7 = 3a$

$\mathbf{a = \frac{1}{3}(T - 14.7)}$

Sub this into ①: $5g - T = 5 \times \frac{1}{3}(T - 14.7)$

$8T = 147 + 73.5$

$T = 27.6\ N$

***Rough Inclined Plane** questions need **Really Good** force diagrams*

You know the routine... resolve forces parallel and perpendicular to the plane... yawn.

Example:

A 3 kg mass is held in equilibrium on a rough ($\mu = 0.4$) plane inclined at 30° to the horizontal. It is attached by a piece of light, inextensible string to a mass M kg hanging vertically beneath a smooth pulley, as shown. Find M if the 3 kg mass is on the point of sliding up the plane.

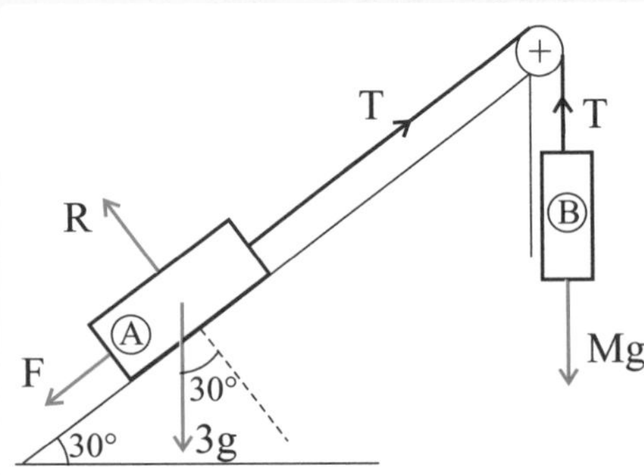

For B: Resolving vertically: $F_{net} = ma$

$Mg - T = M \times 0$

$\mathbf{T = Mg}$

For A: Resolving in ↖ direction: $F_{net} = ma$

$R - 3g\cos30 = 3 \times 0$

$\mathbf{R = 3g\cos30}$

It's limiting friction, so: $F = \mu R$

$F = 0.4 \times 3g\cos30$

$= \mathbf{10.18\ N}$

For A: Resolving in ↗ direction: $F_{net} = ma$

$T - F - 3g\sin30 = 3 \times 0$

$Mg - 10.18 - 3g\sin30 = 0$

$M = 2.54$ kg (to 3 s.f.)

Connected Particles

This page is for AQA M1, OCR M1, Edexcel M1

Example:

Particles A and B of mass 4 kg and 5 kg are connected by a light inextensible string over a smooth pulley as shown. A force of Q acts on A at an angle of 25° to the rough ($\mu = 0.6$) horizontal plane. Find the range of values of the magnitude of Q if equilibrium is to be maintained.

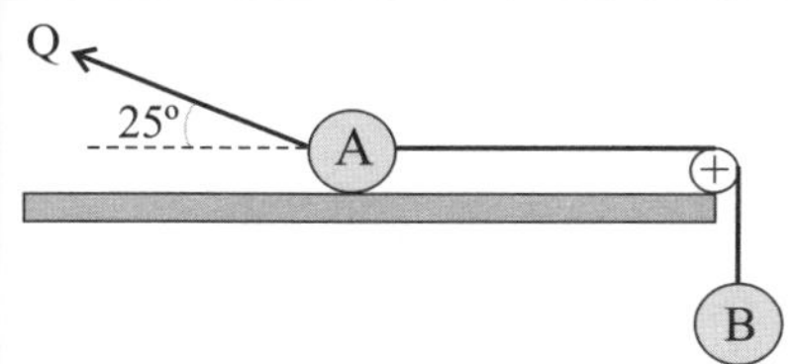

You need to work out Q if A is about to move left or right.

↑ T

↓ 5g

Mass B is the same whichever way the frictional force F acts:

Resolving vertically: **T = 5g** ①

i) Work out Q if A is about to go right:

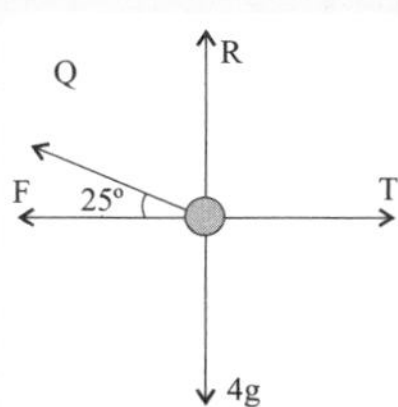

Resolving vertically: $R = 4g - Q\sin25°$

$F = \mu R$

$= 0.6(4g - Q\sin25°)$ ②

Resolving horizontally: $T - Q\cos25° - F = 0$

Using equations 1 and 2. $5g - Q\cos25° - 0.6(4g - Q\sin25°) = 0$

So Q = 39.0 N (to 3 s.f.)

ii) Work out Q if A is about to go left:

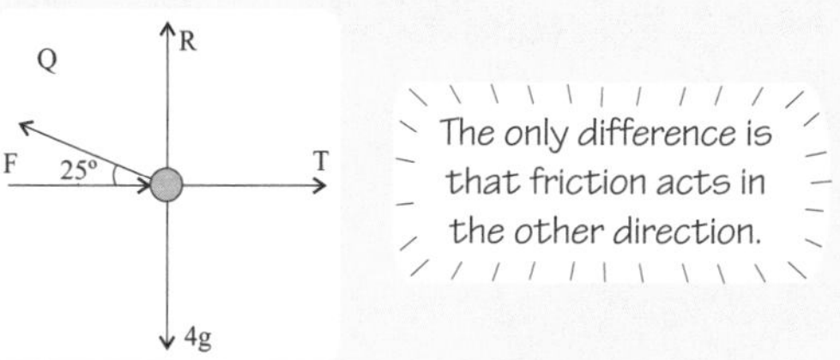

R
Q
F
25°
T
4g
The only difference is that friction acts in the other direction.

Resolving vertically: $R = 4g - Q\sin25°$

$F = \mu R$

$= 0.6(4g - Q\sin25°)$

Resolving horizontally: $Q\cos25° - F - T = 0$

$Q\cos25° - 0.6(4g - Q\sin25°) - 5g = 0$

So Q = 62.5 N (to 3 s.f.)

Practice Questions

1) Particles of mass 3 kg and 4 kg are connected by a light, inextensible string passing over a smooth pulley as shown. The 3 kg mass is on a smooth slope angled at 40° to the horizontal. Find the acceleration of the system if released from rest, and find the tension in the string. What minimum force acting on the 3 kg mass parallel to the plane would be needed to maintain equilibrium?

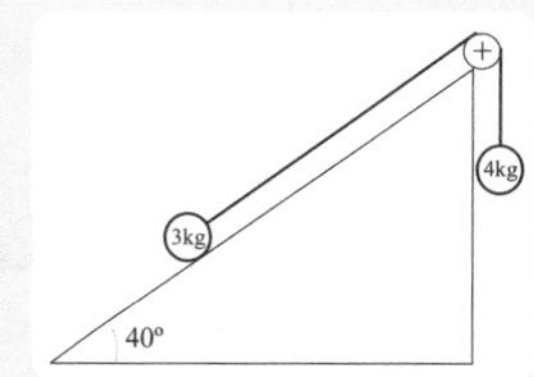

2) Sample exam question:

Two particles P and Q of masses 1 kg and m kg respectively are linked by a light string passing over a smooth pulley as shown. Particle P is on a rough slope inclined at 20° to the horizontal, where the coefficient of friction between P and the plane is 0.1.

a) Given that P is about to slide down the plane, find the mass of Q. [5 marks]

b) Describe the motion of the system if the mass of Q is 1kg. [5 marks]

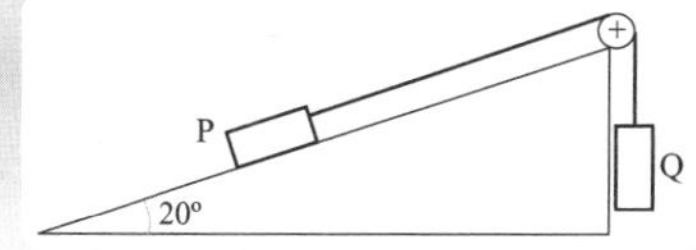

Connected particles — together forever... *isn't it beautiful?*

The key word here is rough. If a question mentions the surface is rough, then cogs should whirr and the word 'friction' should pop into your head. Take your time with force diagrams of rough inclined planes — I had a friend who rushed into drawing a diagram, and he ended up with a broken arm. But that was years later, now that I come to think of it.

Momentum

This page is for AQA M1, OCR M1, Edexcel M1

Momentum is a measure of how much strength a moving object has, due to its mass and velocity.

Momentum has Magnitude and Direction

Total momentum before a collision equals total momentum after a collision.
This idea is called "Conservation of Momentum".

Momentum = Mass × Velocity

Because it's a vector, the sign of the velocity in momentum is important.

Example:

Particles A and B, each of mass 5 kg, move in a straight line with velocities 6 ms^{-1} and 2 ms^{-1} respectively. After collision mass A continues in the same direction with velocity 4.2 ms^{-1}. Find the velocity of B after impact.

Before

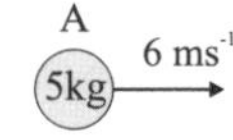

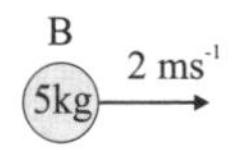

After

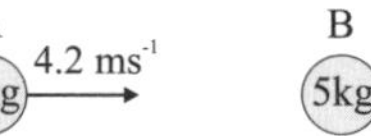

B v 5kg

Draw 'before' and 'after' diagrams to help you see what's going on.

Momentum A + Momentum B = Momentum A + Momentum B

$$(5 \times 6) + (5 \times 2) = (5 \times 4.2) + (5 \times v)$$

$$40 = 21 + 5v$$

So $v = 3.8\ ms^{-1}$ in the same direction as before

Example:

Particles A and B of mass 6 kg and 3 kg are moving towards each other at speeds of 2 ms^{-1} and 1 ms^{-1} respectively. Given that B rebounds with speed 3 ms^{-1} in the opposite direction to its initial velocity, find the velocity of A after the collision.

Before

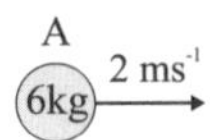

B 1 ms⁻¹ 3kg

After

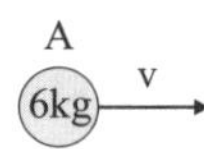

B 3 ms⁻¹ 3kg

$$(6 \times 2) + (3 \times -1) = (6 \times v) + (3 \times 3)$$

$$9 = 6v + 9$$

$$v = 0$$

Masses Joined Together have the Same Velocity

Particles that stick together after impact are said to "coalesce". After that you can treat them as just one object.

Example:

Two particles of mass 40 g and M kg move towards each other with speeds of 6 ms^{-1} and 3 ms^{-1} respectively. Given that the particles coalesce after impact and move with a speed of 2 ms^{-1} in the same direction as that of the 40 g particle's initial velocity, find M.

Before

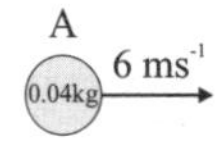

B 3 ms⁻¹ M

After

(M + 0.04) kg 2 ms⁻¹

$$(0.04 \times 6) + (M \times -3) = [(M + 0.04) \times 2]$$

$$0.24 - 3M = 2M + 0.08$$

$$5M = 0.16$$

$$M = 0.032 \text{ kg} = 32 \text{ g}$$

Don't forget to convert all masses to the same units.

Momentum

This page is for AQA M1, OCR M1, Edexcel M1

Example:

A lump of ice of mass 0.1 kg is thrown across the surface of a frozen lake with speed 4 ms^{-1}. It collides with a stationary stone of mass 0.3 kg. The lump of ice and the stone then move in opposite directions to each other with the same speeds. Find their speed.

Before

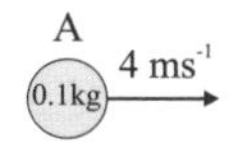

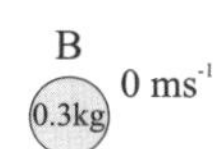

After

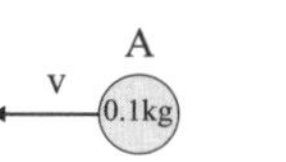

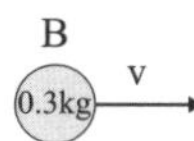

$$(0.1 \times 4) + (0.3 \times 0) = (0.1 \times -v) + 0.3v$$

$$0.4 = -0.1v + 0.3v$$

$$v = 2\ ms^{-1}$$

Practice Questions

Each diagram represents the motion of two particles moving in a straight line. Find the missing mass or velocity. (All masses are in kg and all velocities are in ms^{-1}.)

Before **After**

1) (5) 3 → (4) 1 →   (5) 2 → (4) v →

2) 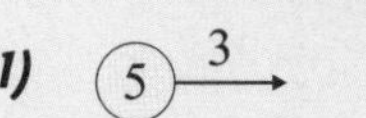 (5) 3 → 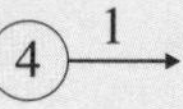 (4) 1 →   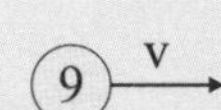 (9) v →

3) (5) 3 → 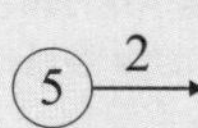 ← 2 (4)   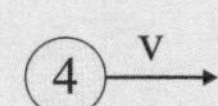 ← v (5) (4) 3 →

4) 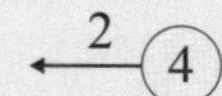 (m) 6 → (8) 2 →   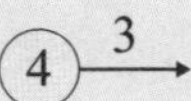 (m) 2 → (8) 4 →

5) ***Sample exam question:***

Two particles of mass 0.8 kg and 1.2 kg are travelling in the same direction along a straight line with speeds of 4 ms^{-1} and 2 ms^{-1} respectively. After collision the 0.8 kg mass has a velocity of 2.5 ms^{-1} in the same direction. The 1.2 kg mass then continues with its new velocity until it collides with a mass m kg travelling with a speed of 4 ms^{-1} in the opposite direction to it. Both particles are brought to rest by this collision. Find the mass m. [4 marks]

Ever heard of Hercules?

Well, he carried out 12 tasks, including: killing the Nemean lion, capturing the Erymanthian boar, acquiring the golden apples of the Hesperides, killing the Stymphalian birds, cleaning the stables of Augeas and capturing the girdle of Hippolyte. Nothing to do with momentum, but next time you're feeling sorry for yourself for doing M1, think on.

Impulse

This page is for Edexcel M1

An impulse changes the momentum of a particle in the direction of motion.

*Impulse is **Change in Momentum***

To work out the impulse that's acted on an object, just subtract the object's initial momentum from its final momentum. Impulse is measured in newton seconds (Ns).

Impulse = mv – mu

Example: A body of mass 500 g is travelling in a straight line. Find the magnitude of the impulse needed to increase its speed from 2 ms^{-1} to 5 ms^{-1}.

Impulse = Change in momentum
= mv – mu
= (0.5 × 5) – (0.5 × 2)
= 1.5 Ns

Momentum = mass x velocity

Example: A 20 g ball is dropped 1 m onto the ground. Immediately after rebounding the ball has a speed of 2 ms^{-1}. Find the impulse given to the ball by the ground. How high does the ball rebound?

First you need to work out the ball's speed as it reaches the ground:

List the variables you're given:
u = 0 ← The ball was dropped, so it started from u = 0.
s = 1
a = 9.8 ← Acceleration due to gravity.
v = ?

Choose an equation containing u, s, a and v:

$v^2 = u^2 + 2as$
$v^2 = 0^2 + (2 \times 9.8 \times 1)$
$\mathbf{v = 4.43\ ms^{-1}}$ ← The sign is really important. Make sure that down is positive in this part of the question.

Now work out the impulse as the ball hits the ground and rebounds:

u = 4.43 ms^{-1} v = 2 ms^{-1} I

Impulse = mv – mu
= (0.02 × 2) – (0.02 × –4.43)
= 0.129 Ns (to 3 s.f.)

Finally you need to use a new equation of motion to find s (the greatest height the ball reaches).

List the variables:
u = 2
v = 0 ← v = 0 at the ball's greatest height.
a = –9.8
s = ?

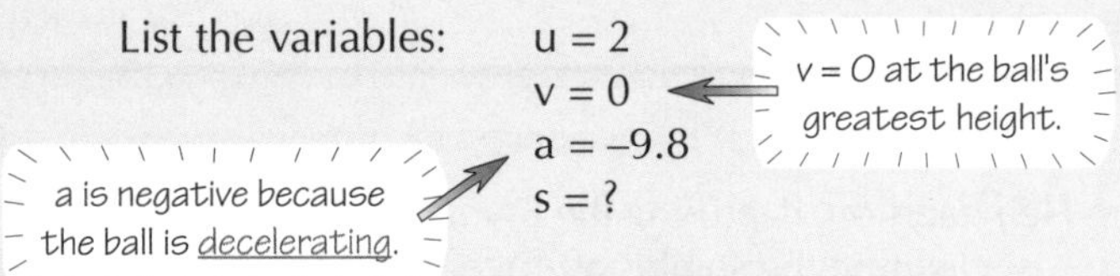
a is negative because
the ball is decelerating.

$v^2 = u^2 + 2as$
$0^2 = 2^2 + (2 \times -9.8 \times s)$
s = 0.204 m (to 3 s.f.)

Impulse

This page is for Edexcel M1

Impulses always **Balance** *in* **Collisions**

During impact between particles A and B, the impulse that A gives to B is the same as the impulse that B gives to A, but in the opposite direction.

Example: A mass of 2 kg moving at 2 ms^{-1} collides with a mass of 3 kg which is moving in the same direction at 1 ms^{-1}. The 2 kg mass continues to move in the same direction at 1 ms^{-1} after impact. Find the impulse given by the 2 kg mass to the other particle.

Using "conservation of momentum":

$$(2 \times 2) + (3 \times 1) = (2 \times 1) + 3v$$

$$\text{So } v = 1\frac{2}{3}$$

Before

A (2kg) 2 ms^{-1} → B (3kg) 1 ms^{-1} →

After

A (2kg) 1 ms^{-1} → B (3kg) v →

Impulse (on B) = mv – mu (for B)

$$= (3 \times 1\frac{2}{3}) - (3 \times 1)$$

$$= \mathbf{2\ Ns}$$

The impulse B gives to A is $(2 \times 1) - (2 \times 2) =$ **–2 Ns**. Aside from the different direction, you can see it's the same — so you didn't actually need to find v for this question.

Impulse *is linked to* ***Force*** *too*

Impulse is also related to the force needed to change the momentum and the time it takes.

Impulse = Force × Time

Example: A 0.9 tonne car increases its speed from 30 kmh^{-1} to 40 kmh^{-1}. Given that the maximum additional forward force the car's engine can produce is 1 kN, find the shortest time it will take to achieve this change in speed.

Impulse = mv – mu

$$= (900 \times \frac{40\,000}{3600}) - (900 \times \frac{30\,000}{3600})$$

$$= 2500 \text{ Ns}$$

Now use Impulse = Force × Time:

$$2500 = 1000 \times t$$

$$t = 2.5 \text{ s}$$

Practice Questions

1) An impulse of 2 Ns acts against a ball of mass 300 g moving with a velocity of 5 ms^{-1}. Find the ball's new velocity.

2) A particle of mass 450 g is dropped 2 m onto a floor. It rebounds to two thirds of its original height. Find the impulse given to the ball by the ground.

3) Sample exam question:

A coal wagon of mass 4 tonnes is rolling along a straight rail track at 2.5 ms^{-1}. It collides with a stationary wagon of mass 1 tonne. During the collision the wagons become coupled and move together along the track.

a) Find their speed after collision. [2 marks]

b) Find the impulse given to the more massive wagon. [2 marks]

c) State two assumptions made in your model. [2 marks]

Still feeling dynamic? Just wait for the projectiles section...

Impulse is change in momentum — remember that and you'll be laughing. Anyway, I know all about impulse. Those orange nylon flares looked great in the shop window, but I really should have tried them on before I bought them.

Projectiles

This page is for AQA M1

A 'projectile' is just any old object that's been lobbed through the air. When you're doing projectile questions you'll have to model the motion of particles in two dimensions whilst ignoring air resistance.

*Split Motion into **Horizontal** and **Vertical** Components*

It's time that connects the two directions.
Remember that the only acceleration is due to gravity — so horizontal acceleration is zero.

Example: A stone is thrown horizontally with speed 10 ms^{-1} from a height of 2 m above the horizontal ground. Find the time taken for the stone to hit the ground and the horizontal distance travelled before impact. Find also the stone's velocity after 0.5 s.

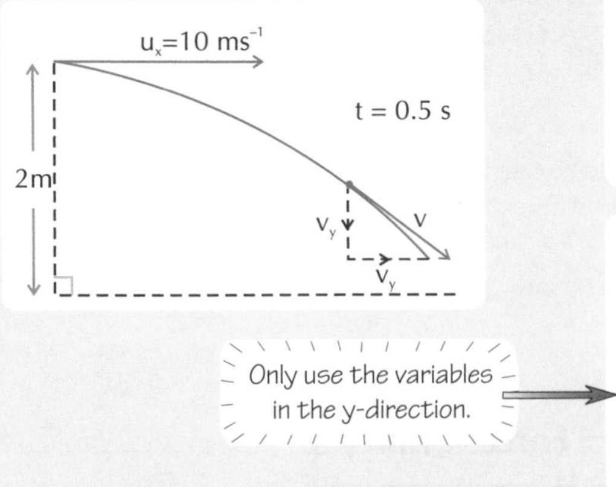

Only use the variables in the y-direction.

Resolving vertically (take down as +ve):

u = 0   s = 2
a = 9.8   t = ?

$s = ut + \frac{1}{2}at^2$

$2 = 0 \times t + \frac{1}{2} \times 9.8 \times t^2$

t = 0.64 s (to 2 s.f.)
i.e. the stone lands after 0.64 seconds

Resolving horizontally (take right as +ve):

u = 10   s = ?
a = 0   t = 0.64

$s = ut + \frac{1}{2}at^2$

$= 10 \times 0.64 + \frac{1}{2} \times 0 \times 0.64^2$

= 6.4 m

i.e. the stone has gone 6.4 m horizontally when it lands.

Now find the velocity after 0.5 s — again, keep the horizontal and vertical bits separate.

v = u + at
$v_y = 0 + 9.8 \times 0.5$
$= \mathbf{4.9\ ms^{-1}}$

v = u + at
$v_x = 10 + 0 \times \frac{1}{2}$
$= \mathbf{10\ ms^{-1}}$

Now you can find the speed and direction if you want to...

v_x=10, α, v_y = 4.9, v

$v = \sqrt{4.9^2 + 10^2} = 11.1\ ms^{-1}$

$\tan\theta = \frac{4.9}{10}$

So $\theta = 26.1°$ below horizontal

*Split **Velocity of Projection** into **Two Components** too*

A particle projected with a speed U at an angle α to the horizontal has two components of initial velocity.

Here's the same information in a diagram:

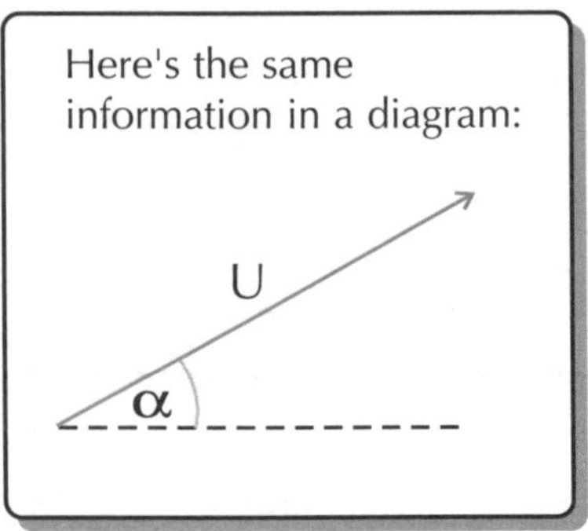

Split the velocity into its x and y components:

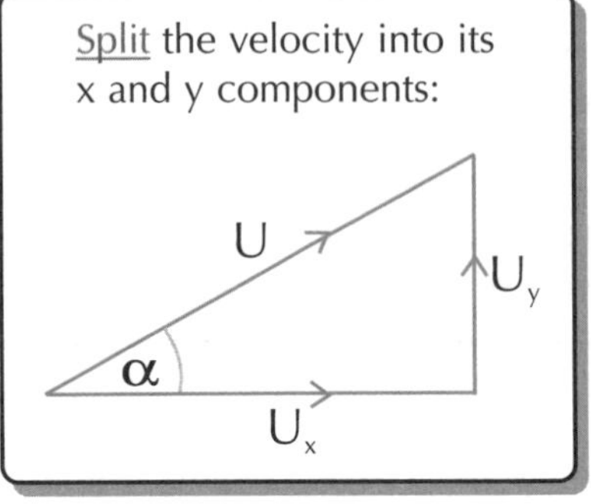

Finally, work out the values of the components using trigonometry:

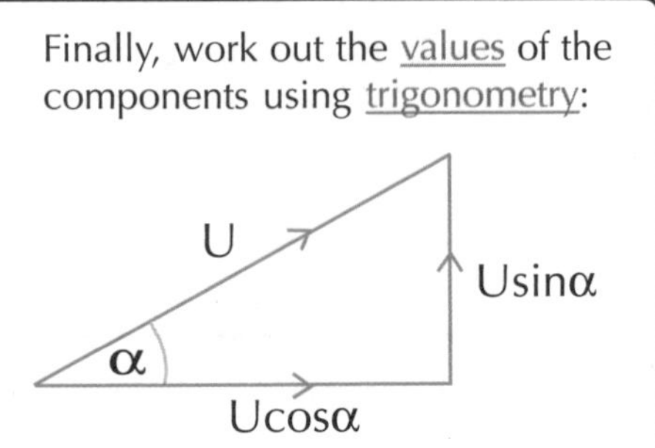

Projectiles

This page is for AQA M1

Example:

A cricket ball is projected with a speed of 30 ms^{-1} at an angle of 25° to the horizontal.

a) Find the maximum height it reaches (h) and the horizontal distance travelled (r). Assume that the ground is horizontal and that the ball is struck at ground level.

b) If instead the ball is struck 1.5 m above the ground, find the new maximum height and new horizontal distance travelled.

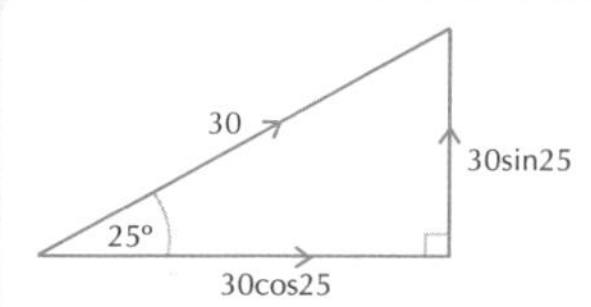

a) **Resolving vertically** (take up as +ve):

u = 30sin25°
v = 0
a = –9.8
s = h

$v^2 = u^2 + 2as$

$0 = (30\sin 25°)^2 + 2(-9.8 \times h)$

h = 8.2 m

v = u + at

0 = 30sin25° – 9.8t

t = 1.294 s

Resolving horizontally (take right as +ve):

u = 30cos25°
s = r
a = 0
t = 2.588

Time to reach max height is 1.294, so total time until landing is 1.294 × 2 = 2.588

$s = ut + \frac{1}{2}at^2$

$r = 30\cos 25° \times 2.588 + \frac{1}{2} \times 0 \times 2.588^2$

= 70.4 m (to 3 s.f.)

b) The first bit's easy: new h is just 8.2 + 1.5 = 9.7 m

Now use an equation of motion to work out the new horizontal distance:

s = –1.5
a = –9.8
u = 30sin25°
t = ?

$s = ut + \frac{1}{2}at^2$

$-1.5 = (30\sin 25°)t - \frac{1}{2}(9.8)t^2$

$t^2 - 2.587t - 0.306 = 0$

$t = -0.11$ or $\mathbf{t = 2.70\ s}$

Time can't be negative, so forget about this answer.

Resolving horizontally (take right as +ve)

s = r
u = 30cos25°
t = 2.70
a = 0

$s = ut + \frac{1}{2}at^2$

$r = 30\cos 25° \times 2.70 + \frac{1}{2} \times 0 \times 2.7^2$

= 73.4 m

Practice Questions

1) A rifle fires a bullet horizontally at 120 ms^{-1}. The target is hit at a horizontal distance of 60 m from the end of the rifle. Find how far the target is vertically below the end of the rifle. Take g = 9.8 ms^{-2}.

2) Sample exam question:

A stationary football is kicked with a speed of 20 ms^{-1}, at an angle of 30° to the horizontal, towards a goal 30 m away. The crossbar is 2.5 m above the level ground. Assuming the path of the ball is not impeded, determine whether the ball passes above or below the crossbar. Take g = 9.8 ms^{-2}. What assumptions does your model make? [6 marks]

Components make the world go round...

Using the equations of motion with projectiles is pretty much the same as before. The thing to remember is that horizontal acceleration is zero — great news because it makes the horizontal calculations as easy as a log-falling beginner's class.

Equations of Path

This page is for AQA M1

There are three simple equations to find the maximum height, time of flight and range on horizontal ground.

*You've got to **Derive** the **Equations** — **Don't** just learn the results*

Each time you use one of these path equations you have to show how the equation is derived.
If you just quote the final equation you'll get no marks.

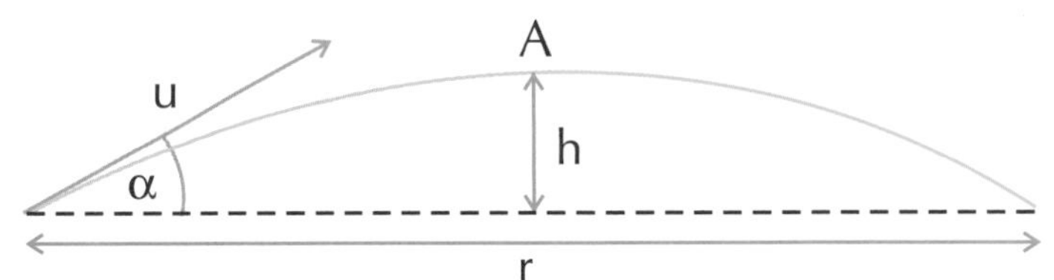

u
A
h
α
r

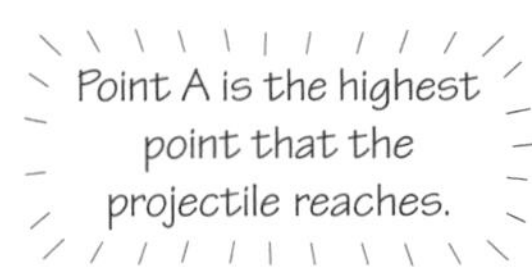

Point A is the highest point that the projectile reaches.

Maximum height, h

Resolving vertically:

$s = h$
$a = -g$
$u = u\sin\alpha$
$v = 0$

$v^2 = u^2 + 2as$
$0 = (u\sin\alpha)^2 + 2(-g)h$

$$h = \frac{u^2 \sin^2 \alpha}{2g}$$

Time of Flight, T

Resolving vertically (to find time to A):

$v = 0$
$u = u\sin\alpha$
$t = ?$
$a = -g$

$v = u + at$
$0 = u\sin\alpha^2 - gt$
$t = \dfrac{u\sin\alpha}{g}$

...but that's only to point A — you need to double t to get total time T:

$$T = \frac{2u\sin\alpha}{g}$$

Range, r

Resolving horizontally:

$s = r$
$u = u\cos\alpha$
$t = T = \dfrac{2u\sin\alpha}{g}$

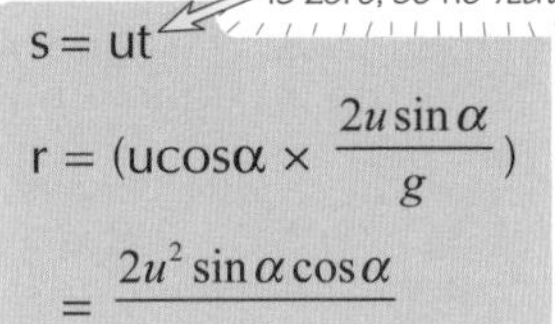

The horizontal acceleration is zero, so no ½at² term.

$s = ut$
$r = (u\cos\alpha \times \dfrac{2u\sin\alpha}{g})$
$= \dfrac{2u^2 \sin\alpha\cos\alpha}{g}$

$$R = \frac{u^2 \sin 2\alpha}{g}$$

Using $\sin 2\alpha = 2\sin\alpha\cos\alpha$.

Example:

A golf ball is struck at 30° to the horizontal with a speed of 40 ms^{-1}. By considering the horizontal and vertical components of the motion of the ball, find the time of flight, the horizontal distance to where the ball first lands and the maximum height reached. Take $g = 10$ ms^{-2}.

Don't forget to derive the equations first (I'll miss out that step but you'd have to write out the derivations for the three equations above).

$$T = \frac{2u\sin\alpha}{g} = \frac{2\times 40\times\sin 30}{10} = 4 \text{ s}$$

$$r = \frac{u^2\sin 2\alpha}{g} = \frac{40^2\sin 60}{10} = 139 \text{ m}$$

$$h = \frac{u^2\sin^2\alpha}{2g} = \frac{40^2\sin^2 30}{2\times 10} = 20 \text{ m}$$

If you ever get asked to work out when the range is largest, here's a handy tip:

r is at a maximum when α = 45°.

Equations of Path

This page is for AQA M1

Example: In the example on the last page, what is the nearest horizontal distance that a 10 m tall tree could be to the golfer if the ball is to go over it?
What assumptions have you made in these examples? (Take $g = 10\ \text{ms}^{-2}$.)

Resolve vertically:

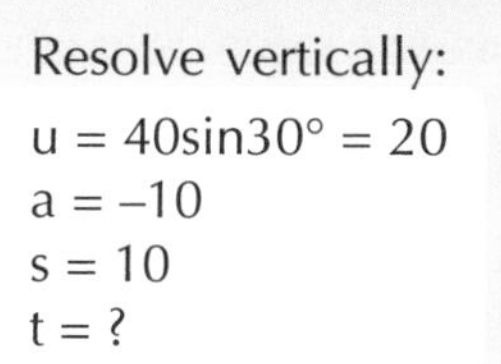

$u = 40\sin 30° = 20$
$a = -10$
$s = 10$
$t = ?$

$s = ut + \frac{1}{2}at^2$
$10 = 20t - 5t^2$
$5t^2 - 20t + 10 = 0$
$t = \dfrac{4 \pm \sqrt{8}}{2}$
$t = 0.586$

Using the quadratic formula.

You get two values for t — the first is when the ball reaches 10 m from the ground on the way up, the second is when the ball reaches the same height on the way down. You need the first one.

Resolve horizontally:

$s = x$
$u = 40\cos 30°$
$t = 0.586$

$s = ut$
$x = 40\cos 30° \times 0.586$
$= 20.3$ m

Assumptions:

- no air resistance
- no wind
- horizontal ground

Wind Speed only changes u_x

A following wind speed should be added to u_x, and you can leave u_y alone. Otherwise the method is the same as before.

Example: For the golfer above, how far does the ball travel before it lands if there is a tail wind of $3\ \text{ms}^{-1}$?

To find the time when the ball lands, put $s = 0$. So $u_y = 20$, $a = -10$, $s = 0$.

$s = u_y t + \frac{1}{2}at^2$
$0 = 20t - 5t^2$
$5t^2 = 20t$
$t = 0$ or $t = 4$.

$t = 0$ is when the ball is hit, so you need $t = 4$ here.

Now put $t = 4$ into your 'distance = speed × time' equation: $s = u_x t = (40\cos 30° + 3) \times 4 = 151$ m

Practice Questions

1) A golf ball takes 4 seconds to land after being hit with a golf club. If it leaves the club with a speed of $22\ \text{ms}^{-1}$, at an angle of α to the horizontal, find α. Take $g = 9.8\ \text{ms}^{-2}$.

2) Sample exam question:

A cannon ball is fired at $50\ \text{ms}^{-1}$ at an angle of 25° to the horizontal.
It hits its target 3 seconds later. Take $g = 9.8\ \text{ms}^{-2}$. Find:

a) The horizontal and vertical components of its initial velocity. [2 marks]
b) The position of the target relative to the cannon. [4 marks]
c) The maximum height reached by the ball. [2 marks]
d) The speed of the ball on impact with the target. [3 marks]

You'll never be a good golfer without path equations under your belt...

...and you'll have a hard time being an M1 mathematician too, come to think of it. OK, one more chorus and then let's call it a day... (to the tune of Auld Lang Syne) 'Derive the equations, don't just learn the results — then you'll get the marks.'

Cows

The stuff on this page isn't strictly on the specification. But I've included it anyway because I reckon it's really important stuff that you ought to know.

There are loads of **Different Types** of Cows

Dairy Cattle

Every day a dairy cow can produce up to 128 pints of milk — which can be used to make 14 lbs of cheese, 5 gallons of ice cream, or 6 lbs butter.

The Jersey
The Jersey is a small breed best suited to pastures in high rainfall areas. It is kept for its creamy milk.

Advantages
1) *Can produce creamy milk until old age.*
2) *Milk is the highest in fat of any dairy breed (5.2%).*
3) *Fairly docile, although bulls can't be trusted.*

Disadvantages
1) *Produces less milk than most other breeds.*

The Holstein-Friesian
This breed can be found in many areas.
It is kept mainly for milk.

Advantages
1) *Produce more milk than any breed.*
2) *The breed is large, so bulls can be sold for beef.*

Disadvantages
1) *Milk is low in fat (3.5%).*

Beef Cattle

Cows are sedentary animals who spend up to 8 hours a day chewing the cud while standing still or lying down to rest after grazing. Getting fat for people to eat.

The Angus
The Angus is best suited to areas where there is moderately high rainfall.

Advantages
1) *Early maturing.*
2) *High ratio of meat to body weight.*
3) *Forages well.*
4) *Adaptable.*

The Hereford
The Hereford matures fairly early, but later than most shorthorn breeds. All Herefords have white faces, and if a Hereford is crossbred with any other breed of cow, all the offspring will have white or partially white faces.

Advantages
1) *Hardy.*
2) *Adaptable to different feeds.*

Disadvantages
1) *Susceptible to eye diseases.*

This is really important — try not to forget it.

Milk comes from **Cows**

Milk is an emulsion of butterfat suspended in a solution of water (roughly 80%), lactose, proteins and salts. Cow's milk has a specific gravity around 1.03.
It's pasteurised by heating it to 63 °C for 30 minutes. It's then rapidly cooled and stored below 10 °C.

Louis Pasteur began his experiments into 'pasteurisation' in 1856.
By 1946, the vacuum pasteurisation method had been perfected, and in 1948, UHT (ultra heat-treated) pasteurisation was introduced.

cow + grass = fat cow

fat cow + milking machine ⇒ milk

You will often see cows with pieces of grass sticking out of their mouths.

SOME IMPORTANT FACTS TO REMEMBER:
- A newborn calf can walk on its own an hour after birth
- A cow's teeth are only on the bottom of her mouth
- While some cows can live up to 40 years, they generally don't live beyond 20.

Cows on the **Internet**

For more information on cows, try these websites:

www.allcows.com (including Cow of the Month)
www.crazyforcows.com (with cow e-postcards)
www.moomilk.com (includes a 'What's the cow thinking?' contest.)
http://www.geocities.com/Hollywood/9317/meowcow.html
(for cow-tipping on the Internet)

The Cow
The cow is of the bovine ilk;
One end is moo, the other, milk.

— Ogden Nash

Famous Cows and Cow Songs

Famous Cows

1) Ermintrude from the Magic Roundabout.
2) Graham Heifer — the Boddingtons cow.
3) Other TV commercial cows — Anchor, Dairylea
4) The cow that jumped over the moon.
5) Greek Mythology was full of gods turning themselves and their girlfriends into cattle.

Cows in Pop Music

1) Size of a Cow — the Wonder Stuff
2) Saturday Night at the Moo-vies — The Drifters
3) What can I do to make you milk me? — The Cows
4) One to an-udder — the Charlatans
5) Milk me baby, one more time — Britney Spears

Where's me Jersey — I'm Friesian...

Cow-milking — an underrated skill, in my opinion. As Shakespeare once wrote, 'Those who can milk cows are likely to get pretty good grades in maths exams, no word of a lie'. Well, he probably would've written something like that if he was into cows. And he would've written it because cows are helpful when you're trying to work out what a question's all about — and once you know that, you can decide the best way forward. And if you don't believe me, remember the saying of the ancient Roman Emperor Julius Caesar, 'If in doubt, draw a cow'.

Answers

Section One — The Riveting Basics of Algebra

Page 8

1) *To remove the roots from $\sqrt{6}-\sqrt{2}$, you multiply it by $\sqrt{6}+\sqrt{2}$ (i.e. use the difference of two squares).*

$$\frac{\sqrt{6}+\sqrt{2}}{\sqrt{6}-\sqrt{2}}=\frac{\sqrt{6}+\sqrt{2}}{\sqrt{6}-\sqrt{2}}\times\frac{\sqrt{6}+\sqrt{2}}{\sqrt{6}+\sqrt{2}}=\frac{(\sqrt{6}+\sqrt{2})(\sqrt{6}+\sqrt{2})}{(\sqrt{6}-\sqrt{2})(\sqrt{6}+\sqrt{2})}$$

$$=\frac{6+2\sqrt{6}\sqrt{2}+2}{6-2}=\frac{8+2\sqrt{12}}{4}$$

$$=\frac{8+2\sqrt{4}\sqrt{3}}{4}=\frac{8+4\sqrt{3}}{4}$$

$$=\mathbf{2+\sqrt{3}}$$

(So a = 2 and b = 1.)

[6 marks available — 1 mark for method of rationalising the denominator, 3 marks for evaluating the expression (allow follow through marks if a mistake has been made) and 1 mark each for correct values of a and b (these do not need to be stated).]

2) a)

$$3^{p-1}=\frac{1}{27}$$

$$\Rightarrow 3^{p-1}=\frac{1}{3^3}$$

$$\Rightarrow 3^{p-1}=3^{-3}$$

$$\Rightarrow p-1=-3$$

$$\Rightarrow \mathbf{p=-2}$$

$$3^q=81$$

$$\Rightarrow 81=3^4$$

$$\Rightarrow \mathbf{q=4}$$

[2 marks available — 1 mark for each correct answer.]

b) Now stick in the values you've found for p and q:

$$(3^q)^{\frac{1}{2}}=(3^4)^{\frac{1}{2}}=3^2=9$$

and $3^{-p}=3^{-(-2)}=3^2=9$

So $(3^q)^{\frac{1}{2}}+3^{-p}=9+9=\mathbf{18}$

[2 marks available — 1 for correctly substituting in p and q and applying the power laws, 1 mark for the correct answer.]

3)

$$\frac{x}{(x-3)(x-2)}+\frac{8}{x^2-4}$$

$$\equiv\frac{x}{(x-3)(x-2)}+\frac{8}{(x+2)(x-2)}\text{ (difference of 2 squares)}$$

$$\equiv\frac{x(x+2)}{(x-3)(x-2)(x+2)}+\frac{8(x-3)}{(x-3)(x-2)(x+2)}$$

(over common denominator)

$$\equiv\frac{x(x+2)+8(x-3)}{(x-3)(x-2)(x+2)}\text{ (adding fractions together)}$$

$$\equiv\frac{x^2+10x-24}{(x-3)(x-2)(x+2)}\text{ (multiplying out the top line)}$$

$$\equiv\frac{(x+12)(x-2)}{(x-3)(x-2)(x+2)}\text{ (factorising the top line)}$$

$$\equiv\frac{\mathbf{x+12}}{\mathbf{(x-3)(x+2)}}\text{ (cancelling common factor)}$$

[4 marks available — 1 mark for a correct common denominator,
1 mark for adding the two fractions correctly, 1 mark for a correct cancellation, 1 mark for the correct answer.]

4)

$$27^x=3^{2y+1}$$

$$\Rightarrow 3^{3x}=3^{2y+1}$$

So $3x=2y+1$

$$\Rightarrow\quad x=\frac{2}{3}y+\frac{1}{3}$$

(So $p=\frac{2}{3}$ and $q=\frac{1}{3}$.)

[2 marks available — 1 for p correct and 1 for q correct.]

5) a) $36^{-\frac{1}{2}}=\frac{1}{\sqrt{36}}=\pm\frac{1}{6}$

b) $\frac{a^6\times a^3}{\sqrt{a^4}}\div a^{\frac{1}{2}}=\frac{a^6\times a^3}{a^2}\div a^{\frac{1}{2}}=a^6\times a^3\times a^{-2}\times a^{-\frac{1}{2}}=a^{6+3-2-\frac{1}{2}}=a^{\frac{13}{2}}$

c) $(5\sqrt{5}+2\sqrt{3})^2=(5\sqrt{5}+2\sqrt{3})\times(5\sqrt{5}+2\sqrt{3})$

The first term is: $(5\sqrt{5})^2=5\sqrt{5}\times5\sqrt{5}$

$$=5\times5\times\sqrt{5}\times\sqrt{5}$$

$$=5\times5\times5$$

$$=125$$

The second term is: $2(5\sqrt{5}\times2\sqrt{3})=2\times5\times2\times\sqrt{5}\times\sqrt{3}$

$$=20\times\sqrt{15}$$

$$=20\sqrt{15}$$

The third term is: $(2\sqrt{3})^2=2\sqrt{3}\times2\sqrt{3}$

$$=2\times2\times\sqrt{3}\times\sqrt{3}$$

$$=2\times2\times3$$

$$=12$$

So the whole thing is: $125+20\sqrt{15}+12$

$$=137+20\sqrt{15}$$

[4 marks — 1 mark each for the first, second and third terms, 1 mark for the correct addition to give the final answer.]

Section Two — Quadratics and the Factor Theorem

Page 23

1) a) $x^2-3px+2p=0$

$a=1,\ b=-3p,\ c=2p$

Now, $b^2-4ac\geq0$

$$\Rightarrow(-3p)^2-4(1)(2p)\geq0$$

$$\Rightarrow 9p^2-8p\geq0$$

$$\Rightarrow p(9p-8)\geq0$$

[3 marks available — 1 mark for stating that $b^2-4ac\geq0$ from the quadratic formula, 1 mark for substituting in the correct values for a, b and c, 1 mark for correctly arriving at answer.]

Answers

b) $x^2 - 3px + 2p = 0$
First of all, stick in the actual value for p:
$x^2 - 6x + 4 = 0$
Now it's a simple quadratic, so just solve it with the quadratic formula.

$a = 1, \quad b = -6 \quad c = 4$

$$x = \frac{-b \pm \sqrt{b^2 - 4ac}}{2a} = \frac{6 \pm \sqrt{36-16}}{2} = \frac{6 \pm \sqrt{20}}{2}$$

$$= \frac{6 \pm \sqrt{4 \times 5}}{2} = \frac{6 \pm \sqrt{4}\sqrt{5}}{2} = \frac{6 \pm 2\sqrt{5}}{2}$$

$$= 3 \pm \sqrt{5}$$

so $\boldsymbol{x = 3 + \sqrt{5}}$ *or* $\boldsymbol{x = 3 - \sqrt{5}}$

[4 marks available — 1 for substituting in p = 2 and finding correct values for a, b and c, 1 for correct use of quadratic formula, 1 mark for each correct root.]

2) $x^2 + 2x - 35 = 0$
$(x - 5)(x + 7) = 0$ *so* $(x - 5) = 0$ *or* $(x + 7) = 0$
i.e. $x = 5$ *or* $x = -7$
The coefficient of x^2 is positive, so the graph of $y = x^2 + 2x - 35$ is u-shaped and crosses the x-axis at $x = -7$ and $x = 5$.
So $x^2 + 2x - 35 < 0$ when $-7 < x < 5$,
i.e. $2x < 35 - x^2$ when $\boldsymbol{-7 < x < 5}$
[3 marks available — 1 mark for attempting to solve as quadratic, 1 mark for correctly solving it, 1 mark for giving correct range for answer.]

3) Factorising gives: $(x + 3)^2 = 0$, so $x = -3$ is the only root.
So the answer is $\boldsymbol{x > -3}$ **or** $\boldsymbol{x < -3}$ (or $x \neq -3$)
[3 marks — 2 for solving the quadratic and 1 for the correct answer.]

4) The Factor Theorem states that if $f(a) = 0$, then $(x - a)$ is a factor of $f(x)$.
$f(-1) = (-1)^5 + (-1)^4 - 19(-1)^3 - 25(-1)^2 + 66(-1) + 72$
$= 0.$
$f(-1) = 0$, so $(x + 1)$ is a factor of $f(x)$.
[2 marks — 1 for quoting the factor theorem and 1 for the correct answer.]

5) a) $4x + 7 > 7x + 4$
$3 > 3x$
$x < 1$

[2 marks for correct answer, lose 1 mark for a mistake in the method.]

b) Basically, you've got to find the minimum value of $(x - 5)(x - 3)$, and make sure k is less than that. The thing to realise is that the graph's symmetrical — so the minimum will be halfway between $x = 3$ and $x = 5$ — i.e. $x = 4$. So just plug that into the equation to find the lowest point the graph reaches: $(4-5)(4-3) = -1 \times 1$
$= -1$

So if $k < -1$, the graph will never be as low as k.
[3 marks — 1 mark for identifying the x=4, 1 mark for minimum value f(x) = -1, 1 mark for final answer.]

c) Rearrange the expression to get: $(x + 3)(x - 2) < 2$
$x^2 + x - 6 < 2$
$x^2 + x - 8 < 0$

You're interested in when the graph of $y = x^2 + x - 8$ is less than 0.

Now, $x^2 + x - 8 = 0$ when $x = \dfrac{-1 \pm \sqrt{1^2 - (4 \times 1 \times -8)}}{2 \times 1}$

$$= \frac{-1 \pm \sqrt{33}}{2}$$

And since this is a u-shaped parabola, the part below the x-axis is the bit between the roots, so the answer is

$\dfrac{-1-\sqrt{33}}{2} < x < \dfrac{-1+\sqrt{33}}{2}$, or $-3.37 < x < 2.37$ (to 2 d.p.).

[3 marks — 1 mark for rearrangement in form $ax^2 + x + c = 0$, 1 mark for correct evaluation of roots, 1 mark for final answer.]

6) a) Completing the square, you get: $x^2 - 7x + 17 = \left(x - \frac{7}{2}\right)^2 + \frac{19}{4}$

The expression in the question will be maximum when the quadratic in the denominator is as small as possible. The minimum value of the quadratic is 19/4 (since the squared part can never be less than zero), and so $f(x)_{max} = \dfrac{1}{\left(19/4\right)} = \dfrac{4}{19}$

[3 marks — 1 mark for completing the square, 1 mark for identifying that f(x) is a maximum when denominator is minimum, 1 mark for final answer.]

b) If there's only one root, then $b^2 - 4ac = 0$. And so

$b^2 - 4 \times 3 \times 12 = 0$
$b^2 = 144$
$b = \pm 12$

[3 marks — 1 mark for identifying that $b^2 - 4ac = 0$, 1 mark b=12, 1 mark for b=-12.]

Answers

Section Three — Simultaneous Equations and Geometry Page 34

1) a) The gradient of AD $= \frac{y_2 - y_1}{x_2 - x_1}$ *(i.e.* $\frac{\text{difference in } y\text{'s}}{\text{difference in } x\text{'s}}$*)*

$$= \frac{5-0}{0-2}$$
$$= -\frac{5}{2}$$

The y-intercept is 5.

So the equation of AD is $y = -\frac{5}{2}x + 5$.

[2 marks for the correct answer. Award 1 mark for correct gradient but incorrect equation or for incorrect gradient but otherwise correct equation.]

b) *Grad. of AD is* $-\frac{5}{2}$ *and since AD is perpendicular to DC,*

gradient of DC $= \frac{-1}{\text{gradient of AD}}$

$$= \frac{-1}{-\frac{5}{2}}$$
$$= \frac{2}{5}$$

The y-intercept is 5, so the equation of line DC is

$$\mathbf{y = \frac{2}{5}x + 5.}$$

[2 marks available — 1 mark for correctly finding the gradient, 1 mark for correct final equation. Lose a mark for a single mistake, e.g. incorrect y- intercept.]

c) By Pythagoras' Theorem:

$$AD^2 = OD^2 + OA^2$$
$$AD^2 = 5^2 + 2^2$$
$$AD = \sqrt{29}$$

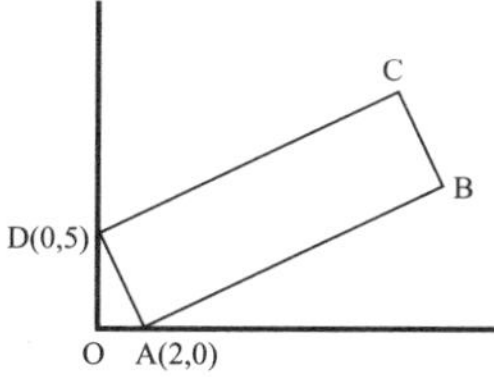

Now to <u>find the length of DC</u>:
First, you need the coordinates of pt C.
You are told the x-coordinate is 10.
Substituting this into the equation of DC from (b):

$$x = 10 \Rightarrow y = \frac{2}{5}x + 5$$
$$\Rightarrow y = \frac{2}{5}(10) + 5$$
$$\Rightarrow y = 9$$

<u>So pt C is (10, 9)</u>.
To find the length DC, make a right-angled triangle CDE.
By Pythagoras' theorem:

$$DC^2 = DE^2 + EC^2$$
$$= 4^2 + 10^2$$
$$DC = \sqrt{116}$$

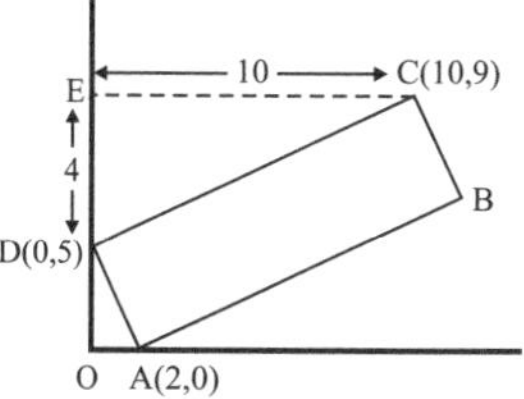

Now just multiply base × height to find the area...

$$\text{Area of rectangle ABCD} = \text{AD} \times \text{DC}$$
$$= \sqrt{29} \times \sqrt{116}$$
$$= \sqrt{29} \times 2\sqrt{29}$$
$$= 29 \times 2 = 58 \text{ units}^2$$

[4 marks available — 1 mark for using correct method to find the distance between two points, 1 mark for each correct side length, 1 mark for correct final answer.]

2) To find the equation of a line, you need its <u>gradient</u> and a <u>point it goes through</u>. The line you want is <u>parallel</u> to $14 - 2y + 5x = 0$, which just means it has the same <u>gradient</u>.
To find the gradient of $14 - 2y + 5x = 0$,
rearrange it into $y = mx + c$ form.
$14 - 2y + 5x = 0$
so $2y = 5x + 14$ and hence $y = \frac{5}{2}x + 7$
So the gradient of $14 - 2y + 5x = 0$ is $\frac{5}{2}$
which means the gradient of the line you want is $\frac{5}{2}$.
The equation of a line, given the gradient and a point is
$(y - y_1) = m(x - x_1)$ (or you could use $y = mx + c$)

Stick in the $m = \frac{5}{2}$, $x_1 = 4$ and $y_1 = 7$: $(y - 7) = \frac{5}{2}(x - 4)$

$2y - 14 = 5x - 20$
$\mathbf{2y - 5x + 6 = 0}$
[3 marks for the correct answer or 1 mark for the correct gradient and 1 mark for a correct method.]

Answers

3) a) Label the equations:
$x + y = -4$ — (1)
$y = -x^2 + 8x - 12$ — (2)
Rewrite eqn (1) as an expression for y in terms of x.
$y = -4 - x$ — (3)
Now substitute this into equation (2):
$-4 - x = -x^2 + 8x - 12$
so $x^2 - 9x + 8 = 0$
It's a quadratic in x. See if it will factorise:
so $(x-1)(x-8) = 0$
which means $x = 1$ or $x = 8$
So you've got two solutions for x. Now substitute these into equation (3) to find the corresponding values for y:
$x = 1$, which means that $y = -5$
$x = 8$, which means that $y = -12$
So the solutions are $\mathbf{x = 1, y = -5}$ and $\mathbf{x = 8, y = -12}$.
[4 marks available — 1 mark for correctly eliminating y (or x), 1 method mark for solving the quadratic to obtain two values for x (or y) and 1 mark for each pair of correct x and y values.]

b) The solutions to part (a) are (1, -5) and (8, -12) which are the coordinates of the two points where the line $x + y = -4$ crosses the curve $y = -x^2 + 8x - 12$.
[1 mark for a correct explanation explaining that the solutions are the points where the line and curve intercept.]

4) a) You need to eliminate one of the variables, so add the two equations together to get rid of the y's:

$$(x+2x)+(-y+y)+(1-8)=0$$
$$3x-7=0$$
$$x=\tfrac{7}{3}$$

Then stick this x-value back into one of the equations to find y:

$$x-y+1=0$$
$$\tfrac{7}{3}-y+1=0$$
$$y=\tfrac{7}{3}+1=\tfrac{10}{3}$$

So A is $\left(\tfrac{7}{3},\tfrac{10}{3}\right)$.

[3 marks — 1 mark for eliminating y, 1 mark for finding x value, 1 mark for finding y value.]

b) Midpoint of AC is: $\left(\dfrac{\frac{7}{3}+\frac{-4}{3}}{2},\dfrac{\frac{10}{3}+\frac{-1}{3}}{2}\right)=\left(\tfrac{1}{2},\tfrac{3}{2}\right)$.

So the gradient of BD is $m_{BD}=\dfrac{y_D-y_B}{x_D-x_B}=\dfrac{\frac{3}{2}--4}{\frac{1}{2}-6}=\dfrac{3+8}{1-12}=-1$.

Now you've got the gradient, you can write the line in the form $y = mx + c$, then put in values for x and y (and m, which you've just found), to find the value of c:

$$y=mx+c$$
$$-4=-6+c$$
$$c=2$$

So equation for BD is: $x + y - 2 = 0$.
[5 marks — 1 mark for finding D, 1 mark for finding gradient of BD, 2 marks for finding value of c, 1 mark for correct final answer.]

c) You've got to show that ABD is right-angled. To do this, just show that two sides of the triangle are <u>perpendicular</u> to each other (so their gradients multiply to give –1).

Gradient of AD: $m_{DA}=\dfrac{y_D-y_A}{x_D-x_A}=\dfrac{\frac{3}{2}-\frac{10}{3}}{\frac{1}{2}-\frac{7}{3}}=\dfrac{9-20}{3-14}=1$.

You know the gradient of BD is –1.
And so these multiply together to give –1, which means that they are perpendicular. And since they form sides of a triangle, it must be a right-angled triangle.
[3 marks — 1 mark for stating requirement of two perpendicular side to give a right-angled triangle, 1 mark for finding gradient of AD, 1 mark for showing that AD and BD are perpendicular]

5) a) *When there's a number added or subtracted <u>inside</u> the brackets — the graph moves <u>sideways</u>.*

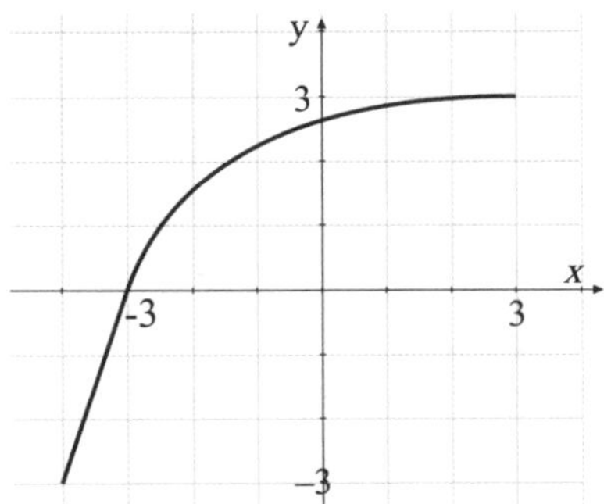

[2 marks — Lose 1 mark for each of these points not correctly marked: (-4,-3), (-3,0), (3,3). Lose 1 mark if general shape not correct.]

Answers

b) When there's a number added or subtracted outside the brackets — the graph moves up or down.

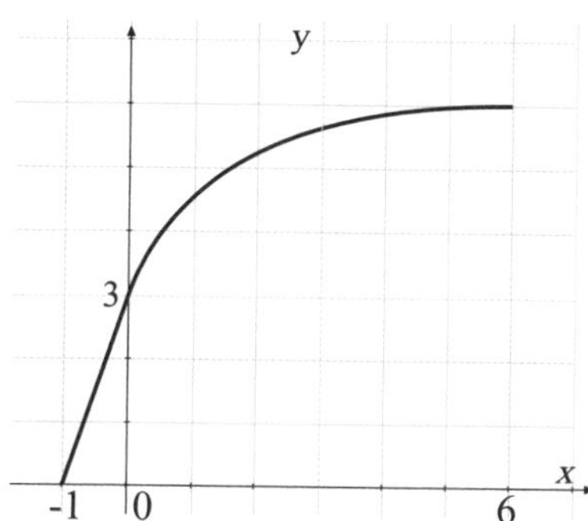

[2 marks — Lose 1 mark for each of these points not correctly marked: (–1,0), (0,3), (6,6). Lose 1 mark if general shape not correct.]

c) When there's a number multiplied inside the brackets — the graph is stretched or squashed horizontally.

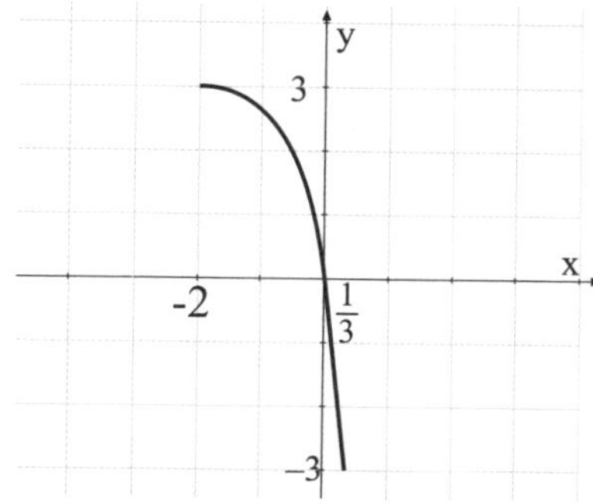

[2 marks — Lose 1 mark for each of these points not correctly marked: (-2,3), (0,0), (1/3,-3). Lose 1 mark if general shape not correct.]

d) When there's a number multiplied outside the brackets — the graph is stretched or squashed vertically.

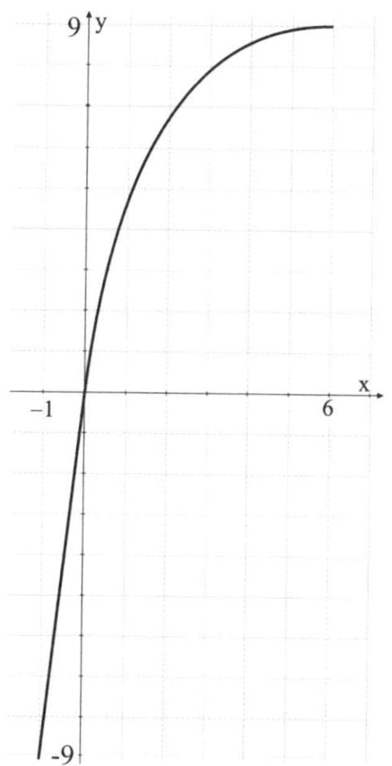

[2 marks — Lose 1 mark for each of these points not correctly marked: (-1,-9), (0,0), (6,9). Lose 1 mark if general shape not correct.]

Section Four — Differentiation

Page 40

1)a) S.A. = (2 × area of circle base) + area of side
$= \underline{2\pi x^2 + 2\pi xy}$

The question says the volume is 450 cm³, so...

$V = \pi r^2 h$ (volume of the cylinder)

$\Rightarrow 450 = \pi x^2 y$

$\Rightarrow \underline{y = \frac{450}{\pi x^2}}$

Surface area $= \underline{2\pi x^2 + (2\pi x)\frac{450}{\pi x^2}}$

$= \underline{\boldsymbol{2\pi x^2 + \frac{900}{x}}}$

[4 marks available for method — 1 mark for each underlined stage above, or similar.]

b) Finding minimums or maximums when you have an expression like this can mean only one thing — differentiation.

$S = 2\pi x^2 + 900x^{-1}$

$\frac{dS}{dx} = 4\pi x - 900x^{-2}$

For maximum and minimum values $\frac{dS}{dx} = 0$

$\Rightarrow 4\pi x - 900x^{-2} = 0$

$\Rightarrow 4\pi x = \frac{900}{x^2}$

$\Rightarrow x^3 = \frac{900}{4\pi}$

$\Rightarrow x = 4.15\text{cm (2 d.p.)}$

So a min or max value of S occurs when $\underline{x = 4.15}$.
To prove that this is a minimum value, you need to look at the sign of the differential either side of this point.
At $x = 4$:

$\frac{dS}{dx} = 4\pi x - 900x^{-2} = 4\pi \times 4 - 900 \times 4^{-2} = -5.98$ (negative)

at $x = 4.2$:

$\frac{dS}{dx} = 4\pi x - 900x^{-2} = 4\pi \times 4.2 - 900 \times 4.2^{-2} = 1.76$ (positive)

So the value of $\frac{dS}{dx}$ goes from negative to zero to positive.
Therefore 4.15 is a minimum.
But, hang on — you're not finished yet. You need to find the surface area. So put $x = 4.15$ back in the surface area formula:

$S = 2\pi x^2 + \frac{900}{x}$

$= 2\pi(4.15^2) + \frac{900}{4.15}$

$= \boldsymbol{325.08\ cm^2}$ to 2 d.p.

[8 marks available in total — 5 method marks and 3 accuracy marks. The 5 method marks are: 1 for the differentiation, 1 for setting the derivative = 0 and solving, 1 for the looking at the sign of the differential either side of 4.15 (or equivalent), 1 for correctly deducing that it is a minimum, 1 for substituting the value of x into the surface area formula. The 3 accuracy marks are: 1 for correct first derivative, 1 for x = 4.15 and 1 for the correct final answer.]

Answers

c) Just stick the value of x at the minimum (4.15 cm) into the formula for y (the height) from earlier.

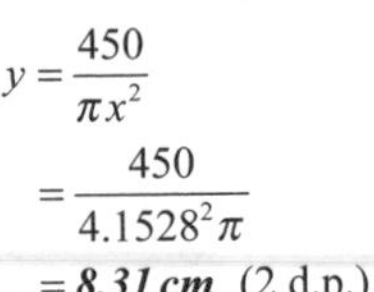

$$y = \frac{450}{\pi x^2} = \frac{450}{4.1528^2 \pi} = \mathbf{8.31\ cm} \text{ (2 d.p.)}$$

[2 marks for correct answer or 1 mark for method.]

2) a) $\frac{dy}{dx} = 3x^2 - 2x - 8$

[1 mark for correctly differentiating.]

b) The turning points occur where the gradient is zero, so solve $\frac{dy}{dx} = 0$.

$3x^2 - 2x - 8 = 0$

$\Rightarrow (3x+4)(x-2) = 0$

So the turning points of $y = x^3 - x^2 - 8x + 24$ occur at $x = -\frac{4}{3}$ and $x = 2$.

[3 marks available – 1 mark for setting the derivative to zero and attempting to solve, 1 mark for each correct value for x.]

c) The tangent at point (1, 16) will be a straight line with the same gradient as the curve at that point.

From part ***a)***, $\frac{dy}{dx} = 3x^2 - 2x - 8$

When $x = 1$, $\frac{dy}{dx} = 3 - 2 - 8 = -7$

So tangent has equation: $y - y_1 = -7(x - x_1)$

Through (1, 16), so: $y - 16 = -7(x - 1)$

Or: $y + 7x = 23$

[1 mark for calculating the gradient of the line, 1 mark for finding a valid equation for the line.]

3) a) Turning points occur when $f'(x) = 0$. So first differentiate $f(x)$, then set the derivative to zero.

$f(x) = -x^2(x-2)$

$= -x^3 + 2x^2$

$f'(x) = -3x^2 + 2(2x^1)$

$= -3x^2 + 4x$

$= x(-3x+4)$

Now put $f'(x)$ equal to 0:

$f'(x) = 0 \Rightarrow x(-3x+4) = 0$

$\Rightarrow x = 0$ or $x = \frac{4}{3}$

So there are turning points at $x = 0$ and $x = \frac{4}{3}$.

But the question asks for the actual points, so you need to find the y-coordinates too, i.e. $f(0)$ and $f\left(\frac{4}{3}\right)$:

$f(0) = 0$, and $f\left(\frac{4}{3}\right) = -\left(\frac{4}{3}\right)^2\left(\frac{4}{3} - 2\right) = -\frac{16}{9}\left(-\frac{2}{3}\right) = \frac{32}{27}$.

So the turning points are (0,0) (i.e. the origin) and $\left(\frac{4}{3}, \frac{32}{27}\right)$.

[3 marks — 1 mark for differentiating f(x), 1 mark for solving f′(x) = 0, 1 mark for putting solutions of f′(x) into f(x) to find final answer.]

b)

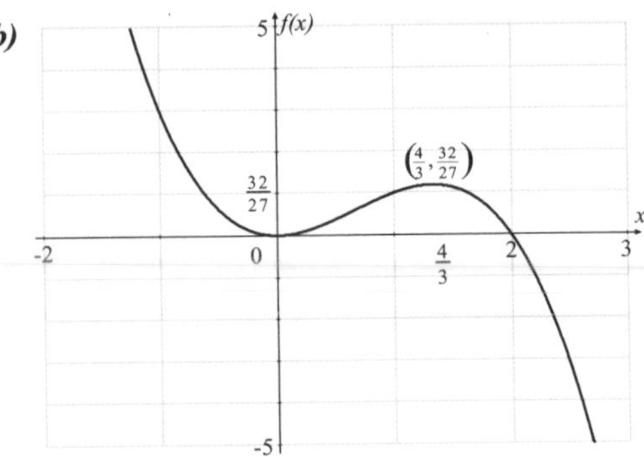

[2 marks — lose 1 mark if general shape is wrong. Lose 1 mark for any of these points not correctly marked: (0, 0), (4/3, 32/27), (2, 0).]

Section Five — Sequences and Series

Page 42

1) a) $u_1 = 1^2 + 3 = 4$; $u_2 = 2^2 + 3 = 7$; $u_3 = 3^2 + 3 = 12$; $u_4 = 4^2 + 3 = 19$

b) $u_{20} = 20^2 + 3 = 403$

2) $a_2 = 3 \times 4 - 2 = 10$; $a_3 = 3 \times 10 - 2 = 28$; $a_4 = 3 \times 28 - 2 = 82$

3) ($u_2 = 3 + 5 = 8$; $u_3 = 8 + 5 = 13$; $u_4 = 13 + 5 = 18$)

n^{th} term: $u_n = 5n - 2$

Page 44

1) $a = 5$, $d = 3$, $l = 65$

$a + (n-1)d = l$

$5 + 3(n-1) = 65$

$3(n-1) = 60$

$n - 1 = 20$

$n = 21$

$S_{21} = 21\left(\frac{5+65}{2}\right) = 735$

2) a) $a + (n-1)d = nth$ term

$7 + (5-1)d = 23$

$4d = 16$

$d = 4$

b) $a + (15-1)d$ is 15th term

$= a + 14d = 7 + 14 \times 4 = 63$

c) $S_{10} = \frac{10}{2}[2 \times 7 + (10-1) \times 4]$

$S_{10} = 5(14 + 36) = 250$

Answers

3) $a + 6d = 36$
$a + 9d = 30$
Subtract one equation from the other:
$-3d = 6$
$\underline{d = -2}$
Plug d into one of the original equations:
$a + 6 \times -2 = 36$
$a - 12 = 36$
$\underline{a = 48}$

$S_5 = \frac{5}{2}[2 \times 48 + (5-1) \times -2]$

$S_5 = \frac{5}{2}(96 - 8) \quad \underline{=220}$

n^{th} *term* $= a + (n-1)d$
$= 48 + (n-1) \times -2$
$= 48 - 2n + 2 \quad = \underline{50 - 2n}$

4) $\sum_{n=1}^{20}(3n - 1) = 2 + 5 + 8 + ... + 59$

$= 20(\frac{2+59}{2}) \quad = 610$

5) $\sum_{n=1}^{10}(48 - 5n)$

$= 43 + 38 + 33 + ... + -2$

$= 10(\frac{43+-2}{2}) \quad = 205$

Page 45

1) a) $l = a + (n-1)d$ *[1 mark]. The average of the terms is* $\frac{a+l}{2}$ *[1mark].*

The sum of the first n terms will be n multiplied by the average.

This is $S_n = n\frac{a+l}{2}$ *[1 mark].*

But $l = a + (n-1)d$*, and so* $S_n = \frac{n}{2}[2a + (n-1)d]$ *[1 mark]*

(You'll get the marks for any suitable method.)

b) $\sum_{n=9}^{32}(2n - 5)$

$= (2 \times 9 - 5) + (2 \times 10 - 5) + (2 \times 11 - 5) + ... + (2 \times 32 - 5)$
$= 13 + 15 + 17 + ... + 59$ *[1 mark]*
So $a = 13$, $l = 59$, $n = 32 - 8 = 24$ *(32 terms minus the first 8)*
[1 mark]

$S_{24} = 24(\frac{13+59}{2})$ *(Using* $S_n = n\frac{(a+l)}{2}$*) [1 mark]*

$S_{24} = 864$ *[1 mark]*

Page 48

1) $a = 2, r = -3$
10th *term,* $u_{10} = ar^9$
$= 2 \times (-3)^9 \quad = -39366$

2) a) r = *2nd term* ÷ 1st *term*
$r = 12 \div 24 \quad = ½$

b) 7th *term* $= ar^6$
$= 24 \times (½)^6$

$= 0.375$ (or $\frac{3}{8}$)

c) $S_\infty = \frac{a}{1-r} \quad = \frac{24}{1-\frac{1}{2}} = 48$

3) **G.P.** — 2, 6, ... $a=2$, $r=3$
5th *term is* $ar^4 = 2 \times 3^4 = \underline{162}$
A.P. — 2, 6, ... $a=2$, $d=4$
You need $a + (n-1)d = 162$
$2 + (n-1)4 = 162$
$4(n-1) = 160$
$n - 1 = 40$
$\underline{n = 41}$
i.e. the 41st *term of the AP is equal to the 5th term of the G.P.*

4) a) *Second term is ar [1 mark]*

Therefore $a \times \frac{-1}{2} = -2$ *[1 mark]*

So $a = 4$ *[1 mark]*

b) $4, -2, 1, \frac{-1}{2}, \frac{1}{4}, \frac{-1}{8}, \frac{1}{16}$

[1 mark for at least 3 terms correct, 2 marks for at least 5 correct, and 3 marks for all 7 correct]

c) $S_7 = \frac{4\left(1-\left(-\frac{1}{2}\right)^7\right)}{1-\frac{-1}{2}} = \frac{4\left(1+\frac{1}{128}\right)}{\frac{3}{2}} = \frac{2}{3} \times 4(\frac{129}{128}) = \frac{43}{16} = 2\frac{11}{16}$

[3 marks for the correct answer, otherwise up to 2 marks for some correct working]

d) $S_\infty = \frac{a}{1-r} = \frac{4}{1-\frac{-1}{2}} = \frac{8}{3} = 2\frac{2}{3}$

[3 marks for the correct answer, otherwise up to 2 marks for some correct working]

Page 50

1) a) $(1 + ax)^{10} = 1 + \frac{10}{1}(ax) + \frac{10\times9}{1\times2}(ax)^2 + \frac{10\times9\times8}{1\times2\times3}(ax)^3 + ...$

[1 mark]

$= 1 + 10ax + 45a^2x^2 + 120a^3x^3 + ...$ *[1 mark]*

b) $(2 + 3x)^5 = 2^5(1 + \frac{3}{2}x)^5$ *[1 mark]*

$= 2^5[1 + \frac{5}{1}\left(\frac{3}{2}x\right) + \frac{5\times4}{1\times2}\left(\frac{3}{2}x\right)^2 + ...]$

x^2 *term is* $2^5 \times \frac{5\times4}{1\times2}\left(\frac{3}{2}\right)^2 x^2$*, so coefficient is* $2^5 \times \frac{5\times4}{1\times2} \times \frac{3^2}{2^2} = 720$

[1 mark]

c) *From a), coefficient of* x^2 *is* $45a^2$
If it's equal to the x^2 *coefficient in part b), then* $45a^2 = 720$
[1 mark]

$a^2 = \frac{720}{45}$

$a^2 = 16$
$a = 4$ *[1 mark]*
(answer can't be -4, as part a) specified a>0)

Answers

Section Six — Trigonometry
Page 60

1) $\cos 315° = \frac{1}{\sqrt{2}}$ (since it's the same as cos 45°).

Now $\frac{1}{\sqrt{2}} \times \frac{\sqrt{2}}{\sqrt{2}} = \frac{\sqrt{2}}{\sqrt{2}\times\sqrt{2}} = \frac{\sqrt{2}}{2}$

$= \frac{1}{2}\sqrt{2}$

[5 marks for the correct answer or 1 mark for correct use of the CAST method (or other suitable method), 1 mark for getting $\cos 315° = \cos 45°$, 1 mark for $\cos 315° = \frac{1}{\sqrt{2}}$, 1 mark for rationalising the denominator.]

2) a) The equation of the curve is $y = \cos 2x + 3$
The +3 shifts the whole graph upwards by 3 units.
$\cos x$ intercepts the y-axis at $y = 1$,
so $\cos 2x + 3$ intercepts the y-axis at $y = 1 + 3 =$ **4**.
[1 mark for correct answer.]

b) Having $\cos 2x$, instead of $\cos x$, means the graph is squashed horizontally (i.e. the period is halved.) On the graph of cos$x$, the equivalent point to point B would occur at $x = 180°$.
So on the graph of $\cos 2x + 3$, pt B must be half the distance along, i.e. at $x = 90°$.
The y-coordinate of pt B will be 3 more than the minimum of $\cos x$, i.e. $-1 + 3 = 2$.
So **B is (90, 2)**.
[2 marks available — 1 mark for the correct x-coordinate and 1 mark for the correct y-coordinate.]

c) $\cos 2x + 3 = 3.5$
$\cos 2x = 0.5$ for $0 \le x < 360°$
$\cos 2x = 0.5$ (Shown as θ in the diagram)

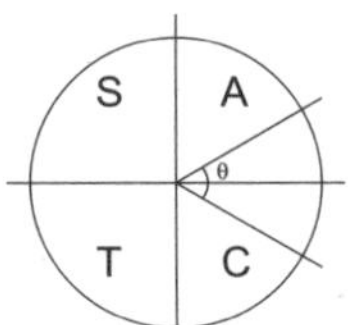

The question says that one solution is $x = 30°$.
But in the CAST diagram, the angle θ is $2x$, so θ must be 60°.
To find another solution, find the other quadrant where cos is positive — the fourth quadrant.
Using the diagram, another possible solution is 360 – θ:
And 360 – θ = 360 – 60 = 300°.
The range for solutions is $0 \le x < 360°$.
But $\theta = 2x$, so the range for θ is $0 \le \theta < 720°$.
This means you get two more solutions by going round again, i.e. adding 360° to the existing solutions:
So the solutions θ = 60°, 300° give θ = 420°, 660° as well.
Now to get the final answer, just give these solutions as values for x, not θ, by dividing by 2:
***x* = 30°, 150°, 210°, 330°**.
[8 marks available in total — 5 marks for method and 1 mark for each of the three new solutions. Lose a method mark for each method mistake, e.g. using x instead of cos 2x as the angle in the CAST diagram.]

3) Arc AB = $r\theta$ where r = 3 cm and $\theta = \frac{\pi}{4}$

So AB = $\frac{3\pi}{4}$ cm
The perimeter of region R is going to be:
major arc AB + OA + OB
Find the major arc:
Circumference of circle = $2\pi r = 2\pi \times 3 =$ **6π cm**

Since the minor arc AB = $\frac{3\pi}{4}$,

the major arc AB = $6\pi - \frac{3\pi}{4} = \mathbf{5\frac{1}{4}\pi}$ **cm**
So Perimeter of R = major arc AB + OA + OB

$= 5\frac{1}{4}\pi + 3 + 3$
$= \mathbf{22.49\,cm}$ (2$d.p.$)

[4 marks available — 2 marks for the correct length of arc AB, otherwise 1 mark for using the correct arc length formula or another correct method, 2 marks for the correct perimeter, otherwise 1 for the correct method. Lose a mark if the correct answer is not given to 2 d.p.]

4) If you use your calculator it'll tell you that the solution of cosθ = 0.5 is 60° ($\cos^{-1}0.5 = 60°$), but you need values between 360° and 720°.
Using the CAST diagram method: cosθ is positive, so you only need the 'cos' and 'all' parts of the diagram.

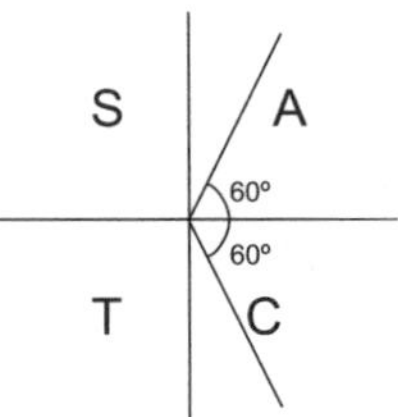

This gives values of 60° and 300°, but you only want values between 360° and 720°. So just add 360° to each value until you get in range:
60° + 360° = 420°, 780° (too big)
300° + 360° = 660°, 1020° (too big)
So cosθ = 0.5 when θ = 420° and 660°.
The question asks for the answers in radians:
180° = π radians, so to convert to radians, divide by 180 and multiply by π.

So θ = 420° and 660° which means $\theta = \frac{7\pi}{3}, \frac{11\pi}{3}$

[4 marks available — 1 mark for a correct method, 1 mark for each of the two correct angles in degrees, 1 mark for converting both angles to radians.]

Answers

Section Seven — Log$_a$

Page 62

1) a) *$3^3 = 27$ so $\log_3 27 = 3$*

b) *to get fractions you need negative powers*
$3^{-3} = {}^1/_{27}$
$\log_3({}^1/_{27}) = -3$

c) *logs are subtracted so divide*
$\log_3 18 - \log_3 2 = \log_3 (18 \div 2)$
$= \log_3 9$
$= 2 \qquad (3^2 = 9)$

2) a) *logs are added so you multiply — remember 2 log 5 = log 5^2*
$\log 3 + 2 \log 5 = \log (3 \times 5^2)$
$= \log 75$

b) *logs are subtracted so you divide and the power half means square root*
$\frac{1}{2} \log 36 - \log 3 = \log (36^{\frac{1}{2}} \div 3)$
$= \log (6 \div 3)$
$= \log 2$

3) *This only looks tricky because of the algebra, just remember the laws.*
$\log_b(\chi^2 - 1) - \log_b(\chi - 1) = \log_b\{(\chi^2 - 1)/(\chi - 1)\}$
using the difference of two squares $(\chi^2 - 1) = (\chi - 1)(\chi + 1)$ and cancelling
$= \log_b(\chi + 1)$

4) a) *Some marks are a give-away*
$\log_3 3 = 1$ *[1 mark]*

b) *You'll get the first mark for showing you can use one of the laws of logarithms and the other for successfully getting $\chi = 32$*
$3 \log_a 2 = \log_a 2^3$
$\log_a 4 + 3 \log_a 2 = \log_a (4 \times 2^3)$
$= \log_a (4 \times 8)$
$= \log_a 32$
[2 marks for the correct answer, otherwise 1 mark for some correct working]

Section Eight — Integration

Page 69

1) a)
$x_0 = 0$: $y_0 = \sqrt{9} = 3$
$x_1 = 1$: $y_1 = \sqrt{8} = 2.8284$
$x_2 = 2$: $y_2 = \sqrt{5} = 2.2361$
$x_3 = 3$: $y_3 = \sqrt{0} = 0$

$$h = \frac{(3-0)}{3} = 1$$

$$\int_a^b y\,dx = \frac{1}{2}[(3+0) + 2(2.8284 + 2.2361)] = 6.5645 \approx 6.56$$

b)
$x_0 = 0.2$: $y_0 = 0.2^3 + 4 = 4.008$
$x_1 = 0.4$: $y_1 = 0.4^3 + 4 = 4.064$
$x_2 = 0.6$: $y_2 = 0.6^3 + 4 = 4.216$
$x_3 = 0.8$: $y_3 = 0.8^3 + 4 = 4.512$
$x_4 = 1.0$: $y_4 = 1.0^3 + 4 = 5.000$
$x_5 = 1.2$: $y_5 = 1.2^3 + 4 = 5.728$
$x_6 = 1.4$: $y_6 = 1.4^3 + 4 = 6.744$

$$h = \frac{(1.4-0.2)}{6} = 0.2$$

$$\int_a^b y\,dx \approx \frac{0.2}{2}[(4.008 + 6.744) + 2(4.064 + 4.216 + 4.512 + 5 + 5.728)]$$
$$= 0.1 \times 57.792 \approx 5.779$$

2) a) $p = 2(1.6)^2 - 2(1.6) = 1.92, \quad q = 2(2)^2 - 2(2) = 4$
[1 mark for each]

b) $$I \approx \frac{0.2}{2}[(0.48 + 4) + 2(1.12 + 1.92 + 2.88)]$$
$$= 1.632$$
[4 marks for the correct answer — otherwise up to 2 marks for substituting the numbers into the trapezium rule correctly, and 1 mark for doing the calculation correctly]

Page 70

1) Integrate $\frac{dy}{dx}$ to get an equation for the curve.
$\frac{dy}{dx} = 6(x^2 - 1) = 6x^2 - 6$
$\int (6x^2 - 6)dx = y = 2x^3 - 6x + C$
To find the value of C, substitute in $x = 0$ and $y = 0$ (because you are told the curve passes through the origin).
$0 = 2(0) - 6(0) + C$, so $C = 0$.
The equation of the curve is therefore: $y = 2x^3 - 6x$
or: $y = 2x(x^2 - 3)$.
The x-intercepts are when $y = 0$, so you know from this second equation that either $2x = 0$ or $x^2 - 3 = 0$. So $x = 0$ or $x = \pm\sqrt{3}$.
The x-intercepts are therefore $(0, 0)$, $(-\sqrt{3}, 0)$ and $(\sqrt{3}, 0)$.
[2 marks for correctly integrating $\frac{dy}{dx}$ to get an equation for the curve, 1 mark for finding the equation of the curve, 1 mark for finding each of the intercepts.]

Answers

2) a) $y = x^3 - 10x^2 + 25x$

$$\Rightarrow \frac{dy}{dx} = 3x^2 - 20x + 25$$

At max/min points, $\frac{dy}{dx} = 0$

$\Rightarrow \; 3x^2 - 20x + 25 = 0$

Factorising:

$(x-5)(3x-5) = 0$

So $x = 5$ or $x = \frac{5}{3}$

But $x = 5$ is the minimum shown on the graph,

so $x = \frac{5}{3} = 1.67$ at point A (to 2 d.p.)

Put this into the curve equation to find y:

$y = x^3 - 10x^2 + 25x$

$y = \left(\frac{5}{3}\right)^3 - 10\left(\frac{5}{3}\right)^2 + 25\left(\frac{5}{3}\right)$

$y = 18.52$ (2 d.p.)

So the maximum turning point is **(1.67, 18.52)**.
But you still need to prove that it is a maximum.
The way to do this is to look at the derivative either side of the turning point:

$\frac{dy}{dx} = 3x^2 - 20x + 25$

at $x = 1.5 \Rightarrow \frac{dy}{dx} = 3(1.5)^2 - 20(1.5) + 25 = 1.75$ (i.e. POSITIVE)

at $x = 2 \Rightarrow \frac{dy}{dx} = 3(2)^2 - 20(2) + 25 = -3$ (i.e. NEGATIVE)

So derivative goes from positive to negative.
Therefore the turning point must be a maximum.
(1.67, 18.52) is a maximum.

[6 marks available — 1 mark for differentiating correctly, 1 mark for setting the derivative equal to 0 and solving for x, 1 mark for the correct value for x, 1 mark for the correct value for y, 1 mark for finding values of the derivative either side of the turning point, 1 mark for correctly proving it is a maximum.]

b) To find the area of the region R, integrate between limits of 0 and 5.

$$\int_0^5 (x^3 - 10x^2 + 25x)dx = \left[\frac{x^4}{4} - \frac{10x^3}{3} + \frac{25x^2}{2}\right]_0^5$$

$$= \left[\frac{(5)^4}{4} - \frac{10(5)^3}{3} + \frac{25(5)^2}{2}\right] - \left[\frac{(0)^4}{4} - \frac{10(0)^3}{3} + \frac{25(0)^2}{2}\right]$$

$$= \left[\frac{625}{4} - \frac{1250}{3} + \frac{625}{2}\right] - 0$$

$$= \frac{625}{12} \text{ units}^2$$

[5 marks available — 1 mark for trying to integrate within the limits, 2 marks for integrating correctly, 1 mark for some correct working with the limits, 1 mark for the correct answer.]

3) a) Solve the equations simultaneously. Substitute $y = x + 8$ into the quadratic equation:

$x + 8 = x(x+3)$ *[1 mark]*

$x + 8 = x^2 + 3x$

$0 = x^2 + 2x - 8$

$(x+4)(x-2) = 0$

$x = -4$ (point A), or $x = 2$ (point B)

Therefore point A: $(-4, 4)$ and point B: $(2, 10)$ *[2 marks]*

b) Region R $= \int_{-4}^{2} (x+8)dx - \int_{-4}^{2} (x^2+3x)dx$ *[2 marks]*

$$= \left[\frac{x^2}{2} + 8x\right]_{-4}^{2} - \left[\frac{x^3}{3} + \frac{3x^2}{2}\right]_{-4}^{2}$$ *[2 marks]*

$$= \left[\left[\frac{(2)^2}{2} + 8(2)\right] - \left[\frac{(-4)^2}{2} + 8(-4)\right]\right] - \left[\left[\frac{(2)^3}{3} + \frac{3(2)^2}{2}\right] - \left[\frac{(-4)^3}{3} + \frac{3(-4)^2}{2}\right]\right]$$ *[2 marks]*

$$= [18 - [-24]] - \left[\frac{26}{3} - \frac{8}{3}\right]$$ *[1 mark]*

$$= 42 - \frac{18}{3} = \frac{108}{3} = 36 \text{ units}^2$$ *[1 mark]*

4) $n = 4$, so $h = \frac{\pi}{12}$ *[1 mark]*

$$\int_0^{\frac{\pi}{3}} \cos x\, dx \approx \frac{\pi}{24}\left[\cos 0 + 2\left(\cos\left(\frac{\pi}{12}\right) + \cos\left(\frac{2\pi}{12}\right) + \cos\left(\frac{3\pi}{12}\right)\right) + \cos\left(\frac{4\pi}{12}\right)\right]$$ *[3 marks]*

$$= \frac{\pi}{24}[1 + 2(2.539...) + 0.5]$$ *[1 mark]*

$= 0.861$ *[1 mark]*

Answers

Section Nine — Data

Page 73

1) *12.8, 13.2, 13.5, 14.3, 14.3, 14.6, 14.8, 15.2, 15.9, 16.1, 16.1, 16.2, 16.3, 17.0, 17.2 (all in cm)*

2)

Boys		Girls
9, 5	1	2
7, 6, 2, 0	2	1, 4, 5, 7, 8, 9
9, 7, 4, 2	3	1, 6, 7, 8, 9

Key 2|1|3 means: Boys 12, Girls 13

3)

Length of call	Lower class boundary (lcb)	Upper class boundary (ucb)	Class width	Frequency	Frequency density = Height of column
0 - 2	0	2.5	2.5	10	4
3 - 5	2.5	5.5	3	6	2
6 - 8	5.5	8.5	3	3	1
9 - 15	8.5	15.5	7	1	0.143

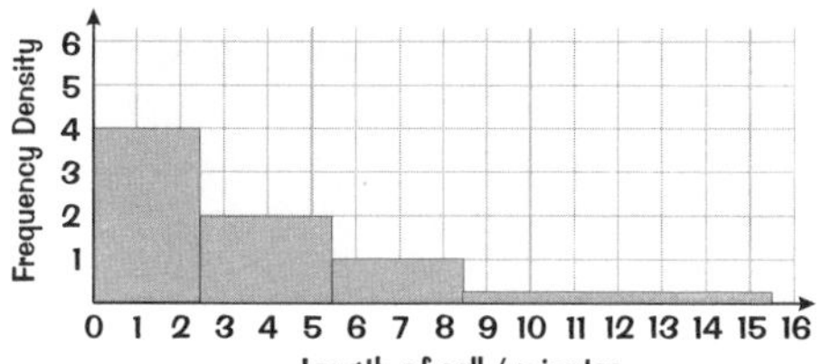

4) a)

Profit	Class width	Frequency	Frequency density = Height of column
4.5 - 5.0	0.5	24	48
5.0 - 5.5	0.5	26	52
5.5 - 6.0	0.5	21	42
6.0 - 6.5	0.5	19	38
6.5 - 8.0	1.5	10	6.67

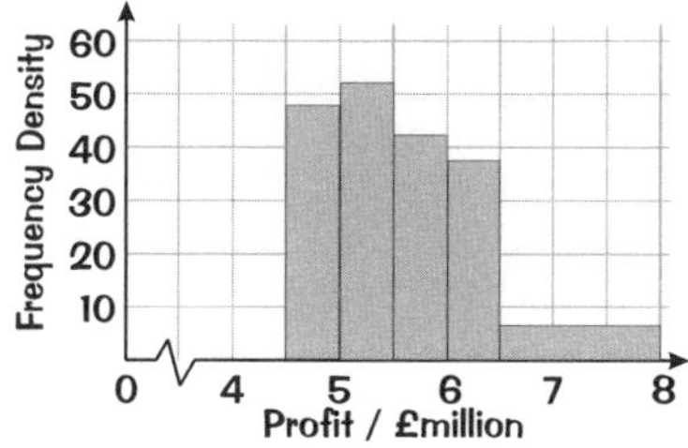

[1 mark for correct axes, then up to 2 marks for the bars drawn correctly]

b) *The distribution is positively skewed — only a few businesses make a high profit. The modal profit is between £5 million and £5.5 million.*
[Up to 2 marks available for any sensible comments]

Page 75

1) *$\Sigma f = 16$, $\Sigma fx = 22$, so mean = 22 ÷ 16 = 1.375*
Median position = 17 ÷ 2 = 8.5, so median = 1
Mode = 0.

2)

Speed	mid-class value x	Number of cars f	fx
30 - 34	32	12 (12)	384
35 - 39	37	37 (49)	1369
40 - 44	42	9	378
45 - 50	47.5	2	95
	Totals	60	2226

Estimated mean = 2226 ÷ 60 = 37.1 mph
Median position is 61 ÷ 2 = 30.5.
This is in class 35 - 39.
30.5 – 12 = 18.5, so median is 18.5th value in class.
Class width = 5, so median is:

$$34.5 + \left(\frac{18.5}{37} \times 5\right) = 37 \text{ mph}$$

Modal class is 35 - 39 mph.

3) a) *There are 30 males, so median is in 31 ÷ 2 = 15.5th position. Take the mean of the 15th and 16th readings to get median = (62 + 65) ÷ 2 = 63.5 [1 mark]*

b) *The female median is 64.5 (halfway between the 8th and 9th readings). Female median is higher than the male median. The females scored better than the males on average.*
Female range = 79 – 55 = 24.
Male range = 79 – 43 = 36
The female range is less than the male range.
Their scores are more consistent than the males'.
[Up to 2 marks available for any sensible comments]

Page 77

1)

Distance	Upper class boundary (ucb)	f	Cumulative frequency (cf)
	0	0	0
0 - 2	2	10	10
2 - 4	4	5	15
4 - 6	6	3	18
6 - 8	8	2	20

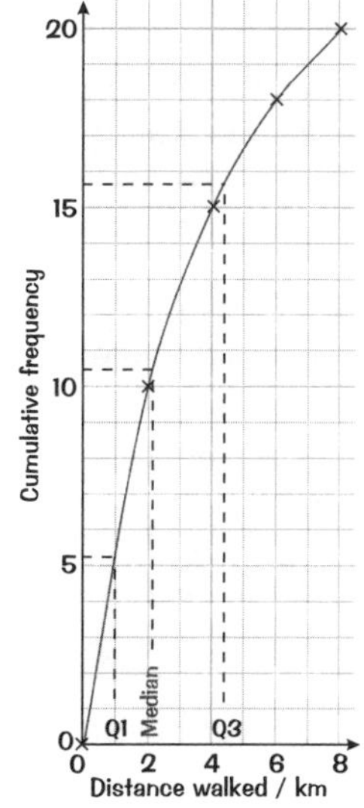

Median = 2.2 km (approximately)
Q_1 = 1 km, Q_3 = 4.4 km, so interquartile range = 3.4 km (approximately)

Answers

2) a)

Age	Upper class boundary (ucb)	f	Cumulative frequency (cf)
Under 5	5	0	0
5 - 10	11	2	2
11 - 15	16	3	5
16 - 20	21	10	15
21 - 30	31	2	17
31 - 40	41	2	19
41 - 70	71	1	20

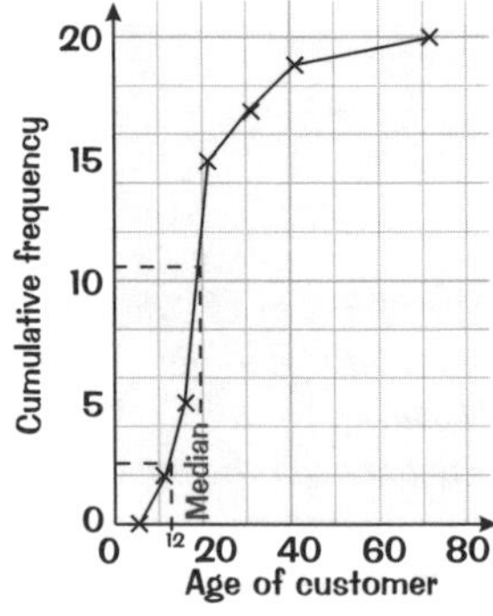

[1 mark for correctly labelled axes, up to 2 marks for calculating the points and plotting them correctly, and 1 mark for joining the points with a suitable line]

b) (i) *Median = 18.5 years (approximately) [1 mark]*

(ii) *Number of customers under 12 = 2 (approximately) [1 mark] So number of customers over 12 = 20 – 2 = 18 (approximately) [1 mark]*

3) a) (i) *Times = 2, 3, 4, 4, 5, 5, 5, 7, 10, 12*

Median position = 5.5, so median = 5 minutes [1 mark]

(ii) *Lower quartile = 4 minutes [1 mark]*

Upper quartile = 7 minutes [1 mark]

b)

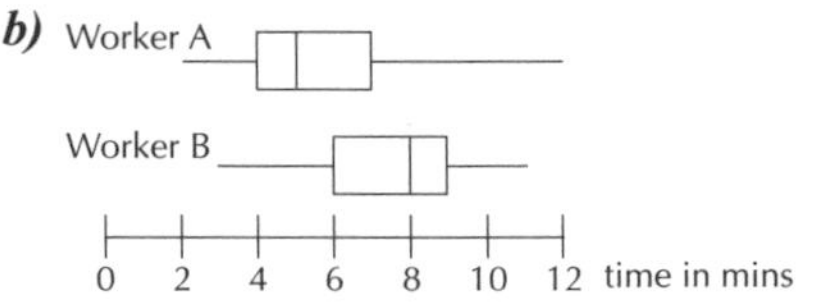

[Up to 3 marks available for each graph — get 1 mark for the median in the right place, 1 mark for both of the quartiles shown correctly, and 1 mark for the lines showing the extremes drawn correctly]

c) *Various statements could be made,*
e.g. the times for Worker B are longer than those for Worker A, on average.
The IQR for both workers is the same — generally they both work with the same consistency.
The range for Worker A is larger than that for Worker B. Worker A has a few items he/she can iron very quickly and a few which take a long time.
[1 mark for any sensible answer]

d) *Worker A would be best to employ. The median time is less than for Worker B, and the Upper Quartile is less than the median of Worker B. Worker A would generally iron more items in a given time than worker B.*
[1 mark for any sensible answer]

Page 79

1) $$\text{Mean} = \frac{11+12+14+17+21+23+27}{7} = \frac{125}{7} = 17.9 \text{ to 3 sig. fig.}$$

$$\text{s.d.} = \sqrt{\frac{11^2+12^2+14^2+17^2+21^2+23^2+27^2}{7} - \left(\frac{125}{7}\right)^2} = \sqrt{30.98} = 5.57 \text{ to 3 sig. fig.}$$

2)

Score	Mid-class value, x	x^2	f	fx	fx^2
100 - 106	103	10609	6	618	63654
107 - 113	110	12100	11	1210	133100
114 - 120	117	13689	22	2574	301158
121 - 127	124	15376	9	1116	138384
128 - 134	131	17161	2	262	34322
		Totals	50 (= n)	5780 (= Σx)	670618 (= Σx^2)

$$\text{Mean} = \frac{5780}{50} = 115.6$$

$$s^2 = \frac{670618}{50} - 115.6^2 = 49$$

So $s = 7$

3) a) $\bar{A} = \frac{60.3}{20} = 3.015 \text{ g}$ *[1 mark]*

b) $s_A^2 = \frac{219}{20} - 3.015^2 = 1.86 \text{ g}^2$

So $s_A = 1.36$ g to 3 sig. fig.

[3 marks for the correct answer — otherwise 1 mark for a correct method to find the variance, and 1 mark for taking the square root to find the s.d.]

c) *Brand A chocolate drops are heavier on average than brand B. Brand B chocolate drops are much closer to the mean of 2.95 g.*
[1 mark for each of 2 sensible statements]

d)

$$\text{Mean of } A \text{ and } B = \frac{\sum A + \sum B}{50} = \frac{60.3 + (30 \times 2.95)}{50} = 2.976 \text{ g}$$

$$\frac{\sum B^2}{30} - 2.95^2 = 1, \text{ and so } \sum B^2 = 291.075$$

$$\text{Variance of } A \text{ and } B = \frac{\sum A^2 + \sum B^2}{50} - 2.976^2 = \frac{219 + 291.075}{50} - 2.976^2$$

$$= 1.3449$$

[4 marks for the correct answer — otherwise 1 mark for the correct total mean, 1 mark for the sum of the B^2, 1 mark for the correct method to find the total variance/standard deviation]

Answers

Page 81

1) *Let $y = x - 20$.*

Then $\bar{y} = \bar{x} - 20$ or $\bar{x} = \bar{y} + 20$

$\sum y = 125$ and $\sum y^2 = 221$

So $\bar{y} = \frac{125}{100} = 1.25$ and $\bar{x} = 1.25 + 20 = \underline{21.25}$

$s_y^2 = \frac{221}{100} - 1.25^2 = 0.6475$ and so $s_y = 0.805$ to 3 sig. fig.

Therefore $s_x = 0.805$ to 3 sig. fig.

2)

Time	Mid-class x	$y = x - 35.5$	f	fy	fy^2
30 - 33	31.5	-4	3	-12	48
34 - 37	35.5	0	6	0	0
38 - 41	39.5	4	7	28	112
42 - 45	43.5	8	4	32	256
		Totals	20 (= n)	48 (= Σy)	416 (= Σy^2)

$\bar{y} = \frac{48}{20} = 2.4$

So $\bar{x} = \bar{y} + 35.5 = 2.4 + 35.5 = \underline{37.9 \text{ minutes}}$

$s_y^2 = \frac{416}{20} - 2.4^2 = 15.04$, and so $s_y = 3.88$ minutes, to 3 sig. fig.

But $s_x = s_y$, and so $\underline{s_x = 3.88 \text{ minutes, to 3 sig. fig.}}$

3) a) *Let $y = x - 30$.*

$\bar{y} = \frac{228}{19} = 12$ and so $\underline{\bar{x} = \bar{y} + 30 = 42}$

$s_y^2 = \frac{3040}{19} - 12^2 = 16$ and so $s_y = 4$

But $s_x = s_y$ and so $\underline{s_x = 4}$

[3 marks for the correct answers — otherwise 1 mark for the correct mean and 1 mark for the correct s.d. or variance of the y]

b) $\bar{x} = \frac{\sum x}{19} = 42$

And so $\sum x = 42 \times 19 = \underline{798}$

$s_x^2 = \frac{\sum x^2}{19} - \bar{x}^2 = \frac{\sum x^2}{19} - 42^2 = 16$

And so $\sum x^2 = (16 + 42^2) \times 19 = \underline{33820}$

[3 marks for both correct answers — otherwise 1 mark for either correct]

c) New $\sum x = 798 + 32 = 830$.

So new $\bar{x} = \frac{830}{20} = \underline{41.5}$

New $\sum x^2 = 33820 + 32^2 = 34844$.

So new $s_x^2 = \frac{34844}{20} - 41.5^2 = 19.95$ and new $\underline{s_x = 4.47}$ to 3 sig. fig.

[2 marks for each correct answer — otherwise 1 mark for some correct working for each part]

Page 83

1) *IQR = 88 – 62 = 26, so 3 × IQR = 78.*
So upper fence = 88 + 78 = 166.
This means that: ***a)*** *161 is not an outlier.*
b) *176 is an outlier.*
Lower fence = 62 – 78 = -16.
This means that: ***c)*** *0 is not an outlier.*

2) *Put the 20 items of data in order:*
1, 4, 5, 5, 5, 5, 5, 6, 6, 7, 7, 8, 10, 10, 12, 15, 20, 20, 30, 50
Then the median position is 10.5, and since the 10th and the 11th items are both 7, the median = $\underline{7}$.
Lower quartile = $\underline{5}$.
Upper quartile = (12 + 15) ÷ 2 = $\underline{13.5}$.

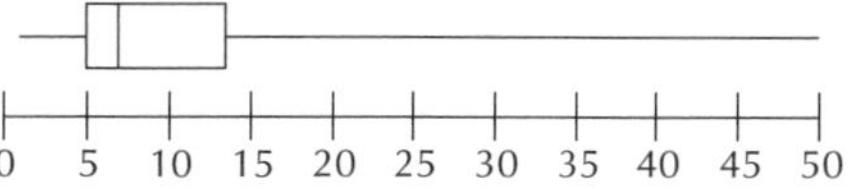

This data is positively skewed. Most 15-year-olds earned a small amount of pocket money. A few got very large amounts.

3) Pearson's coefficient of skewness $= \frac{10.3 - 10}{1.5} = 0.2$

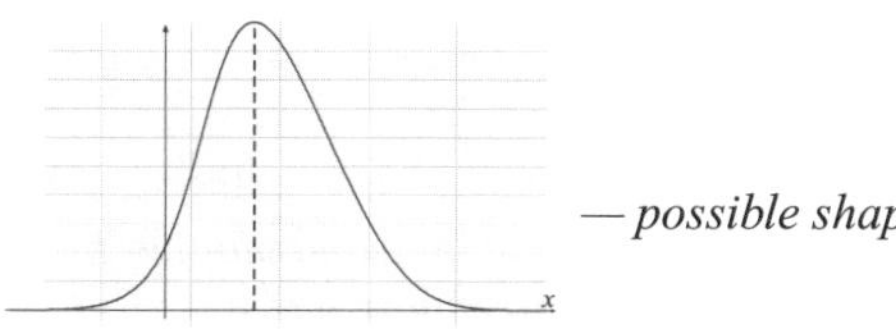

— possible shape

This is positively skewed, but not by much.

4) a) *Total number of people = 38.*
Median position = (38 + 1) ÷ 2 = 19.5.
19th value = 15; 20th value = 16, so median = $\underline{15.5 \text{ hits}}$.
Mode = $\underline{15 \text{ hits}}$.
[2 marks for the correct median, otherwise 1 mark for some correct working, plus 1 mark for the correct mode]

b) *Lower quartile = 10th value = 14*
Upper quartile = 29th value = 17. [1 mark for both]
So interquartile range = 17 – 14 = 3,
and upper fence = 17 + (3 × 3) = 26.
This means that 25 is not an outlier. [1 mark]

c)

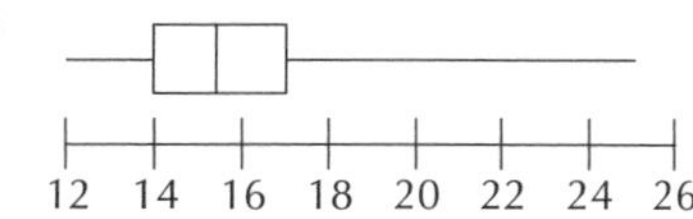

[1 mark]
The distribution is positively skewed. [1 mark]
(Different kinds of sketch would be allowed.)

d) *If 25 was removed then the right-hand tail of the box plot would be much shorter, and the distribution would be less positively skewed. [1 mark]*

Answers

5) a)

mm of rain	Upper class boundary (ucb)	*f*	Cumulative frequency (cf)
Under 5	5	0	0
5 - 10	10	2	2
10 - 15	15	3	5
15 - 20	20	5	10
20 - 25	25	7	17
25 - 30	30	10	27
30 - 35	35	3	30

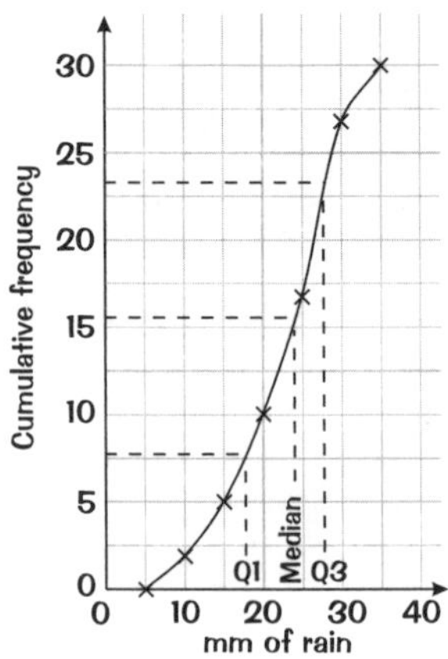

[Up to 2 marks for correctly calculating and plotting the points, plus 1 mark for correct axes/labels etc.]

b) *From the diagram, median = 24.*
[1 mark for answer in range 23.5-24.5]
Lower quartile = 17.5. [1 mark for answer in the range 17-18]
Upper quartile = 27.5. [1 mark for answer in the range 27-28]

c) *Quartile coefficient of skewness* $= \dfrac{27.5-(2\times 24)+17.5}{27.5-17.5} = -0.3$

[1 mark — answers may differ slightly, depending on answers to part b)]
The graph is negatively skewed — most of the days tend to have higher rainfall. [1 mark]

Page 85

1) *Number each item with a 3-digit number from 000 to 999. Then use a random number table (or the Ran# button on a calculator) to generate 100 different numbers. If a number is repeated, discount it and choose another.*

2) Mean $= \bar{x} = \dfrac{\sum x}{n} = \dfrac{90}{15} = 6$ *[2 marks]*

Variance $= s^2 = \dfrac{\sum (x-\bar{x})^2}{n-1}$

$= \dfrac{16+36+4+1+4+16+25+4+9+16+4+4+16+0+9}{14}$

$= \dfrac{164}{14} = 11\frac{5}{7} (=11.7)$ *[2 marks]*

Section Ten — Probability

Page 87

1) a) *The sample space would be as below:*

Coin \ Dice	1	2	3	4	5	6
H	2	4	6	8	10	12
T	5	6	7	8	9	10

b) *There are 12 outcomes in total, and 9 of these are more than 5, so P(score >5) = 9/12 = 3/4*

c) *There are 6 outcomes which have a tail showing, and 3 of these are even, so P(even score given that you throw a tail) = 3/6 = 1/2*

2) a) *20% of the people eat chips, and 10% of these is 2% — so 2% eat both chips and sausages.*
Now you can draw the Venn diagram:

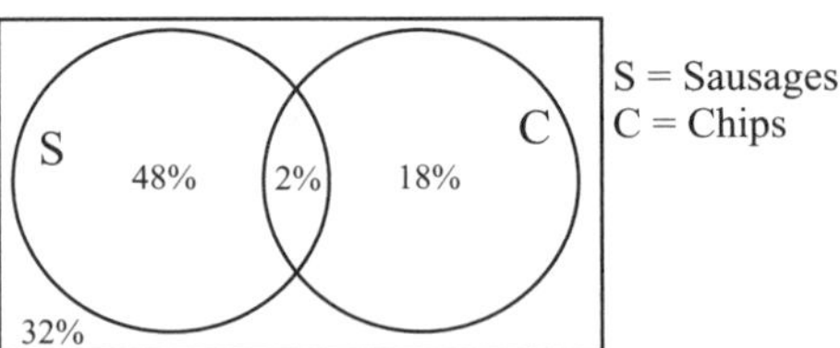

By reading the numbers in the appropriate sets from the diagram you can see

b) *18% eat chips but not sausages.*

c) *18% + 48% = 66% eat chips or sausages, but not both.*

3) a) *The Venn diagram would look something like this:*

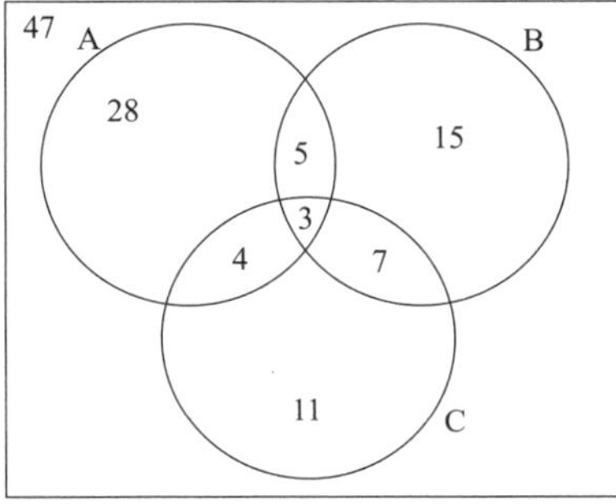

[1 mark for the central figure correct, 2 marks for '5', '7' and '4' correct (get 1 mark for 2 correct), plus 2 marks for '28', '15' and '11' correct (get 1 mark for 2 correct)]

b) i) *Add up the numbers in all the circles to get 73 people out of 120 buy at least 1 type of soap. So the probability = 73/120 [2 marks]*

ii) *Add up the numbers in the intersections to get 5 + 3 + 4 + 7 = 19, meaning that 19 people buy at least two soaps, so the probability a person buys at least two types = 19/120 [2 marks]*

iii) *28 + 11 + 15 = 54 people buy only 1 soap, and of these 15 buy soap B [1 mark]. So probability of a person who only buys one type of soap buying type B is 15/54 = 5/18 [2 marks]*

Answers

Page 89

1) *Draw a sample space diagram*

2nd Dice						
6	7	8	9	10	11	12
5	6	7	8	9	10	11
4	5	6	7	8	9	10
3	4	5	6	7	8	9
2	3	4	5	6	7	8
1	2	3	4	5	6	7
1st Dice	1	2	3	4	5	6

There are 36 outcomes altogether.

a) *15 outcomes are prime (since 2, 3, 5, 7 and 11 are prime), so P(prime) = 15/36 = 5/12*

b) *7 outcomes are square numbers (4 and 9), so P(square) = 7/36*

c) *Being prime and a square number are exclusive events, so P(prime or square) = 15/36 + 7/36 = 22/36 = 11/18*

2) a)

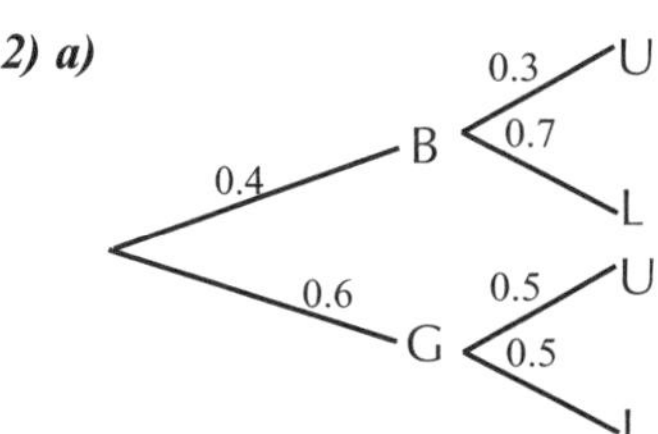

b) *Choosing an upper school pupil means either 'boy and upper' or 'girl and upper'. P(boy and upper) = 0.4 × 0.3 = 0.12. P(girl and upper) = 0.6 × 0.5 = 0.30. So P(Upper) = 0.12 + 0.30 = 0.42.*

3) a)

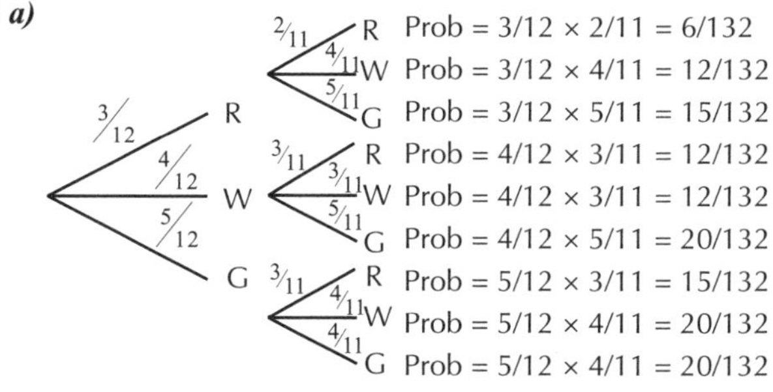

[4 marks available — 1 mark for each set of branches correct]

b) *The second counter is green means one of three outcomes 'red then green'* ***or*** *'white then green'* ***or*** *'green then green'. So Prob(2nd is green) = 15/132 + 20/32 + 20/132 = 55/132 = 5/12 [2 marks]*

c) *For both to be red there's only one outcome: 'red then red' Prob(both red) = 6/132 = 1/22 [2 marks]*

d) *'Both same colour' is the complementary event of 'not both same colour'.*
So Prob (not same colour) = 1 – P(both same colour)
Both same colour is either R and R or W and W or G and G
Prob(not same colour) = 1 – [6/132 + 12/132 + 20/132]
= 1 – 38/132 =94/132 = 47/66
[3 marks for the correct answer — otherwise up to 2 marks available for using a suitable method]

Page 91

1) *You could draw a Venn diagram — but you don't have to, it just makes things easier.*

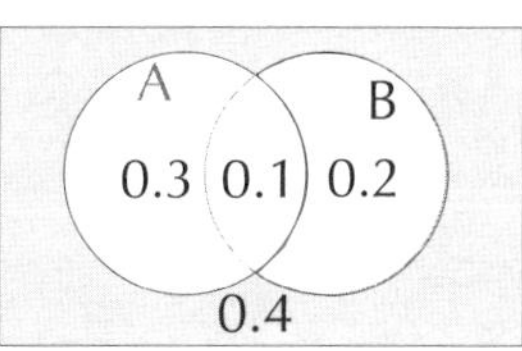

a) *$P(B|A) = P(A \cap B) \div P(A) = 0.1 \div 0.4 = 0.25$*

b) *$P(B|A') = P(A' \cap B) \div P(A') = 0.2 \div 0.6 = 1/3$*

c) *P(B|A) = 0.25, but P(B) = 0.3 so they're not independent.*

2) *Draw a tree diagram:*

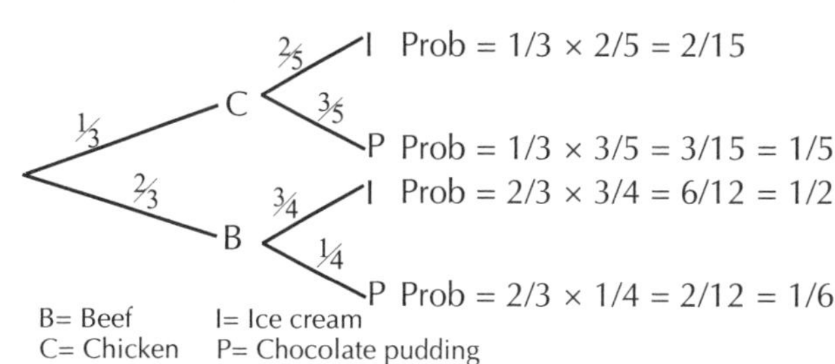

a) *$P(\text{chicken or ice cream but not both}) = P(C \cap P) + P(B \cap I) = 1/5 + 1/2 = 7/10$*

b) *$P(\text{ice cream}) = P(C \cap I) + P(B \cap I) = 2/15 + 1/2 = 19/30$*

c) *$P(\text{chicken}|\text{ice cream}) = P(C \cap I) \div P(I) = (2/15) \div (19/30) = 4/19$*

3) a) (i) *V and W are independent, so*
$P(V \cap W) = P(V) \times P(W) = 0.2 \times 0.6 = 0.12$ [1 mark]
(ii) *$P(V \cup W) = P(V) + P(W) - P(V \cap W) = 0.2 + 0.6 - 0.12 = 0.68$ [2 marks]*

b) *$P(U|V') = P(U \cap V') \div P(V')$*
Now $U = V' \cap W'$, so $U \cap V' = U$ — think about it — all of U is contained in V', so $U \cap V'$ (the 'bits in both U and V') are just the bits in U. Therefore $P(U \cap V') = P(U) = 1 - P(V \cup W) = 1 - 0.68 = 0.32$
And so $P(U|V') = 0.32 \div 0.8 = 0.4$
[3 marks for the correct answer — otherwise up to 2 marks available for correct working]

4) *Draw a tree diagram*

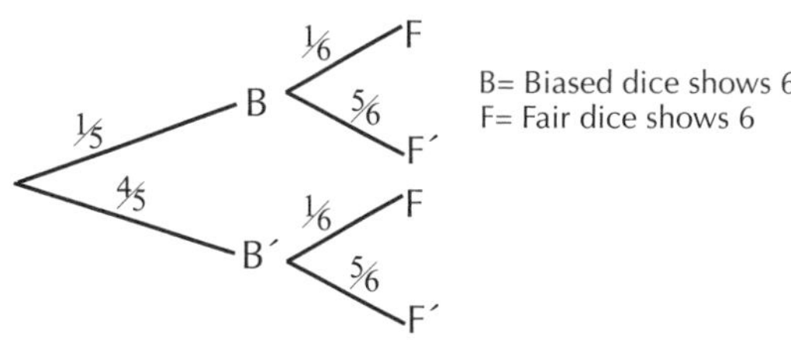

a) *P(B') = 0.8 [1 mark]*

b) *Either at least one of the dice shows a 6 or neither of them do, so these are complementary events. Call F the event 'the fair dice shows a 6'.*
Then $P(F \cup B) = 1 - P(F' \cap B') = 1 - (4/5 \times 5/6) = 1 - 2/3 = 1/3$ [2 marks]

Answers

c) *P(exactly one 6 | at least one 6)*
= P(exactly one 6 $\cap$ at least one 6) ÷ P(at least one 6).
This next step might be a bit easier to get your head round if you draw a Venn diagram.

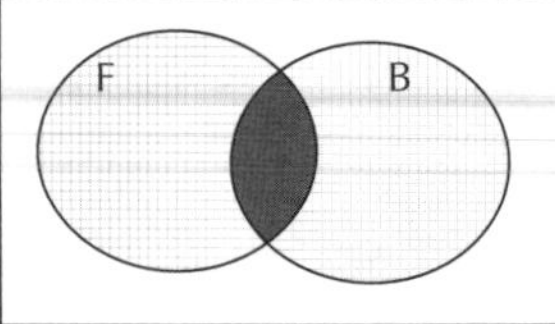

'exactly one 6' $\cap$ 'at least one 6' = 'exactly one 6'
(Look at the diagram — 'exactly one 6' is the cross-hatched area, and 'at least one 6' is the cross-hatched area plus the grey bit. So the bit in common to both is just the cross-hatched area.)
Now, that means P(exactly one 6 $\cap$ at least one 6) = P(B$\cap$F') + P(B'$\cap$F) — this is the cross-hatched area in the Venn diagram, i.e. P(exactly one 6 $\cap$ at least one 6)
= (1/5 × 5/6) + (4/5 × 1/6)
= 9/30 = 3/10 (using the fact that B and F are independent).
P(at least one 6) = 1/5 + 1/6 – (1/5 × 1/6) = 10/30 = 1/3
And all of this means P(exactly one 6 | at least one 6)
= 3/10 ÷ 1/3 = 9/10
[3 marks for the correct answer — otherwise up to 2 marks available for correct working]

Page 93

1) *STATISTICS has 10 letters, 3 repeated S's, 3 repeated T's and 2 repeated I's. This gives 10!÷(3!×3!×2!) = 50400 arrangements.*

2) a) *the 6 people can sit in 6! = 720 ways.*
b) *Put Mr & Mrs Brown together as one object — you can do this in two ways. This means there are 5 objects to rearrange (which can be done in 5! ways) and for each of these, there are 2 "Brown arrangements".*
This means there are 2×5! = 240 ways to seat the Browns together.
c) *The number of ways the Browns sit apart is 720 – 240 = 480.*
So P(Browns sit apart) = 480 ÷ 720 = 2/3

3) a) *$^{26}C_6$ = 26×25×24×23×22×21÷(6×5×4×3×2×1)*
= 230230 ways
b) *We can choose the 5 vowels in one way, leaving 21 choices for the last letter. This means there are 21 ways.*
So P(5 vowels) = 21 ÷ 230230 = 3/32890
c) *Choosing no vowels is the complement of at least one vowel.*
No vowels can be chosen in $^{21}C_6$ = 54264 possible ways.
P(no vowels) = 54264÷230230 = 0.236 (to 3 sig. fig.)
So P(at least 1 vowel) = 1 – 0.236 = 0.764 (to 3 sig. fig.)
d) *There are 6×5×4×3 = 360 ways to arrange 4 discs.*
You can start with a vowel in 3 ways, leaving 5 letters for 3 places so there are 3 × 5 × 4 × 3 =180 arrangements that start with a vowel.
So P(starts with a vowel) = 180/360 = 1/2
(Or you could say that it just depends on the first tile, whether the arrangement starts with a vowel or not. There are 6 tiles, and 3 of these are vowels, so the probability of a random arrangement starting with a vowel is 1/2.)

4) a) *10 people can be arranged in 10! = 3 628 800 ways. [1 mark]*
b) *You can stand the 5 women in 5! = 120 ways.*
This leaves 6 gaps for the 5 men to stand in (including the places at the start and end of the line). If you number these gaps 1-6, then the men have to be standing in either gaps 1-5 or gaps 2-6 (otherwise two women stand together).
Ignoring where they stand for now, the men can be arranged in 5! = 120 ways. So for each of the 120 arrangements of women, there are 2 × 120 = 240 arrangements of the men (since they could be in gaps 1-5 or 2-6).
This means the total number of ways of them all standing apart is 120 × 240 = 28800, so the probability of this happening is 28800 ÷ 3628800 = 1/126
[3 marks for the correct answer — otherwise up to 2 marks available for correct working]
c) *You can have 2, 3, 4 or 5 men (corresponding to 4, 3, 2 or 1 women).*
Suppose 2 men and 4 women are to be chosen.
There are $^5C_2$ = 10 ways to choose 2 men, and $^5C_4$ = 5 ways to choose the 4 women. This means there would be 10 × 5 = 50 ways to choose the 2 men and 4 women.
Similarly 3 men and 3 women could be chosen in $^5C_3 \times {}^5C_3$ = 100 ways.
4 men and 2 women could be chosen in $^5C_4 \times {}^5C_2$ = 50 ways.
5 men and 1 woman could be chosen in $^5C_5 \times {}^5C_1$ = 5 ways.
So the total number of ways in which the people could be chosen is:
50 + 100 + 50 + 5 = 205.
[3 marks for the correct answer — otherwise up to 2 marks available for correct working]

Section Eleven — Probability Distributions

Page 95

1) a) *All the probabilities have to add up to 1.*
So 0.5 + k + k + 3k = 0.5 + 5k = 1, i.e. 5k = 0.5, i.e. k = 0.1.
b) *P(Y < 2) = P(Y = 0) + P(Y = 1) = 0.5 + 0.1 = 0.6.*

Answers

2) *Make a table showing the possible values of X, i.e. total scores on the dice (though there are other ways to do this):*

Score on dice 1 (columns); Score on dice 2 (rows)

+	1	1	1	2	2	3
1	2	2	2	3	3	4
1	2	2	2	3	3	4
1	2	2	2	3	3	4
2	3	3	3	4	4	5
2	3	3	3	4	4	5
3	4	4	4	5	5	6

There are 36 outcomes in total, and of these 10 have X = 4.

So $P(X = 4) = \frac{10}{36} = \frac{5}{18}$.

Altogether there are 5 possible values for X, and the rest of the pdf is found in the same way that you found P(X = 4). The pdf is summarised in this table:

x	2	3	4	5	6
$P(X = x)$	$\frac{1}{4}$	$\frac{1}{3}$	$\frac{5}{18}$	$\frac{1}{9}$	$\frac{1}{36}$

3) a) *The probability of getting 3 heads is:* $\frac{1}{2}\times\frac{1}{2}\times\frac{1}{2}=\frac{1}{8}$ *[1 mark]*

The probability of getting 2 heads is: $3\times\frac{1}{2}\times\frac{1}{2}\times\frac{1}{2}=\frac{3}{8}$ *(multiply by 3 because any of the three coins could be the tail — the order in which the heads and the tail occur isn't important). [1 mark]*

Similarly the probability of getting 1 head is: $3\times\frac{1}{2}\times\frac{1}{2}\times\frac{1}{2}=\frac{3}{8}$. *[1 mark]*

And the probability of getting no heads is $\frac{1}{2}\times\frac{1}{2}\times\frac{1}{2}=\frac{1}{8}$. *[1 mark]*

Hence the pdf of X is:

x	20p	10p	nothing
$P(X = x)$	$\frac{1}{8}$	$\frac{3}{8}$	$\frac{1}{2}$

b) *You need the probability that X >10p [1 mark]. This is just* $P(X = 20p) = \frac{1}{8}$ *[1 mark]*

Page 97

1) $P(W \le 0.2) = P(W = 0.2) = 0.2$

$P(W \le 0.3) = P(W = 0.2) + P(W = 0.3) = 0.4$

$P(W \le 0.4) = P(W = 0.2) + P(W = 0.3) + P(W = 0.4) = 0.7$

$P(W \le 0.5) = P(W = 0.2) + P(W = 0.3) + P(W = 0.4) + P(W = 0.5) = 1$

So the cumulative distribution function of W is:

w	0.2	0.3	0.4	0.5
$P(W \leqslant w)$	0.2	0.4	0.7	1

2) $P(R = 0) = P(R \le 0) = F(0) = 0.1$

$P(R = 1) = P(R \le 1) - P(R \le 0) = 0.5 - 0.1 = 0.4$

$P(R = 2) = P(R \le 2) - P(R \le 1) = 1 - 0.5 = 0.5$

So the pdf of R is:

r	0	1	2
$P(R = r)$	0.1	0.4	0.5

$P(0 \le R \le 1) = 0.5$

3) *There are 5 possible outcomes, and the probability of each of them is k, so k = 1 ÷ 5 = 0.2.*

Mean of X $= \frac{0+4}{2} = 2$. *Variance of X* $= \frac{(4-0+1)^2-1}{12} = \frac{24}{12} = 2$.

4) a) *All the probabilities must add up to 1, so 2k + 3k + k + k = 1, i.e. 7k = 1, and so* $k = \frac{1}{7}$. *[1 mark]*

b) $P(X \le 0) = P(X = 0) = \frac{2}{7}$ *[1 mark]*

$P(X \le 1) = P(X = 0) + P(X = 1) = \frac{5}{7}$ *[1 mark]*

$P(X \le 2) = P(X = 0) + P(X = 1) + P(X = 2) = \frac{6}{7}$ *[1 mark]*

$P(X \le 3) = P(X = 0) + P(X = 1) + P(X = 2) + P(X = 3) = 1$ *[1 mark]*

So the distribution function is as in the following table:

x	0	1	2	3
$P(X \leqslant x)$	$\frac{2}{7}$	$\frac{5}{7}$	$\frac{6}{7}$	1

c) $P(X > 2) = 1 - P(X \le 2) = 1 - \frac{6}{7} = \frac{1}{7}$ *[1 mark]*

(Or $P(X > 2) = P(X = 3) = \frac{1}{7}$*, using part a).)*

5) a)

x	0	1	2	3	4	5	6	7	8	9
$P(X = x)$	0.1	0.1	0.1	0.1	0.1	0.1	0.1	0.1	0.1	0.1

[1 mark]

b) $\text{Mean} = \frac{0+9}{2} = 4.5$ *[1 mark]* $\text{Variance} = \frac{(9-0+1)^2-1}{12} = \frac{99}{12} = 8.25$ *[2 marks]*

c) $P(X < 4.5) = P(X = 0) + P(X = 1) + P(X = 2) + P(X = 3) + P(X = 4) = 0.5$ *[2 marks]*

Page 99

1) a) *As always, the probabilities have to add up to 1, so*

$k = 1 - \left(\frac{1}{6}+\frac{1}{2}+\frac{5}{24}\right) = 1 - \frac{21}{24} = \frac{3}{24} = \frac{1}{8}$

b) $E(X) = \left(1\times\frac{1}{6}\right)+\left(2\times\frac{1}{2}\right)+\left(3\times\frac{1}{8}\right)+\left(4\times\frac{5}{24}\right) = \frac{4+24+9+20}{24} = \frac{57}{24} = \frac{19}{8}$

$E(X^2) = \left(1^2\times\frac{1}{6}\right)+\left(2^2\times\frac{1}{2}\right)+\left(3^2\times\frac{1}{8}\right)+\left(4^2\times\frac{5}{24}\right) = \frac{4+48+27+80}{24} = \frac{159}{24}$

$Var(X) = E(X^2) - [E(X)]^2 = \frac{159}{24} - \left(\frac{57}{24}\right)^2 = \frac{3816-3249}{576} = \frac{567}{576} = \frac{63}{64}$

c) $E(2X-1) = 2E(X) - 1 = 2\times\frac{19}{8} - 1 = \frac{30}{8} = \frac{15}{4}$

$Var(2X-1) = 2^2Var(X) = 4\times\frac{63}{64} = \frac{63}{16}$

2) a) *P(X = 1) = a, P(X = 2) = 2a, P(X = 3) = 3a. Therefore the total probability is 3a + 2a + a = 6a. This must equal 1, so* $a = \frac{1}{6}$.

[1 mark]

b) $E(X) = \left(1\times\frac{1}{6}\right)+\left(2\times\frac{2}{6}\right)+\left(3\times\frac{3}{6}\right) = \frac{1+4+9}{6} = \frac{14}{6} = \frac{7}{3}$ *[2 marks]*

c) $E(X^2) = Var(X) + [E(X)]^2 = \frac{5}{9} + \left(\frac{7}{3}\right)^2 = \frac{5+49}{9} = \frac{54}{9} = 6$ *[2 marks]*

d) $E(3X+4) = 3E(X) + 4 = 3\times\frac{7}{3} + 4 = 11$ *[1 mark]*

$Var(3X+4) = 3^2Var(X) = 9\times\frac{5}{9} = 5$ *[2 marks]*

Answers

Page 101

1) a) $P(X=2)=\binom{5}{2}0.3^2(1-0.3)^3=\frac{5!}{2!3!}\times0.09\times0.343=0.3087$

b) $P(X\le3)=P(X=0)+P(X=1)+P(X=2)+P(X=3)$

$$=\binom{5}{0}\times0.3^0\times0.7^5+\binom{5}{1}\times0.3^1\times0.7^4+\binom{5}{2}\times0.3^2\times0.7^3+\binom{5}{3}\times0.3^3\times0.7^2$$

$=0.96922$

c) $P(X<2)=P(X=0)+P(X=1)$

$$=\binom{5}{0}\times0.3^0\times0.7^5+\binom{5}{1}\times0.3^1\times0.7^4$$

$=0.52822$

d) $E(X)=np=5\times0.3=1.5$

e) $Var(X)=npq=5\times0.3\times(1-0.3)=1.05$

2) a) *This is a binomial distribution (call the random variable X) where n = 20 and p = $\frac{1}{6}$ ('success' here is rolling a 6).*

$P(X=10)=\binom{20}{10}\times\left(\frac{1}{6}\right)^{10}\times\left(\frac{5}{6}\right)^{10}=4.93\times10^{-4}$ (= 0.000493) *to 3 sig. fig.*

(It's quickest to use tables for this really.)

b) *It's definitely quicker to use tables for this one:*
$P(X\ge10)=1-P(X\le9)=1-0.999401=0.000599$

3) *This is another binomial distribution, where X is 'the number of people eating salad', n = 10 and p = 0.3. You need P(X < 3).*

$P(X<3)=P(X=0)+P(X=1)+P(X=2)$

$$=\binom{10}{0}\times0.3^0\times0.7^{10}+\binom{10}{1}\times0.3^1\times0.7^9+\binom{10}{2}\times0.3^2\times0.7^8$$

$=0.3828$ *to 4 sig. fig.*

4) a) *This is a binomial distribution. Here, n = 30 and p = $\frac{1}{7}$, so if X is the number of people born on a Saturday,* $X\sim B\left(30,\frac{1}{7}\right)$.

[1 mark]
The conditions needed for a binomial distribution are:
(i) there are a fixed number of trials,
(ii) there are just two possible outcomes of each trial,
(iii) each of the trials is independent of all the others,
(iv) the variable is the total number of successes.
[4 marks available — 1 mark for each condition]

b) *This is P(X = 5), and is given by:*

$P(X=5)=\binom{30}{5}\left(\frac{1}{7}\right)^5\left(\frac{6}{7}\right)^{25}$

$=\frac{30!}{5!25!}\left(\frac{1}{7}\right)^5\left(\frac{6}{7}\right)^{25}=0.1798$ *to 4 sig. fig.* *[2 marks]*

Page 103

1) $P(X=15)=0.9^{14}\times0.1=0.0229$ *to 3 sig. fig.*
The expected value of X is 1 ÷ 0.1 = 10.

2) a) *When you throw two dice, there are 36 possible outcomes and 6 of these will be doubles. So if X is the random variable 'number of throws till you get a double', then X ~ Geo($\frac{1}{6}$).*
[1 mark]
Then the mean number of throws needed to start the game is $1\div\frac{1}{6}=6$.
[1 mark]

b) (i) $P(X=4)=\left(\frac{5}{6}\right)^3\times\frac{1}{6}=0.0965$ *[2 marks]*

(ii) *If it takes at least 5 throws, then that means that the first four throws were unsuccessful [1 mark]. Since the probability of 'not a success' is $\frac{5}{6}$, then the required probability must be*

$\left(\frac{5}{6}\right)^4=0.482$. *[2 marks]*

c) *The probability that a player <u>has</u> started by the time he/she's had four throws is $1-\left(\frac{5}{6}\right)^4$ (using the above). Since all throws are independent, the probability that both players <u>have</u> started is $\left[1-\left(\frac{5}{6}\right)^4\right]^2$ [1 mark]. So the probability that at least one of them <u>hasn't</u> started is $1-\left[1-\left(\frac{5}{6}\right)^4\right]^2=0.732$ to 3 sig. fig. [1 mark]*

Page 105

1) a) *Normalise the probabilities and then use tables to answer these. Remember, X ~ N(50, 16) (and 16 is the variance, not the standard deviation).*

$P(X<55)=P\left(Z<\frac{55-50}{\sqrt{16}}\right)=P(Z<1.25)=0.8944$

b) $P(X<42)=P\left(Z<\frac{42-50}{\sqrt{16}}\right)=P(Z<-2)$
$=1-P(Z<2)=1-0.9772=0.0228$

c) $P(X>56)=1-P(X<56)=1-P\left(Z<\frac{56-50}{\sqrt{16}}\right)=$
$1-P(Z<1.5)=1-0.9332=0.0668$

d) $P(47<X<57)=P(X<57)-P(X<47)=P\left(Z<\frac{57-50}{\sqrt{16}}\right)-P\left(Z<\frac{47-50}{\sqrt{16}}\right)$

$=P(Z<1.75)-P(Z<-0.75)$
$=P(Z<1.75)-(1-P(Z<0.75))$
$=0.9599-(1-0.7734)$
$=0.7333$

Answers

2) a) *Here X ~ N(600, 202). You need to use your 'percentage points' table for these.*

If $P(X<a)=0.95$, *then* $P\left(Z<\frac{a-600}{\sqrt{202}}\right)=0.95$.

So $\frac{a-600}{\sqrt{202}}=1.645$ *(using the table).*

Rearrange this to get $a=600+1.645\times\sqrt{202}=623.38$

b) *|X – 600| < b means that X is 'within b' of 600, i.e. 600 – b < X < 600 + b. Since 600 is the mean of X, and since a normal distribution is symmetrical,*

$P(600-b<X<600+b)=0.8$ *means that* $P(600<X<600+b)=0.4$

i.e. $P\left(\frac{600-600}{\sqrt{202}}<Z<\frac{600+b-600}{\sqrt{202}}\right)=0.4$, *i.e.* $P\left(0<Z<\frac{b}{\sqrt{202}}\right)=0.4$

This means that $P\left(Z<\frac{b}{\sqrt{202}}\right)-P(Z<0)=0.4$

i.e. $P\left(Z<\frac{b}{\sqrt{202}}\right)-0.5=0.4$, *or* $P\left(Z<\frac{b}{\sqrt{202}}\right)=0.9$

Use your percentage points table to find that

$\frac{b}{\sqrt{202}}=1.2816$, *or* $b=1.2816\times\sqrt{202}=18.21$

3) a) *Here, the random variable X (the distribution of the marks) is distributed X ~ N(50, 30²). First, you need P(X > 41) — this will tell you the fraction of marks that are above 41.*

$P(X>41)=P\left(Z>\frac{41-50}{30}\right)=P(Z>-0.3)$

This is equal to $P(Z<0.3)=0.6179$ *(from tables). [2 marks]*

Then to estimate the number of candidates that passed the exam, multiply this by 1000 — so roughly 618 students are likely to have passed [1 mark].

b) *Let the mark required for an A-grade be k. Then since 90% of students don't get an A, P(X < k) = 0.9 [1 mark]*

Normalise this to get $P\left(Z<\frac{k-50}{30}\right)=0.9$ *[1 mark]*

Now you can use your percentage points table to get that

$\frac{k-50}{30}=1.282$, *or* $k=30\times1.282+50=88.45$ *[1 mark]. So the mark needed for an A-grade will be around 88-89 marks.*

4) *Assume that the lives of the batteries are distributed as N(μ, σ²). Then P(X < 20) = 0.4 [1 mark] and P(X < 30) = 0.8. Normalise these 2 equations to get*

$P\left(Z<\frac{20-\mu}{\sigma}\right)=0.4$ *and* $P\left(Z<\frac{30-\mu}{\sigma}\right)=0.8$ *[1 mark]*

Now you need to use your percentage points table to get:

$\frac{20-\mu}{\sigma}=-0.2533$ *and* $\frac{30-\mu}{\sigma}=0.8416$ *[2 marks]*

Now rewrite these as:

$20-\mu=-0.2533\sigma$ *and* $30-\mu=0.8416\sigma$. *[2 marks]*

Subtract these two equations to get:

$10=(0.8416+0.2533)\sigma$

i.e. $\sigma=\frac{10}{0.8416+0.2533}=9.1333$ *[1 mark]*

Now use this value of σ in one of the equations above:

$\mu=20+0.2533\times9.1333=22.31$ *[1 mark]*

So $X\sim N(22.31,\ 9.13^2)$ *i.e.* $X\sim N(22.31, 83.4)$

Page 107

1) *This is a binomial distribution, but we can use the normal approximation here, since n is very large (or np = 10 and nq = 190, both of which are bigger than 5). To find the correct normal distribution, we need the mean (= np). This is 200 × 0.05 = 10. The variance (= npq) is given by 200 × 0.05 × 0.95 = 9.5. So we need to use the approximation A ~ N(10, 9.5). We need P(A < 11) — using a continuity correction means you have to actually find P(A < 10.5). Now you can normalise as usual...*

$P(A<10.5)=P\left(Z<\frac{10.5-10}{\sqrt{9.5}}\right)=P(Z<0.1622)$

Using tables this is 0.5644

So the probability that A is less than 11 is 0.5644.

2) *This is another binomial question — here, if X is the number of plants that are short of water, X ~ B(500, 0.3). We can use a normal approximation. This is justified, since n is large (or np = 150 and nq = 350 are both very large). The mean (= np) is 500 × 0.3 = 150. The variance = npq = 500 × 0.3 × 0.7 = 105. So approximately, X ~ N(150, 105). We need P(X < 149), which, with the continuity correction, means finding P(X < 148.5). Now you can normalise as usual...*

$P(X<148.5)=P\left(Z<\frac{148.5-150}{\sqrt{105}}\right)=P(Z<-0.1464)=1-P(Z<0.1464)$

Using tables, this is $1-0.5582=0.4418$

Answers

3) *The probability of getting 2 sixes with two throws is* $\frac{1}{6}\times\frac{1}{6}=\frac{1}{36}$ *[1 mark]. So if X is the number of people who get 2 sixes,* $X \sim B\left(200, \frac{1}{36}\right)$ *[1 mark]. Once again, you can use a normal approximation, since n is so large (or because* $200\times\frac{1}{36}=5.6$ *and* $200\times\frac{35}{36}=194.4$ *are both bigger than 5) [1 mark]. The mean of X is np = 5.556, and the variance is* $200\times\frac{1}{36}\times\frac{35}{36}=5.401$.

So the right approximation is X ~ N(5.556, 5.401) [1 mark]. You need $P(X\geq 10)$*, so with the continuity correction this is* $P(X>9.5) = 1 - P(X<9.5)$ *[1 mark].*
Now normalise as usual...

$$1-P(X<9.5)=1-P\left(Z<\tfrac{9.5-5.556}{\sqrt{5.401}}\right)=1-P(Z<1.697)$$

Using tables, this is $1-0.9552=0.0448$ *[1 mark]*

4) a) *If X is the random variable distributed as the weight of the bags of sweets, then* $X \sim N(30, 5^2) = N(30, 25)$*, and so for a sample of 50 bags, the mean is distributed as* $\bar{X}\sim N\left(30, \frac{25}{50}\right)=N(30, 0.5)$. *[1 mark]*

b) *Once again, just normalise...*

$$P(\bar{X}<29)=P\left(Z<\tfrac{29-30}{\sqrt{0.5}}\right)=P(Z<-1.414)$$
$$=1-P(Z<1.414)=1-0.9213=0.0787 \text{ [2 marks]}$$

c) *This time, we have a sample of size n, so the distribution of the sample mean is* $\bar{X}\sim N\left(30,\frac{25}{n}\right)$ *[1 mark]. We need n big enough that* $P(\bar{X}<29)=0.05$ *[1 mark]*

Normalise again...

$$P(\bar{X}<29)=P\left(Z<\frac{29-30}{\sqrt{25/n}}\right)=0.05$$

Use your tables to find that $\frac{29-30}{\sqrt{25/n}}=-1.645$ *[1 mark]*

So $\sqrt{\frac{25}{n}}=0.6079$*, and* $\frac{25}{n}=0.3695$,

which means that $n=67.7$ *[1 mark]*

So if we choose n to be 68 or more, we should be all right [1 mark].

Page 109

1) a) $\bar{X}$ = *(35.7 + 38.5 + 40.3 + 39.4 + 37.2 + 32.1 + 30.4 + 36.3) ÷ 8 = 289.9 ÷ 8 = 36.2375*

For 95% confidence interval, z = 1.9600 (as on p.110), hence the confidence interval is given by:

$$\left(\bar{X}-z\left(\frac{\sigma}{\sqrt{n}}\right), \bar{X}+z\left(\frac{\sigma}{\sqrt{n}}\right)\right)$$
$$=\left(36.2375-1.96\left(\frac{3.4}{\sqrt{8}}\right), 36.2375+1.96\left(\frac{3.4}{\sqrt{8}}\right)\right)$$

= (33.88, 38.59) each to 2 d.p.

b) *The value of 35 grams for the population mean is within this confidence interval, so there is insufficient evidence to reject the claim.*

2) a) *For 90% confidence interval, P(-z < Z < z) = 0.90, hence P(Z < z) = 0.95, and from Z-tables z = 1.645) [1 mark], hence the confidence interval is given by:*

$$\left(\bar{X}-z\left(\frac{s}{\sqrt{n}}\right), \bar{X}+z\left(\frac{s}{\sqrt{n}}\right)\right) \text{ [1 mark]}$$
$$=\left(4.5-1.6449\left(\frac{3.2}{\sqrt{50}}\right), 4.5+1.6449\left(\frac{3.2}{\sqrt{50}}\right)\right) \text{ [1 mark]}$$

= (3.76, 5.24) each to 2 d.p. [1 mark]

b) *The sample size is large [1 mark], so the Central Limit Theorem applies, i.e. the distribution of the sample mean is approximately normal [1 mark].*

Section Twelve — Correlation and Regression

Page 111

1)

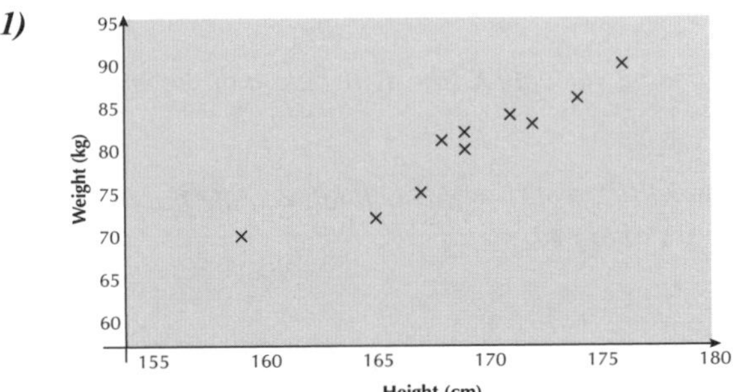

I've called the height x and the weight y.
You need to work out these sums:

$\sum x=1690$, $\sum y=803$, $\sum x^2=285818$, $\sum y^2=64835$, $\sum xy=135965$.

$$\text{Then } r=\frac{135965-\frac{1690\times803}{10}}{\sqrt{\left(285818-\frac{1690^2}{10}\right)\left(64835-\frac{803^2}{10}\right)}}=\frac{258}{\sqrt{208\times354.1}}=0.951$$

This value is very close to 1, which tells you that the variables 'height' and 'weight' have a high positive correlation — as the height increases, generally the weight increases as well.

Answers

2)

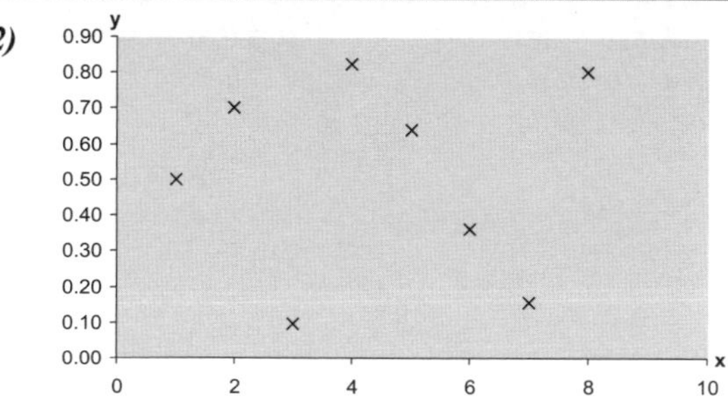

[Up to 2 marks available]
Again, you need to work out these sums:

$\sum x = 36$, $\sum y = 4.08$, $\sum x^2 = 204$, $\sum y^2 = 2.6272$, $\sum xy = 18.36$.
[5 marks available — 1 for each correct sum]

$$\text{Then } r = \frac{18.36 - \frac{36 \times 4.08}{8}}{\sqrt{\left(204 - \frac{36^2}{8}\right)\left(2.6272 - \frac{4.08^2}{8}\right)}} = \frac{0}{\sqrt{42 \times 0.5464}} = 0 \quad \text{[1 mark]}$$

This value of zero for the correlation tells you that there appears to be no linear relationship between the two variables [1 mark].

Page 113

1) *It's best to draw a table like the one below:*

Physics	54	34	23	58	52	58	13	65	69	52
English	16	73	89	81	23	81	56	62	61	37
Physics rank	5	8	9	3.5	6.5	3.5	10	2	1	6.5
English rank	10	4	1	2.5	9	2.5	7	5	6	8
d	-5	4	8	1	-2.5	1	3	-3	-5	-1.5
d^2	25	16	64	1	6.25	1	9	9	25	2.25

$\sum d^2 = 158.5$

$$\text{Then } r_s = 1 - \frac{6\sum d^2}{n(n^2-1)} = 1 - \frac{6 \times 158.5}{10 \times (10^2 - 1)} = 1 - \frac{951}{990} = 0.0394 \text{ to 3 sig. fig.}$$

This is quite low, so it appears that doing well in Physics is no indication as to how well you will do in English and vice versa.

2) a) *First work out S_{xy}, S_{xx} and S_{yy}:*

$$S_{xy} = 26161 - \frac{386 \times 460}{8} = 3966 \quad \text{[1 mark]}$$

$$S_{xx} = 25426 - \frac{386^2}{8} = 6801.5 \quad \text{[1 mark]}$$

$$S_{yy} = 28867 - \frac{460^2}{8} = 2417 \quad \text{[1 mark]}$$

$$\text{Then } r = \frac{S_{xy}}{\sqrt{S_{xx}S_{yy}}} = \frac{3966}{\sqrt{6801.5 \times 2417}} = 0.978 \text{ to 3 sig. fig. [2 marks]}$$

b) *Since r is very close to 1, the quantities x and y are very closely positively correlated. This means that the more money spent on advertising one of the products, the higher the sales tend to be. [2 marks]*

Page 115

1) a)

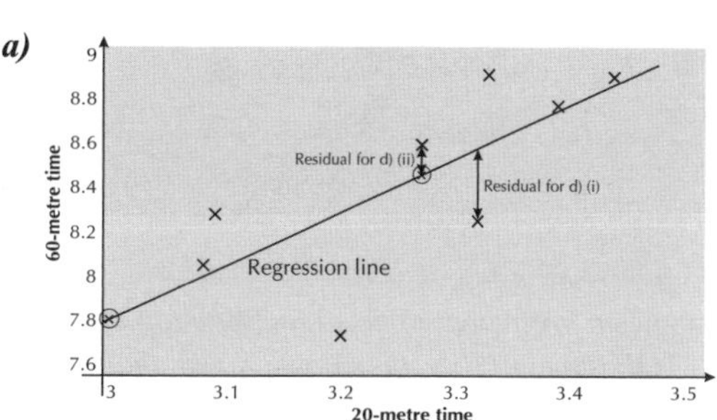

[3 marks available for the scatter diagram]

b) *It's best to make a table like this one, first:*

									Totals
20-metre time, *x*	3.39	3.2	3.09	3.32	3.33	3.27	3.44	3.08	26.12
60-metre time, *y*	8.78	7.73	8.28	8.25	8.91	8.59	8.9	8.05	67.49
x^2	11.4921	10.24	9.5481	11.0224	11.0889	10.6929	11.8336	9.4864	85.4044
y^2	77.0884	59.7529	68.5584	68.0625	79.3881	73.7881	79.21	64.8025	570.6509
xy	29.7642	24.736	25.5852	27.39	29.6703	28.0893	30.616	24.794	220.645

[2 marks available for at least 3 correct totals on the right-hand side]

$$\text{Then: } S_{xy} = 220.645 - \frac{26.12 \times 67.49}{8} = 0.29015 \quad \text{[1 mark]}$$

$$S_{xx} = 85.4044 - \frac{26.12^2}{8} = 0.1226 \quad \text{[1 mark]}$$

Then the gradient b is given by: $b = \frac{S_{xy}}{S_{xx}} = \frac{0.29015}{0.1226} = 2.3666$

[1 mark]
And the intercept a is given by:

$$a = \bar{y} - b\bar{x} = \frac{\sum y}{n} - b\frac{\sum x}{n} = \frac{67.49}{8} - 2.3666 \times \frac{26.12}{8} = 0.709 \quad \text{[1 mark]}$$

So the regression line has equation: y = 2.367x + 0.709
[1 mark]
To plot the line, find two points that the line passes through. A regression line always passes through $(\bar{x}, \bar{y})$, *which here is equal to (3.27, 8.44). Then put x = 3 (say) to find that the line also passes through (3, 7.81).*
Now plot these points (in circles) on your scatter diagram, and draw the regression line through them [1 mark for plotting the line correctly]

c) **(i)** *x = 3.15, y = 2.367 × 3.15 + 0.709 = 8.17 (to 3 sig. fig.)*
[2 marks] — this should be reliable, since we are using interpolation in a known region
[1 mark].
(i) *x = 3.88, y = 2.367 × 3.88 + 0.709 = 9.89 (to 3 sig. fig.)*
[2 marks] — this could be unreliable, since we are using extrapolation [1 mark].

d) **(i)** *x = 3.32, residual = 8.25 – (2.367 × 3.32 + 0.709)*
= –0.317 (3 sig. fig.)
[1 mark for calculation, 1 mark for plotting residual correctly]
(ii) *x = 3.27, residual = 8.59 – (2.367 × 3.27 + 0.709)*
= 0.141 (3 sig. fig.)
[1 mark for calculation, 1 mark for plotting residual correctly]

Answers

Section Thirteen — Kinematics

Page 117

1) $u = 3$; $v = 9$; $a = a$; $s = s$; $t = 2$

Use $s = ½(u + v)t$

$s = ½(3 + 9) \times 2$

$s = ½(12) \times 2 = 12$ m

2) $u = 3$; $v = 0$; $a = -9.8$; $s = s$; $t = t$

[v = 0 because when projected objects reach the top of their motion, they STOP momentarily, then come down again.]

[$a = -9.8$ ms^{-2} because gravity will SLOW the ball down, so it's negative.]

Use $v = u + at$

$0 = 3 + (-9.8)t$

$0 = 3 - 9.8t$

$9.8t = 3$

$$t = \frac{3}{9.8} = 0.31 \text{ s}$$

3) a) $u = u$; $v = v$; $a = 10$; $s = 4h$; $t = 1.2$

You need an equation with u and s because they're in the equation you have to show is true. You know t and a, so the equation to use is one with u, a, s and t in it – that's $s = ut + ½ at^2$.

$4h = 1.2u + ½ \times 10 \times 1.2^2$ [1 mark]

$4h = 1.2u + 7.2$ [1 mark]

b) i) *You have to think of the entire fall through 8 floors, because the question demands we have u involved.*

$u = u$; $v = v_1$; $a = 10$; $s = 8h$; $t = 1.8$

[s = 8h because it's 4 floors + 4 floors]

[t = 1.8 because it's 1.2 + 0.6]

Use $s = ut + ½ at^2$ [1 mark]

$8h = 1.8u + ½ \times 10 \times 1.8^2$

$8h = 1.8u + 16.2$ [1 mark]

ii) *You now have two equations, both with h and u in, so solve simultaneously.*

The easiest way to do this is to double the equation from a):

$8h = 2.4u + 14.4$

and subtract the equation from b)i) from this:

$0 = 0.6u - 1.8$

$0.6u = 1.8$

$u = 3$ ms^{-1} [1 mark]

Substitute this value back into the equation from a):

$4h = 1.2 \times 3 + 7.2$

$4h = 10.8$

$h = 2.7$ m [1 mark]

c) Modelling assumption made is that the sandwich is a particle, so there's no air resistance on it. [1 mark]

Page 119

1)

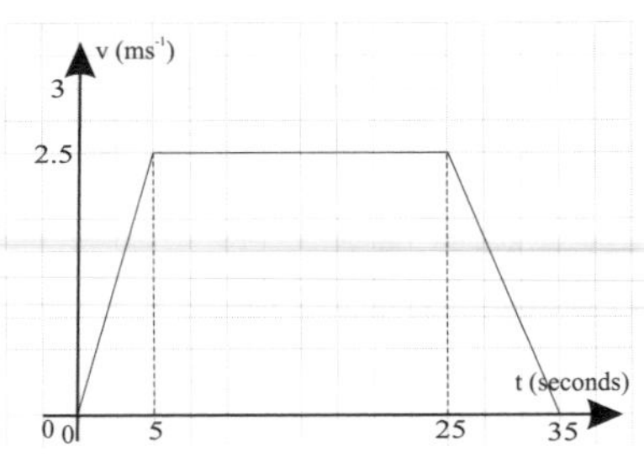

distance = $(5 \times 2.5) \div 2 + (20 \times 2.5) + (10 \times 2.5) \div 2$

$= 68.75$ m

2) velocity = area under (t,a) graph

a) $t = 3$, area $= (3 \times 5) \div 2 = 7.5$ ms^{-1}

b) $t = 5$, area $= 7.5 + (2 \times 5) = 17.5$ ms^{-1}

c) $t = 6$, area $= 17.5 + (1 \times 5) \div 2 = 20$ ms^{-1}

3) a) Between $t = 10$ and $t = 15$. The graph is horizontal at this point. [1 mark]

b) $50 \div 10 = 5$ ms^{-1} [2 marks]

c) 50 m ÷ 20 s = 2.5 ms^{-1} back towards the starting point. [2 marks]

Page 121

1) a) Multiplying out gives $v = 15t^2 - 2t^3$

Differentiating v with respect to t gives $a = 30t - 6t^2$ [1 mark]

If $a = 0$:

$30t - 6t^2 = 0$

$6t(5 - t) = 0$

So $t = 0$ or $t = 5$ [2 marks]

b) Integrating v with respect to t between $t = 0$ and $t = 4$ gives r, the distance the particle has moved.

$$r = \int_0^4 (15t^2 - 2t^3)dt = \left[5t^3 - \frac{1}{2}t^4\right]_0^4 \quad \text{[2 marks]}$$

so $r = 320 - 128 = 192$ [2 marks]

2) a) Integrating v gives the position vector $\mathbf{r}$:

$$\mathbf{r} = \begin{bmatrix} 3t^2 - 2t + c_1 \\ 5t + c_2 \end{bmatrix} \quad \text{[2 marks]}$$

When $t = 0$, $\mathbf{r} = \begin{bmatrix} 4 \\ 3 \end{bmatrix}$ [1 mark]

so $c_1 = 4$ and $c_2 = 3$, giving $\mathbf{r} = \begin{bmatrix} 3t^2 - 2t + 4 \\ 5t + 3 \end{bmatrix}$ [2 marks]

b) Differentiating v with respect to t gives $\mathbf{a} = \begin{bmatrix} 6 \\ 0 \end{bmatrix}$ [2 marks]

Answers

Section Fourteen — Vectors

Page 123

1) *Displacement = (15 × 0.25) – (10 × 0.75) = -3.75 km*
Time taken = 1 hour
Average velocity = -3.75 km/h (i.e. 3.75 kmh^{-1} south)

2) $(3\mathbf{i} + 7\mathbf{j}) + 2 \times (-2\mathbf{i} + 2\mathbf{j}) - 3 \times (\mathbf{i} - 3\mathbf{j})$
$= (3 - 4 - 3)\mathbf{i} + (7 + 4 + 9)\mathbf{j}$
$= -4\mathbf{i} + 20\mathbf{j}$

3) *Horizontally:* $0 + 5\cos 30 = 4.33$ N *[1 mark]*
Vertically: $4 - 5\sin 30 = 1.5$ N *[1 mark]*

$\theta = \tan^{-1}\left(\frac{1.5}{4.33}\right) = 19.1°$

i.e. θ = 19.1° above the horizontal [1 mark]

Magnitude $= \sqrt{1.5^2 + 4.33^2} = 4.58$ N *[2 marks]*

Page 125

1)

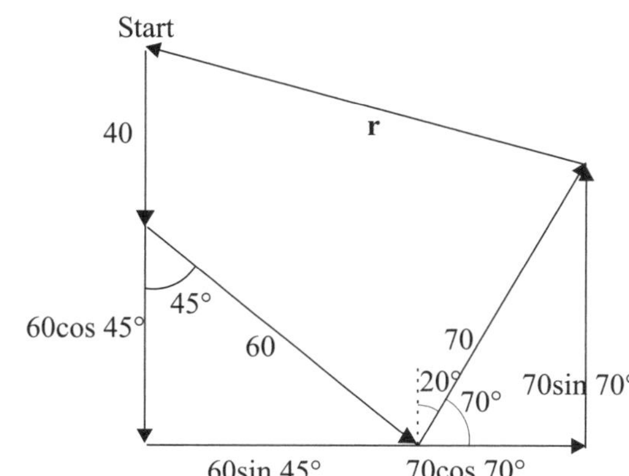

Resolving East: $60\sin 45° + 70\cos 70° = 66.4$ *miles*
Resolving North: $-40 - 60\cos 45° + 70\sin 70°$
$= -16.6$ *miles*

Magnitude of **r** $= \sqrt{66.4^2 + 16.6^2} = 68.4$ *miles*

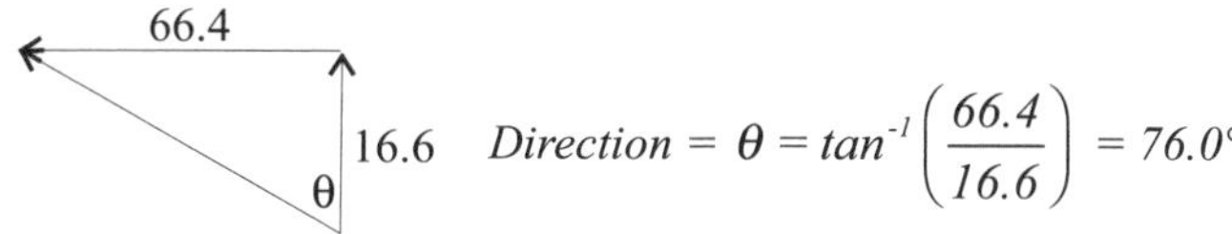

Direction $= \theta = \tan^{-1}\left(\frac{66.4}{16.6}\right) = 76.0°$

Bearing is 360° – 71.8° = 284°

2)

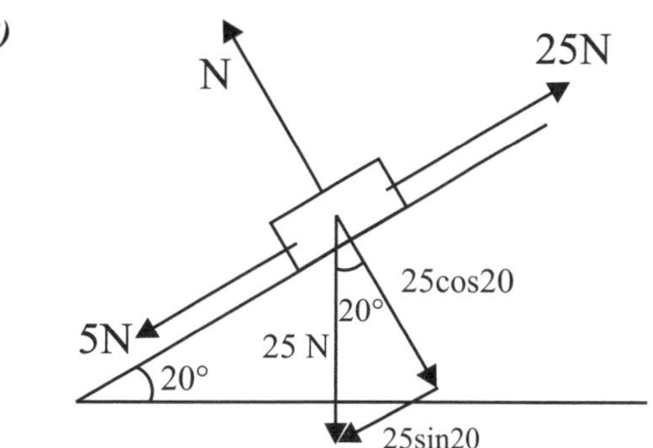

Force perpendicular to the slope: $N = 25\cos 20° = 23.5$ N
Force parallel to the slope: $25 - 25\sin 20° - 5 = 11.4$ N.
So the resultant force is 11.4 N up the slope.

3)

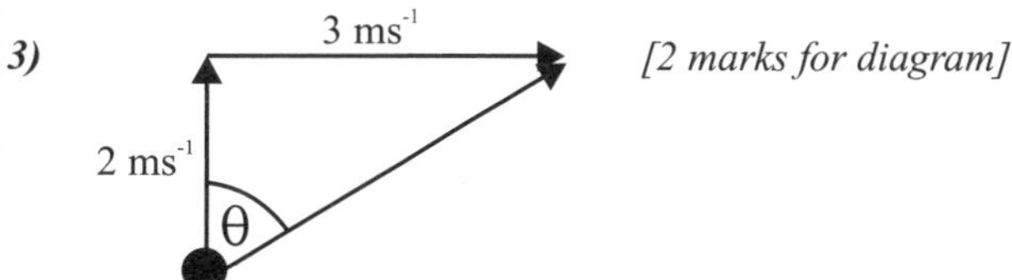

[2 marks for diagram]

Magnitude $= \sqrt{2^2 + 3^2} = \sqrt{13} = 3.61$ ms^{-1} *[2 marks]*

$\theta = \tan^{-1}\left(\frac{3}{2}\right) = 56.3°$ *[1 mark]*

So angle to river bank is 90° – 56.3° = 33.7° [1 mark]

Section Fifteen — Statics

Page 127

1) a) W *Small point mass, no air resistance, no wind, released from rest*

b) W *Small point mass, no air resistance, no wind, released from rest*
The drinks can is very smooth so ignoring air resistance is OK.

c) W *Same assumptions as in a) and b), (although realistically it might not be safe to ignore wind if you're outside).*

2) D R F W *Assumptions: Point mass, one point of contact to ground, constant driving force D from engine, constant friction, F includes road resistance and air resistance, acceleration = 0 as it's moving at 25mph.*

3) f W *Assumptions: Point mass, can ignore friction (air) upwards as parachute isn't open yet.*

Answers

Page 129

1) a)

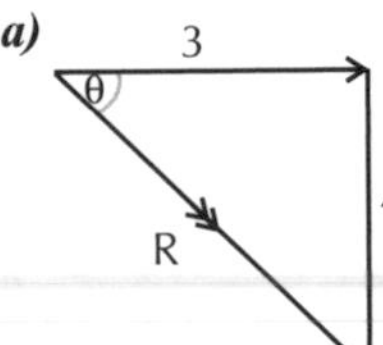

$R = \sqrt{4^2 + 3^3} = 5 \text{ N}$

$\tan\theta = \dfrac{4}{3}$

$\theta = 53.1^\circ$

b)

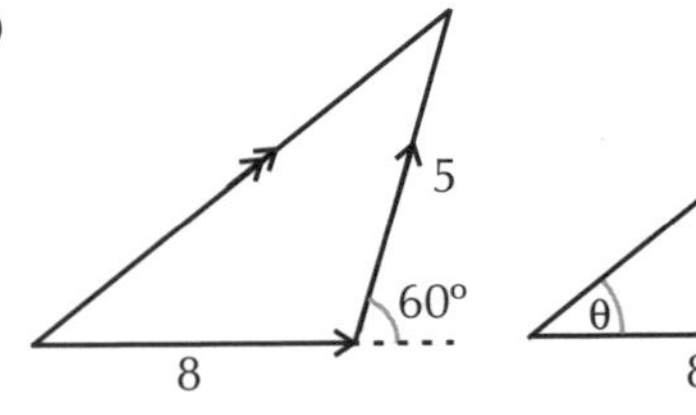

$R = \sqrt{(8 + 5\cos 60)^2 + (5\sin 60)^2} = 11.4 \text{ N}$

$\tan\theta = \dfrac{5\sin 60}{8 + 5\cos 60} = 0.412 \quad \text{so} \quad \theta = 22.4^\circ$

c) *Total force up = 6 – 4sin10° – 10sin20°*
= 1.885 N
Total force left = 10cos20° – 4cos10°
= 5.458 N

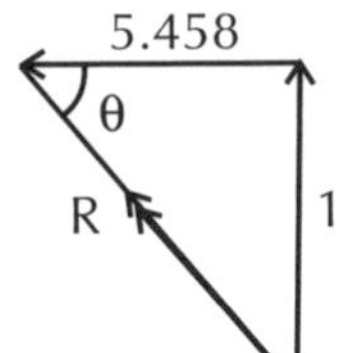

$R = \sqrt{1.885^2 + 5.458^2} = 5.77 \text{ N}$

$\theta = \tan^{-1}\dfrac{1.885}{5.458} = 19.1^\circ$

2) a) *[1 mark]*

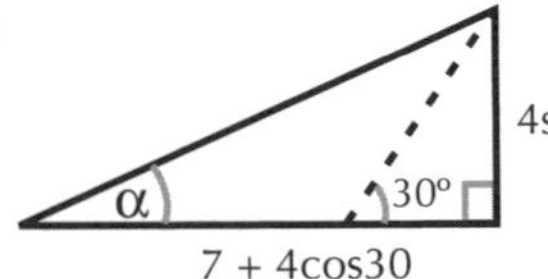

$R = \sqrt{(4\sin 30^\circ)^2 + (7 + 4\cos 30^\circ)^2}$ *[1 mark]*

$= 10.7 \text{ N}$ *[1 mark]*

b) $\tan\alpha = \dfrac{4\sin 30^\circ}{7 + 4\cos 30^\circ} = 0.1911$ *[1 mark for tanα, 1 mark for the rest]*

$\alpha = 10.8^\circ$ *[1 mark]*

3)

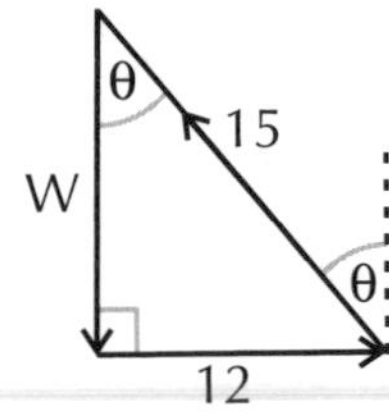

a) $\sin\theta = \dfrac{12}{15}$ *[1 mark]*

so $\theta = 53.1^\circ$ *[1 mark]*

b) $15^2 = W^2 + 12^2$ *[1 mark]*

$W = \sqrt{15^2 - 12^2} = 9 \text{ N}$ *[1 mark]*

Remove W and the particle moves in the opposite direction to W, i.e. upwards. This resultant of the two remaining forces is 9 N [1 mark] upwards [1 mark].

Page 131

1)

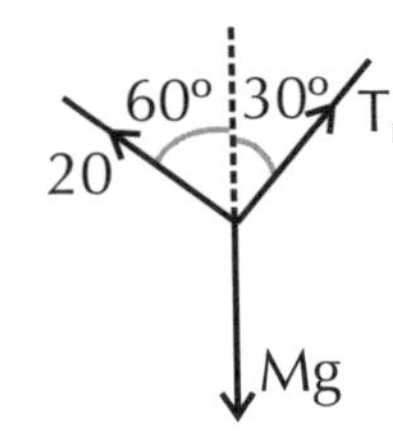

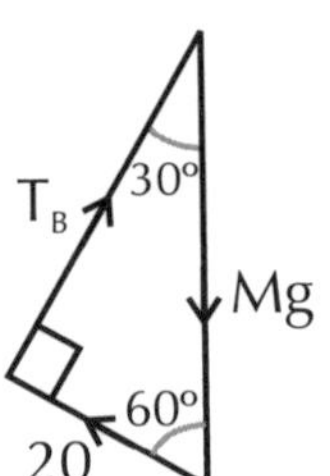

a) $\tan 30 = \dfrac{20}{T_B}$

$T_B = \dfrac{20}{\tan 30}$

$= 34.6 \text{ N}$

b) $\sin 30 = \dfrac{20}{Mg}$

$Mg = \dfrac{20}{\sin 30}$

$M = 4.08 \text{ kg}$

2)

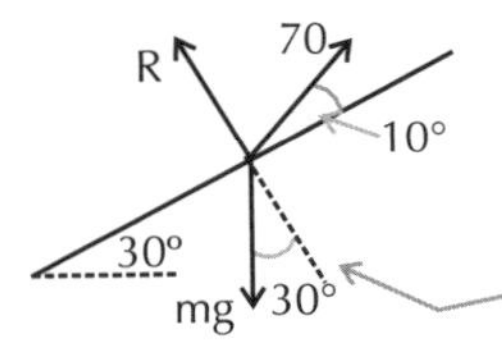

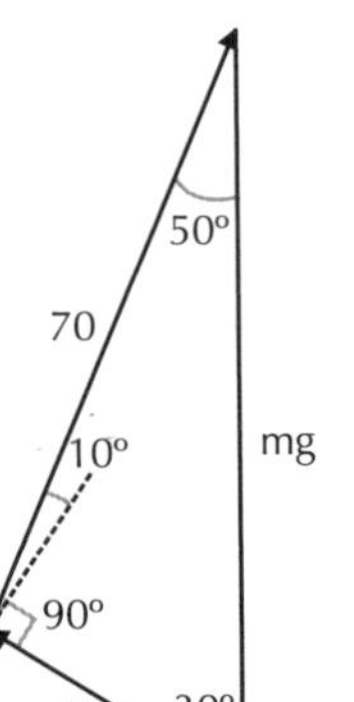

Sine rule: $\dfrac{mg}{\sin 100} = \dfrac{70}{\sin 30}$

So $mg = \dfrac{70\sin 100}{\sin 30}$

$m = 14.1 \text{ kg}$

$\dfrac{R}{\sin 50} = \dfrac{70}{\sin 30}$

So $R = \dfrac{70\sin 50}{\sin 30} = 107 \text{ N}$

Answers

3)

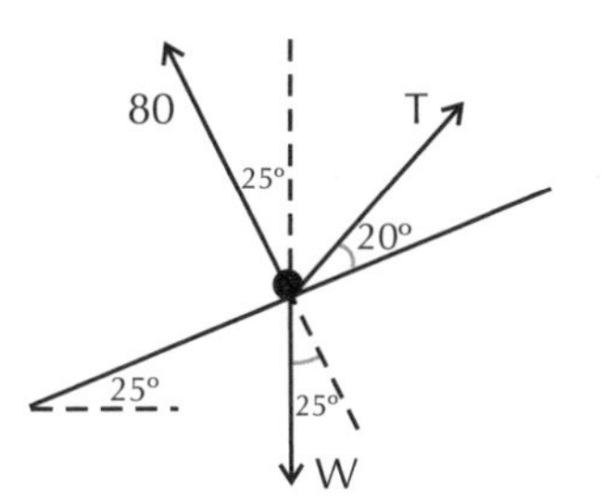

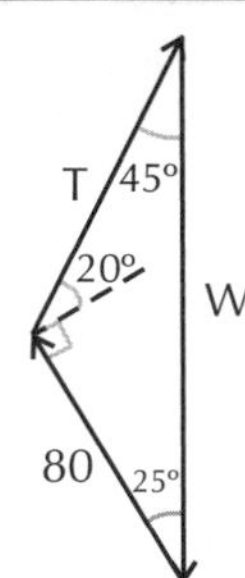

[2 marks for method and diagram]

a) $\dfrac{T}{\sin 25°} = \dfrac{80}{\sin 45°}$ *[1 mark]*

So $T = \dfrac{80 \sin 25°}{\sin 45°} = 47.8 \text{ N}$ *[1 mark]*

b) $\dfrac{W}{\sin 110°} = \dfrac{80}{\sin 45°}$ *[1 mark]*

So $W = \dfrac{80 \sin 110°}{\sin 45°} = 106 \text{ N}$ *[1 mark]*

Page 133

1)

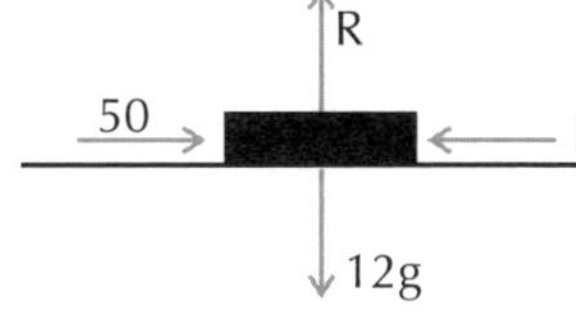

Resolve vertically: $R = 12g$

Use formula: $F \leq \mu R$

$F \leq \frac{1}{2}(12g)$

$F \leq 58.8 \text{ N}$

50 N isn't big enough to overcome friction — so it doesn't move.

2) *Force would have to be > 58.8 N*

3) tan α = μ = 0.2 *[1 mark]*
So α = $\tan^{-1}$ 0.2 = 11.3° *[1 mark]*

4) *Resolve horizontally*: $S\cos 40° = F$ *[1 mark]*

Resolve vertically: $R = 2g + S \sin 40°$ *[1 mark]*

It's limiting friction so $F = \mu R$ *[1 mark]*

So $S\cos 40° = \frac{3}{10}(2g + S\sin 40°)$

$S\cos 40° = 0.6g + 0.3S\sin 40°$

$S\cos 40° - 0.3S\sin 40° = 0.6g$

$S(\cos 40° - 0.3\sin 40°) = 0.6g$ *[1 mark]*

$S = 10.3 \text{ N}$ *[1 mark]*

Page 135

1) *Moments about B:* $60g \times 3 = T_2 \times 8$

So $T_2 = \dfrac{180g}{8} = 220.5 \text{ N}$

Vertically balanced forces, so $T_1 + T_2 = 60g$

$T_1 = 367.5 \text{ N}$

2)

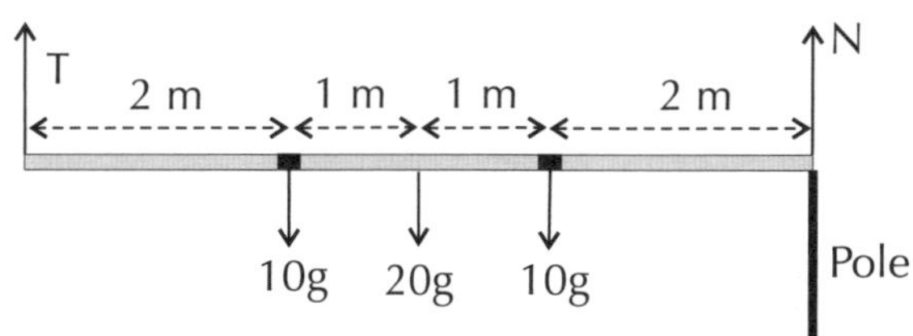

a) Take moments about where the beam is resting on the pole:
6T = (2 × 10g) + (3 × 20g) + (4 × 10g) *[2 marks]*
6T = 20g + 60g + 40g = 120g ⇒ T = 20g *[1 mark]*

b) Resolve vertically:
T + N = 10g + 20g + 10g *[1 mark]*
20g + N = 40g ⇒ N = 20g *[1 mark]*

Section Sixteen — Dynamics

Page 137

1)

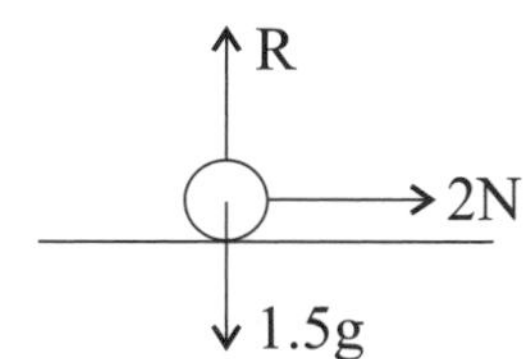

Resolve horizontally: $F_{net} = ma$

$2 = 1.5a$ *so* $a = 1\frac{1}{3}$ ms^{-2}

$v = u + at$

$v = 0 + (1\frac{1}{3} \times 3) = 4$ ms^{-1}

Answers

2) $(24\mathbf{i} + 18\mathbf{j}) = \begin{pmatrix} 24 \\ 18 \end{pmatrix}$

$(6\mathbf{i} + 22\mathbf{j}) = \begin{pmatrix} 6 \\ 22 \end{pmatrix}$

$$F_{net} = \begin{pmatrix} 24 \\ 18 \end{pmatrix} + \begin{pmatrix} 6 \\ 22 \end{pmatrix} = \begin{pmatrix} 30 \\ 40 \end{pmatrix} = 30\mathbf{i} + 40\mathbf{j}$$

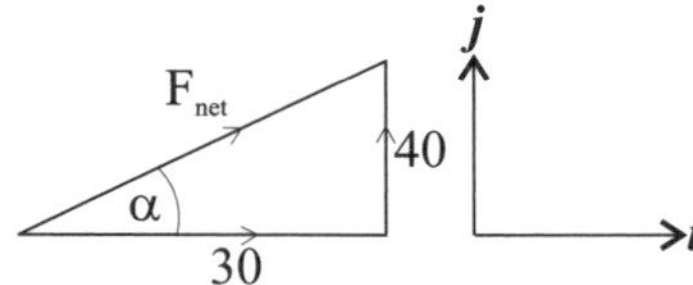

$\tan\alpha = \frac{40}{30}$

So $\alpha = 53.1°$

$F_{net} = \sqrt{30^2 + 40^2} = 50$ N

Resolve in direction of F_{net}:

$F_{net} = ma$

$50 = 8a$ so $a = 6.25$ ms^{-2}

$s = ut + \frac{1}{2}at^2$

$s = 0 \times 3 + \frac{1}{2} \times 6.25 \times 3^2 = 28.1$ m

3)

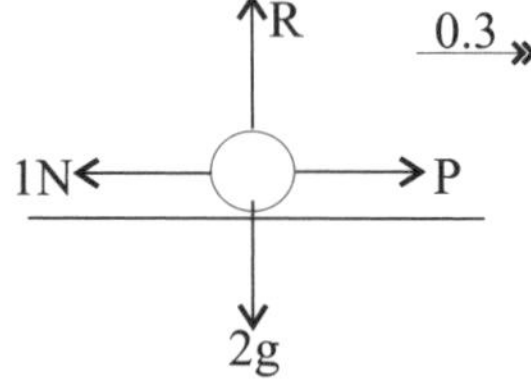

Resolve horizontally: $F_{net} = ma$

$P - 1 = 2 \times 0.3$

$P = 1.6$ N

Resolve vertically: $R = 2g$

Limiting friction: $F = \mu R$

$1 = \mu \times 2g$

So $\mu = 0.05$ (to 2 d.p.)

4) a)

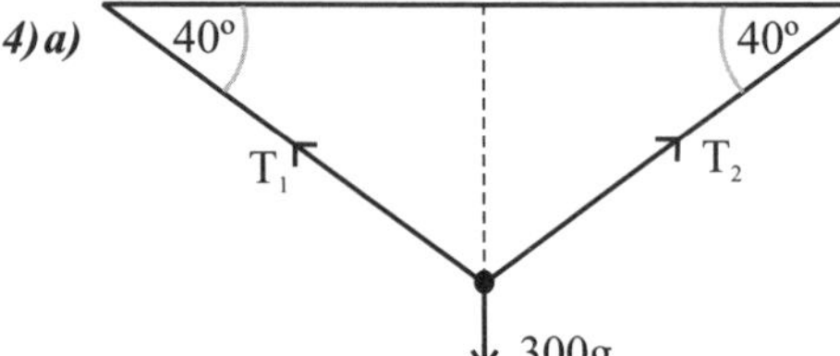

Constant velocity, so $a = 0$

Resolve horizontally: $F_{net} = ma$

$T_2\cos40 - T_1\cos40 = 300 \times 0$

$T_2\cos40 = T_1\cos40$

$T_2 = T_1$ *(1 mark)*

Resolve vertically: $F_{net} = ma$

$T_1\sin40 + T_2\sin40 - 300g = 300 \times 0$ *(1 mark)*

Let $T_1 = T_2 = T$: $2T\sin40 = 300g$ *(1 mark)*

$T = 2290$ N (to 3 s.f.) *(1 mark)*

b) *Resolve horizontally:* $F_{net} = ma$

$T_2\cos40 - T_1\cos40 = 300 \times 0.4$ *(1 mark)*

$T_2 - T_1 = 156.65$ N ① *(1 mark)*

Resolve vertically: $F_{net} = ma$

$T_1\sin40 + T_2\sin40 - 300g = 300 \times 0$ *(1 mark)*

$T_1\sin40 + T_2\sin40 = 300g$

So $T_1 + T_2 = 4573.83$ N ② *(1 mark)*

from ①: $T_2 = T_1 + 156.65$

into ②: $T_1 + (T_1 + 156.65) = 4573.83$

so $2T_1 = 4417.17$

$T_1 = 2210$ N (to 3 s.f.) *(1 mark)*

So $T_2 = 2370$ N (to 3 s.f.) *(1 mark)*

c) *Modelling assumptions: particle is considered as a point mass, there's no air resistance, it's a constant acceleration, etc.*
(1 mark each for any 2)

Page 139

1)

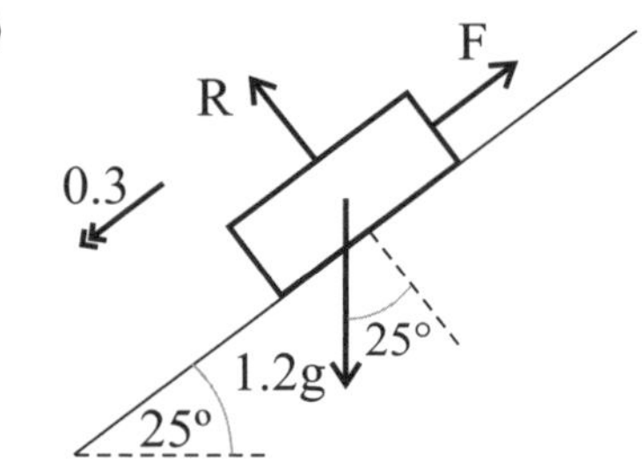

Resolving in ↖ *direction:* $F_{net} = ma$

$R - 1.2g\cos25° = 1.2 \times 0$

$R = 1.2g\cos25°$

$R = 10.66$ N

Resolving in ↙ *direction:* $F_{net} = ma$

$1.2g\sin25° - F = 1.2 \times 0.3$

So $F = 1.2g\sin25 - 1.2 \times 0.3$

$= 4.61$ N

Limiting friction, so: $F = \mu R$

$4.61 = \mu \times 10.66$

$\mu = 0.43$ (to 2 d.p.)

Assumptions: any two from: i) brick slides down line of greatest slope; ii) acceleration is constant; iii) no air resistance; iv) point mass / particle

Answers

2)

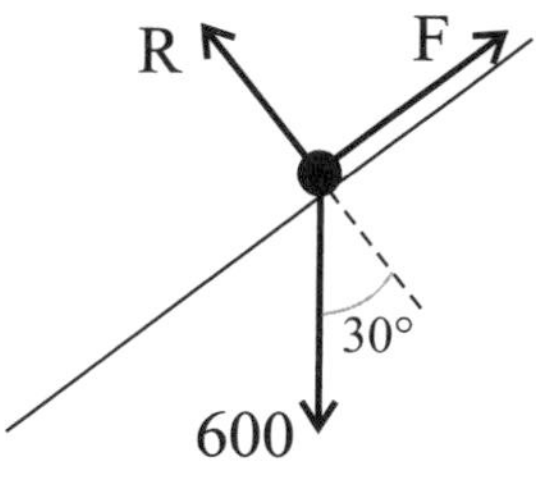

m = 0.5

Resolving in ↖ *direction:* $F_{net} = ma$

$$R - 600\cos 30° = \left(\frac{600}{g}\right) \times 0$$

$$R = 600\cos 30°$$

Sliding, so $F = \mu R$

$$F = 0.5 \times 600\cos 30° = 259.8$$

Resolving in ↙ *direction:* $600\sin 30° - F = \left(\frac{600}{g}\right)a$

$$600\sin 30 - 259.8 = 61.22 \times a$$

$$a = 0.656\ ms^{-2}$$

$u = 0$, $s = 20$, $a = 0.656$, $v = ?$

$$v^2 = u^2 + 2as$$

$$v^2 = 0^2 + 2 \times 0.656 \times 20$$

$$v^2 = 26.26$$

$$v = 5.12\ ms^{-1}$$

3) a)

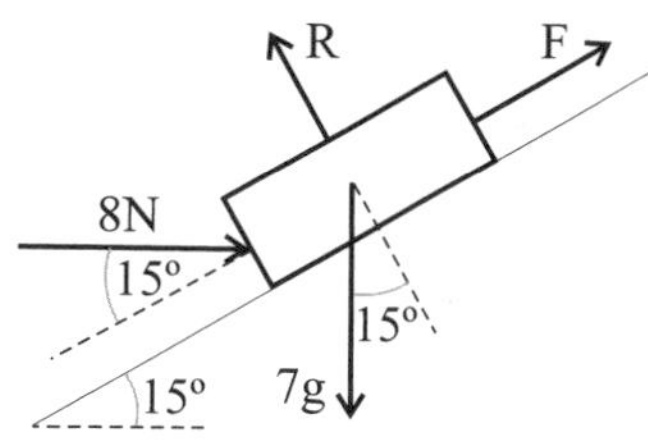

$$F_{net} = ma$$

Resolving in ↗ *direction:* $8\cos 15 + F - 7g\sin 15 = 7 \times 0$

$$F = 7g\sin 15 - 8\cos 15$$

$$= 10.03\ N \quad [2\ marks]$$

Resolving in ↖ *direction:* $F_{net} = ma$

$$R - 8\sin 15 - 7g\cos 15 = 7 \times 0 \quad [1\ mark]$$

$$R = 8\sin 15 + 7g\cos 15$$

$$= 68.33\ N \quad [1\ mark]$$

Limiting friction: $F = \mu R$

$$10.03 = \mu \times 68.33$$

$$\mu = 0.15 \quad [2\ marks]$$

b) *8 N removed:*

Resolving in ↙ *direction:* $7g\sin 15 - F = 7a$ ① *[1 mark]*

Resolving in ↖ *direction:* $R - 7g\cos 15 = 7 \times 0$

$$R = 7g\cos 15$$

$$= 66.26\ N \quad [1\ mark]$$

$$F = \mu R$$

$$F = 0.15 \times 66.26 = 9.939\ N \quad [2\ marks]$$

①: $7g\sin 15° - 9.939 = 7a$

$a = \frac{7.82}{7} = 1.12\ ms^{-2}$ (to 2d.p.) *[1 mark]*

$s = 3$; $u = 0$; $a = 1.12$; $t = ?$

$s = ut + \frac{1}{2}at^2$

$3 = 0 + \frac{1}{2} \times 1.12 \times t^2$

$t = \sqrt{\frac{6}{1.12}} = 2.3\ s$ *[2 marks]*

Answers

Page 141

1) a) $\mathbf{u} = \begin{pmatrix} 0 \\ -6 \end{pmatrix}$; $\mathbf{v} = \begin{pmatrix} 8 \\ 0 \end{pmatrix}$; $t = 20$; $\mathbf{a} = ?$

$\mathbf{v} = \mathbf{u} + \mathbf{a}t$

$\begin{pmatrix} 8 \\ 0 \end{pmatrix} = \begin{pmatrix} 0 \\ -6 \end{pmatrix} + 20\mathbf{a}$

$\mathbf{a} = \begin{pmatrix} 0.4 \\ 0.3 \end{pmatrix} = 0.4\mathbf{i} + 0.3\mathbf{j}$

$a = \sqrt{0.3^2 + 0.4^2} = 0.5 \text{ ms}^{-2}$

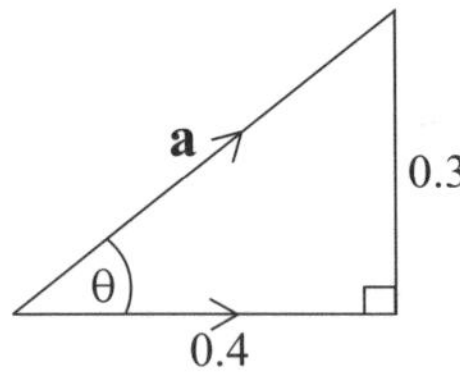

$\theta = \tan^{-1}\left(\frac{0.3}{0.4}\right) = 36.9°$

Bearing = 90° – 36.9° = 053.1°
So $\mathbf{a} = 0.5 \text{ ms}^{-2}$ on bearing 053.1°

b) $\mathbf{s} = \mathbf{u}t + \frac{1}{2}\mathbf{a}t^2$

$\begin{pmatrix} x \\ y \end{pmatrix} = \begin{pmatrix} 0 \\ -6 \end{pmatrix} 5 + \frac{1}{2}\begin{pmatrix} 0.4 \\ 0.3 \end{pmatrix} 5^2 = \begin{pmatrix} 5 \\ -26 \end{pmatrix}$ (to 2 s.f.)

At $t = 5$, $P(5, -26)$

2) a) $\mathbf{r} = \begin{pmatrix} 15t + 10 \\ 15\sqrt{3}t - 5t^2 \end{pmatrix}$ $t > 1$

Plug in $t = 3$: $\mathbf{r} = \begin{pmatrix} 55 \\ 32.9 \end{pmatrix}$ [1 mark]

So $d = \sqrt{32.9^2 + 55^2} = 64$ m (to 2 s.f.) [1 mark]

b)

t	1	2	3
x	25	40	55
y	21	32	33

(values to 2 s.f.) [2 marks]

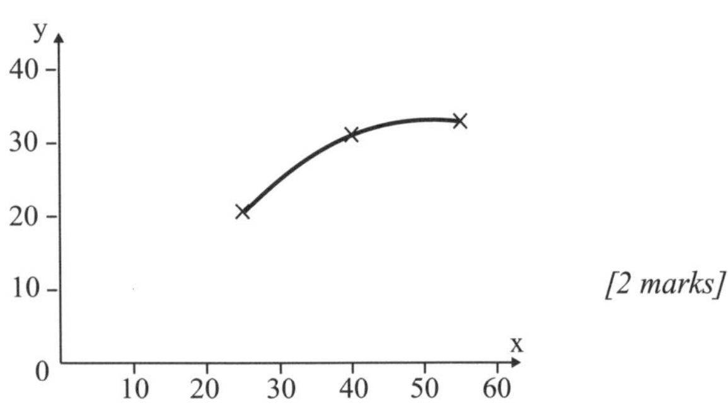

[2 marks]

c) $\begin{pmatrix} x \\ y \end{pmatrix} = \begin{pmatrix} 15t + 10 \\ 15\sqrt{3}t - 5t^2 \end{pmatrix}$ so $x = 15t + 10$

So $t = \left(\frac{x - 10}{15}\right)$ [1 mark]

Plug into y: $y = 15\sqrt{3}\left(\frac{x-10}{15}\right) - 5\left(\frac{x-10}{15}\right)^2$ [1 mark]

$y = \sqrt{3}x - 10\sqrt{3} - \frac{1}{45}(x^2 - 20x + 100)$

$y = -\frac{1}{45}x^2 + (\sqrt{3} + \frac{4}{9})x - (\frac{20}{9} + 10\sqrt{3})$ [1 mark]

$y = 2.2x - 20 - 0.02x^2$ [1 mark]

d) $\mathbf{r} = \begin{pmatrix} 15t + 10 \\ 15\sqrt{3}t - 5t^2 \end{pmatrix}$ $\mathbf{v} = \frac{d\mathbf{r}}{dt} = \begin{pmatrix} 15 \\ 15\sqrt{3} - 10t \end{pmatrix}$ [2 marks]

$\mathbf{a} = \frac{d\mathbf{v}}{dt} = \begin{pmatrix} 0 \\ -10 \end{pmatrix}$ [1 mark]

e) For large t, $y < 0$ — i.e. the aircraft crashes!

Page 143

1) Taking tractor and trailer together (and calling the resistance force on the trailer R):

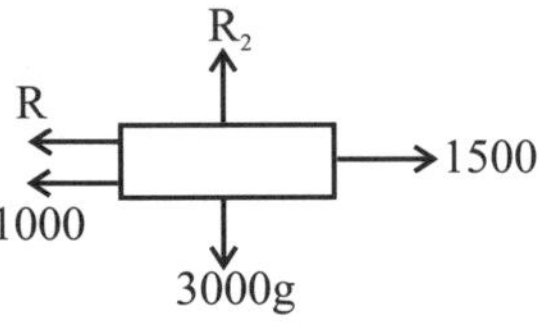

Resolving horizontally: $F_{net} = ma$

$1500 - R - 1000 = 3000 \times 0$

$R = 500$ N

For trailer alone:

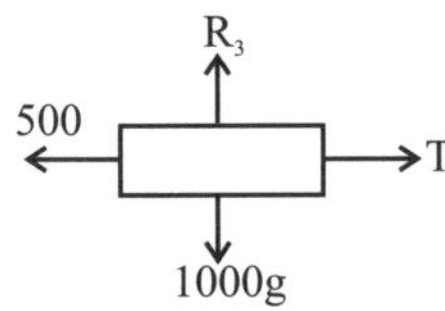

Resolving horizontally: $F_{net} = ma$

$T - 500 = 1000 \times 0$

$T = 500$ N

T could be found instead by looking at tractor forces horizontally.

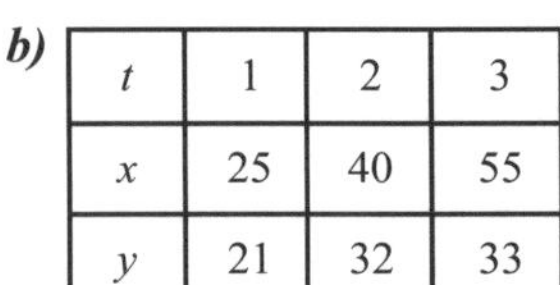

Answers

2)

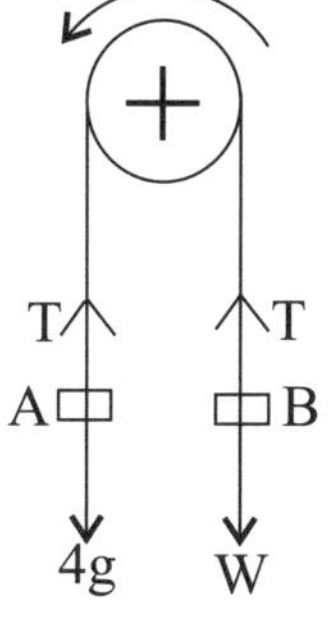

Resolving downwards for A: $F_{net} = ma$

$4g - T = 4 \times 1.2$

$T = 4g - 4.8$ ①

Resolving upwards for B: $F_{net} = ma$

$T - W = \frac{W}{g} \times 1.2$ ②

Sub ① into ② : $(4g - 4.8) - W = \frac{W}{g}(1.2)$

$4g - 4.8 = W(1 + \frac{1.2}{g})$

So $W = 30.6\ N$

3)a)

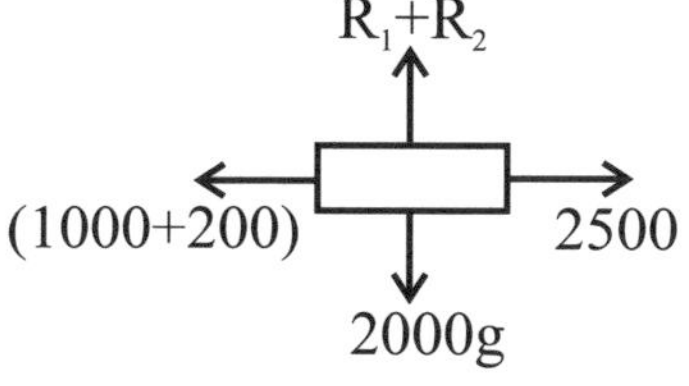

Resolving horizontally: $F_{net} = ma$

$2500 - 1200 = 2000a$

$a = 0.65\ ms^{-2}$

b) *Either:*

Caravan

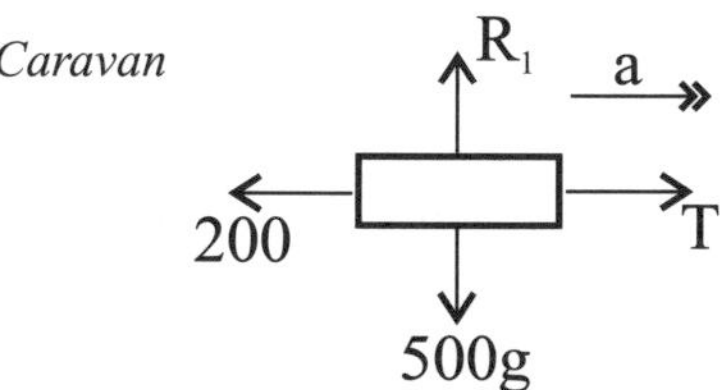

Resolving horizontally: $F_{net} = ma$

$T - 200 = 500 \times 0.65$ *[2 marks]*

$T = 525\ N$ *[1 mark]*

Or:

Car

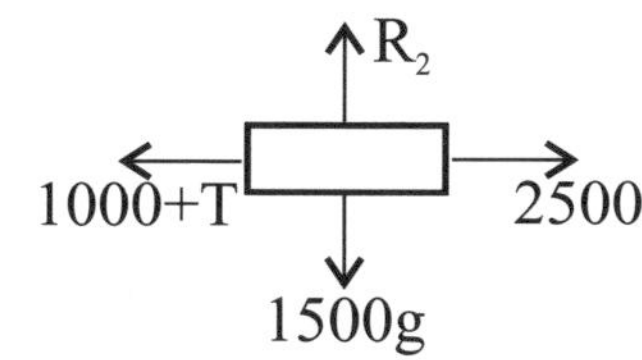

Resolving horizontally: $F_{net} = ma$

$2500 - (1000 + T) = 1500 \times 0.65$

$2500 - 1000 - T = 975$ *[2 marks]*

$1500 - 975 = T$

$T = 525\ N$ *[1 mark]*

<u>*Page 145*</u>

1)

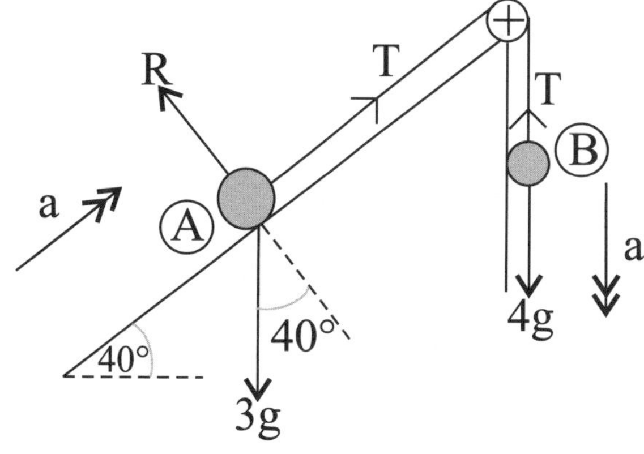

For B, resolving vertically: $F_{net} = ma$

$4g - T = 4a$

$T = 4g - 4a$ ①

For A , resolving in ↗ direction: $F_{net} = ma$

$T - 3g\sin 40 = 3a$ ②

Sub ① into ② : $4g - 4a - 3g\sin 40 = 3a$

$4g - 3g\sin 40 = 7a$

$a = 2.9\ ms^{-2}$

Sub into ① : $T = 4g - (4 \times 2.9) = 27.6\ N$ *(to 3 s.f.)*

If equilibrium, then for B: $T = 4g$

Then for Ⓐ

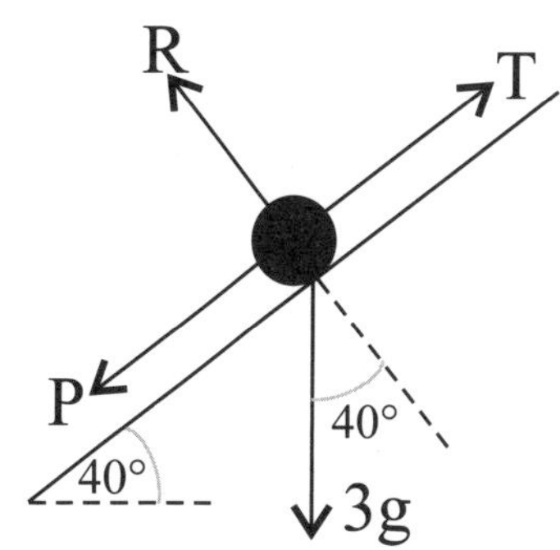

Resolving in ↗ direction: $F_{net} = ma$

$T - 3g\sin 40 - P = 0$

$4g - 3g \sin 40 = P$

$P = 20.3N$

Answers

2) a)

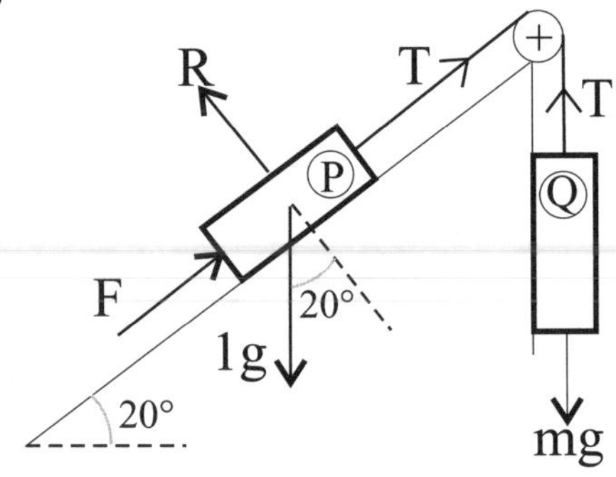

For Q: $F_{net} = ma$

Resolving vertically: $mg - T = 0$

$T = mg$ *[1 mark]*

For P: Resolving in ↖ direction:

$F_{net} = ma$

$R - 1g\cos 20° = 1 \times 0$

$R = g\cos 20°$ *[1 mark]*

Limiting friction: $F = \mu R$

$F = 0.1 \times g\cos 20°$ *[1 mark]*

Resolving in ↙ direction:

$1g\sin 20° - F - T = 1 \times 0$ *[1 mark]*

$1g\sin 20° - 0.1g\cos 20° - mg = 0$

$\sin 20° - 0.1\cos 20° = m$

$m = 0.248$ kg *[1 mark]*

b) *If Q = 1kg:*

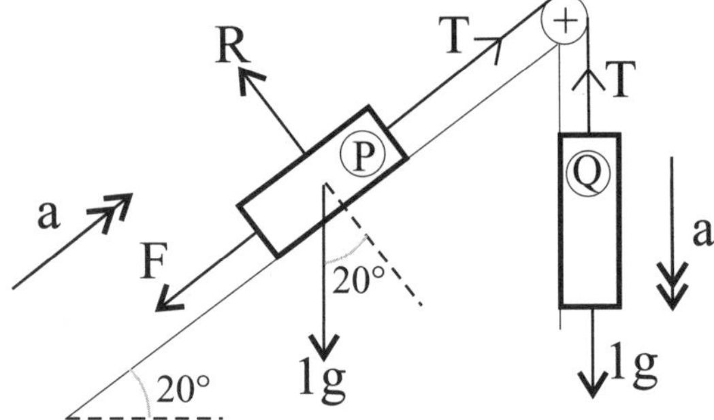

$F_{net} = ma$

For Q: resolving vertically: $1g - T = 1a$

$T = g - a$ ① *[1 mark]*

For P: resolving in ↖ direction: $R = g\cos 20°$

$F = \mu R = 0.1g\cos 20°$ *[1 mark]*

Resolving in ↗ direction: $T - 1g\sin 20° - F = 1a$

$T - 1g\sin 20° - 0.1g\cos 20° = 1a$ ② *[1 mark]*

Sub ① into ②:

$(g - a) - g\sin 20° - 0.1g\cos 20° = a$ *[1 mark]*

$g - g\sin 20° - 0.1g\cos 20° = 2a$

$5.527 = 2a$

$a = 2.76\ ms^{-2}$ *[1 mark]*

i.e. the masses move with an acceleration of $2.76\ ms^{-2}$

Page 147

1) $(5 \times 3) + (4 \times 1) = (5 \times 2) + (4 \times v)$

$19 = 10 + 4v$

$v = 2\frac{1}{4}\ ms^{-1}$ *to the right*

2) $(5 \times 3) + (4 \times 1) = 9v$

$19 = 9v$

$v = 2\frac{1}{9}\ ms^{-1}$ *to the right*

3) $(5 \times 3) + (4 \times -2) = (5 \times -v) + (4 \times 3)$

$7 = -5v + 12$

$5v = 5$

$v = 1\ ms^{-1}$ *to the left*

4) $(m \times 6) + (8 \times 2) = (m \times 2) + (8 \times 4)$

$6m + 16 = 2m + 32$

$4m = 16$

$m = 4$ kg

5) ***Before*** ***After***

(0.8) →4 (1.2) →2 (0.8) →2.5 (1.2) →v

$(0.8 \times 4) + (1.2 \times 2) = (0.8 \times 2.5) + 1.2v$ *[1 mark]*

$3.2 + 2.4 = 2.0 + 1.2v$

$v = 3\ ms^{-1}$ *[1 mark]*

Before ***After***

(1.2) →3 4← (m) (1.2+m) 0

$(1.2 \times 3) + (m \times -4) = (1.2 + m) \times 0$ *[1 mark]*

$3.6 = 4m$

$m = 0.9$kg *[1 mark]*

Page 149

1) *Impulse acts against motion, so* $I = -2$ Ns

$I = mv - mu$

$-2 = 0.3v - (0.3 \times 5)$

$v = -1\frac{2}{3}\ ms^{-1}$

2) *You need to find the particle's velocities just before and just after impact.*

Falling (down = +ve):

$u = 0,\ s = 2,\ a = 9.8,\ v = ?$

$v^2 = u^2 + 2as$

$v = \sqrt{2 \times 9.8 \times 2}$

$= 6.261\ ms^{-1}$

Rebound (this time, let up = +ve):

$v = 0,\ u = ?,\ a = -9.8,\ s = 1\frac{1}{3}$

$v^2 = u^2 + 2as$

$0 = u^2 + 2 \times -9.8 \times 1\frac{1}{3}$

$u = 5.112\ ms^{-1}$

Taking up = +ve:

Impulse $= mv - mu$

$= (0.45 \times 5.112) - (0.45 \times -6.261)$

$= 5.12$ Ns

Answers

3) a) Before — After

(4000) 2.5→ (1000) 0 — (5000) v→

$(4000 \times 2.5) + (1000 \times 0) = 5000v$ [1 mark]

$v = 2\ ms^{-1}$ [1 mark]

b) Impulse $= mv - mu$

$= (4000 \times 2) - (4000 \times 2.5)$ [1 mark]

$= -2000\ Ns$ [1 mark]

c) Track horizontal; no resistance (e.g. friction) to motion; wagons can be modelled as particles — or any other valid assumption.
[1 mark each for any two]

Section Seventeen — Projectiles

Page 151

1) Resolving horizontally:
$u = 120;\ s = 60;\ a = 0;\ t = ?$

$s = ut + \frac{1}{2}at^2$

$60 = 120t + \frac{1}{2} \times 0 \times t^2$

$t = 0.5\ s$

Resolving vertically: $u = 0;\ s = ?;\ a = 9.8;\ t = 0.5$

$s = ut + \frac{1}{2}at^2$

$= (0 \times 0.5) + (0.5 \times 9.8 \times 0.5^2)$

$= 1.23\ m$ (to 3 s.f.)

2) Resolving horizontally:
$u = 20\cos30°;\ s = 30;\ a = 0;\ t = ?$

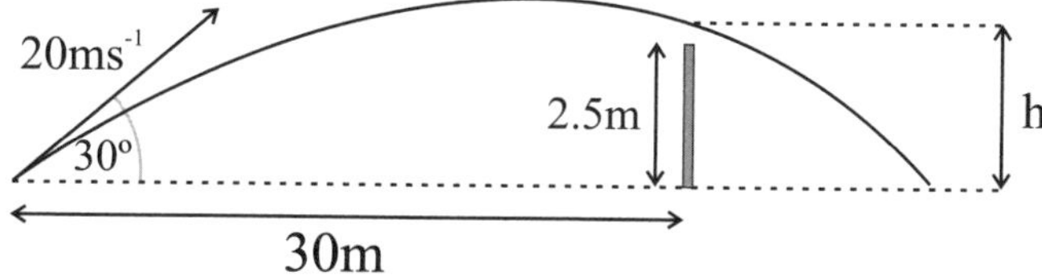

$s = ut + \frac{1}{2}at^2$ [1 mark]

$30 = (20\cos30° \times t) + (\frac{1}{2} \times 0 \times t^2)$

$t = 1.732\ s$ [1 mark]

Resolving vertically:
$s = h;\ u = 20\sin30°;\ t = 1.732;\ a = -9.8$

$s = ut + \frac{1}{2}at^2$ [1 mark]

$h = (20\sin30° \times 1.732) + (\frac{1}{2} \times -9.8 \times 1.732^2)$

$= 2.62\ m$ (to 3 s.f.) [1 mark]

Therefore the ball goes over the crossbar. [1 mark]
Assumptions: ball is a point mass, no air resistance/wind.
[1 mark]

Page 153

1) First find time of flight to top:
$v = 0;\ u = u\sin\alpha;\ a = -g,\ \alpha = ?$

$v = u + at$

$0 = u\sin\alpha - gt$

$t = \frac{u\sin\alpha}{g}$

(this is time taken to reach highest point — i.e. half way)

So $T = \frac{2u\sin\alpha}{g}$

Now plug in the values to get α.
$u = 22;\ g = 9.8;\ T = 4;\ \alpha = ?$

$4 = \frac{2 \times 22 \times \sin\alpha}{9.8}$

$\sin\alpha = 0.89$

$\alpha = 63.0°$ (to 3 s.f.)

2) a) $u_x = 50\cos25° = 45.3\ ms^{-1}$ [1 mark]

$u_y = 50\sin25° = 21.1\ ms^{-1}$ [1 mark]

b) Resolving vertically:
$u = 50\sin25°;\ a = -9.8;\ t = 3;\ s = ?$

$s = ut + \frac{1}{2}at^2$ [1 mark]

$= (50\sin25° \times 3) + (\frac{1}{2} \times -9.8 \times 9)$

$= 19.3\ m$ [1 mark]

Resolving horizontally:
$u = 50\cos25°;\ a = 0;\ t = 3;\ s = ?$

$s = ut$ [1 mark]

$= 50\cos25° \times 3$

$= 136\ m$ [1 mark]

So the target is 136 m away (horizontally) and is 19.3 m above the cannon.

c) $s = h;\ a = -9.8;\ u = 50\sin25°;\ v = 0$

$v^2 = u^2 + 2as$ [1 mark]

$0 = (50\sin25°)^2 + (2 \times -9.8 \times h)$

$h = \frac{(50\sin25°)^2}{2(9.8)}$

$= 22.8\ m$ [1 mark]

d) Resolving vertically:
$u = 50\sin25°;\ a = -9.8;\ t = 3;\ v = ?$

$v = u + at$

$v = 50\sin25° - (9.8 \times 3)$

$= -8.269\ ms^{-1}$ [1 mark]

Resolving horizontally:
$u = 50\cos25°;\ t = 3;\ v = ?;\ a = 0$

$v = u + at$

$= 50\cos25° + (0 \times 3)$

$= 45.315$ [1 mark]

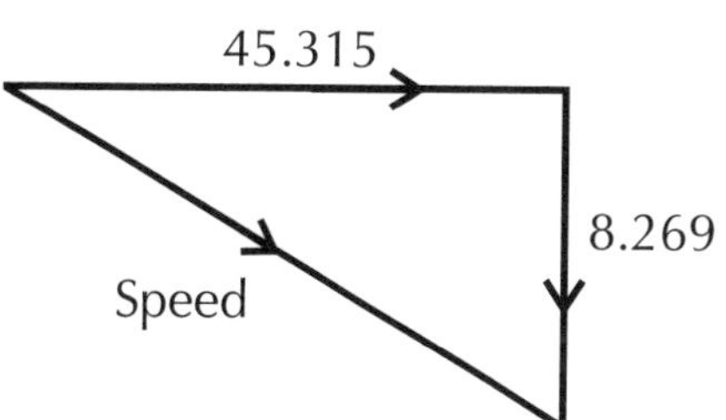

Speed $= \sqrt{45.315^2 + 8.269^2}$

$= 46.1\ ms^{-1}$ [1 mark]

Index

A

a real stinker 110
a right pain 106
acceleration 116-121, 123
acceleration due to gravity 116-117
acceleration-time graphs 118
adding integrals 67
air resistance 126
algebraic division 20
algebraic fractions 4
'almost' quadratic equations 19
arc lengths 53
areas 52, 53, 66-68
area under a motion graph 118
arithmetic progressions 43-45
arrows 122
assumptions 117, 126-127, 136, 142, 153
averages 74-75
averaging 28
$ax^2 + bx + c$ 10

B

$b^2 - 4ac$ 15
back-to-back 73
balance 127
balancing impulses 149
bar charts 72
beam 126
before and after diagrams 146-147
bias 85
binomial distribution 100-101, 103, 106
binomial expansions 49-50
binomial tables 101
book on a table example 127
box and whisker plots 77, 82
box of biscuits 89
brackets 2-3, 5, 10-12

C

casino 97
CAST diagrams 56-58
catwalks 126
Central Limit Theorem 107
change in momentum 148-149
circles 30-31
classic probability machine 86
coalescing particles 146
coding 80-81
coefficient of friction 132
coefficient of skewness 82
coefficients 14
collisions 146-149
combinations 92, 93
common denominator 4
common factors 3
complementary events 88-89
completing the square 12-13, 16
complicated questions 95, 119
components 120, 122-123, 128, 150
conditional probability 90-91
confidence intervals 108-109
connected particles 142-145
conservation of momentum 146-149
constant acceleration 116-117, 136, 140-141
constants 1
constants of integration 64-65, 121
converging 47
continuity correction 106
continuous random variable 104
coordinate geometry 27-29
coordinates 65, 141
correlation 110-113
correlation coefficient 111-113
cos 51
cosine rule 51-52, 128
cows 154
cubic graphs 32
cubics 22, 32
cumulative binomial tables 101
cumulative distribution function 96-97
cumulative frequency 76
cumulative frequency diagram 76-77, 83
curve sketching 32, 38
customers at least 12 years old 77

D

damned lies 73
definite integrals 66
degrees 51
denominators 4, 7
dependent variable 114
derivatives 35, 37, 65
deriving equations 152
differentiation 35-39, 120
direction 123-124, 128-129, 140
discrete random variable 94, 96-97, 99
discrete uniform distribution 97
disguised quadratics 19
displacement 116, 118, 124
distance 116-119, 151
distance-time graphs 118
distribution function 96-97
diverging 47-48
drag 126
dynamics 136-149

E

elimination 24
equations 1, 10, 19
equation of a circle 30-31
equations of motion 116-117, 120-121
equilibrium 126, 129, 132-134, 144-145
estimator 85, 107
events 86
exclusive 88
expected values 98-99, 103-104
experiment 86
experimental error 115
explanatory variable 114
exponentials 48
extrapolation 115

F

factor theorem 20-21
factorising 5, 10-11, 21-22
factors 3, 12, 21, 32
fence 83
finite sequences 42
forces 128-139, 149
fractional powers 6-7
fractions 4
frequency 71-72, 74, 79
frequency density 71
frequency polygons 72
friction 125, 130-133, 138-139
functions 1

G

geometric distribution 102
geometric interpretation 26
geometric progressions 46-48
gerbils 87
golf balls 152-153
gradients 27, 29 36-39, 118
graph sketching 32, 38
graph transformations 33
graphs of motion 118, 119, 141
gravity 116
grouped data 81

Index

H

height 151-153
hidden values 117
histograms 71-72
Horace 90

I

ice hockey 126
identities 1
impact 146-149
impulse 148-149
inclined planes 125, 131, 137-139, 144- 145
indefinite integrals 64
independent events 91
independent variable 114
indices laws 6
inequalities 17-18
inextensible bodies 126, 142, 145
infinite sequences 42
infinity 47-48, 66
integrals 67
integration 64-69, 120-121
interpercentile range 77
interpolation 115
interquartile range (IQR) 76, 83
intersection of lines 34

J

jargon 100

K

kinematics 116-121

L

labradors 113
laminas 126
laws of indices 6
least squares regression 115
less than or equal to 17
lies 73
light bodies 126, 142
limits 64, 66
limiting friction 132-133
limiting motion 139
linear equations 25, 27
linear inequalities 17
linear regression 114-115
linear transformation 112
line segment 28
lines of best fit 114
lobbing things through the air 150
logarithms (logs) 61-62
lotto 92, 103
lower class boundary 71, 75
lower fence 83
lower quartile (Q_1) 76-77, 82-83

M

magnitude 122, 128, 139-140
make life much easier 81
Mark Twain 89, 95
mathematical modelling 126-127
maximum (of a function) 37
maximum height 151-153
mean 74-75, 78-82, 97-101, 103-105, 107
median 74-77, 82-83
mid-class value 75, 79, 81
midpoint 28
minimum (of a function) 37
modal class 75
mode 74-75, 82-83, 94
modelling assumptions 117, 126-127, 136, 153
moments 134-135
momentum 146-149
more than one moving object 117
motion graphs 118-119
motion under gravity 116
multiplying out brackets 2
mutually exclusive 88

N

n-shaped graphs 9, 32
nasty-looking equations 19
natural numbers 45
negative correlation 110
negative skew 82-83
Newton's laws 136-137
non-uniform mass 126
normal approximation 106
normal distribution 104-107
normal distribution tables 104
normalising 104-106
normal reaction 125, 130, 132-133
normals (perpendicular lines) 29, 39
nose to tail arrows 122-125, 128
n^{th} terms 41, 43
numerical integration 68-69

O

odd functions 54
one of those days 102
outcomes 86, 93, 95, 97, 100
outliers 83

P

paint ball 83
parabola 32
parachutes 118
parallel lines 29
parameter 85
paranormal distribution 105
particles 117, 126-128, 136, 140, 144-147, 149, 150
Pascal's triangle 49
path equations 141, 152-153
Pearson's Coefficient of Skewness 82-83
pegs 142
percentiles 77
periodic functions 54
periodic sequences 42
permutations 92-93
perpendicular directions 123, 125, 136-139
perpendicular lines 29, 39
pie charts 72
planes 126
plastic aeroplane kits 126
PMCC 110-113
points 26-27, 37
point masses 126-127, 136
polygons of forces 129
polynomial division 20
polynomials 1, 21
population 84-85
population mean 85, 109
population variance 85
positive correlation 110-111
positive skew 82-83
powers 6, 35, 49, 64
presenting data 72
probability 86-91, 93-95, 100, 104
probability density function (pdf) 94-99, 104
probability distributions 94-95, 98-99
Product-Moment Correlation Coefficient (PMCC) 110-113
projectiles 150-153
pulleys 142
Pythagoras 28, 51, 123, 128

Index

Q

Q1 (lower quartile) 76-77, 82-83
Q3 (upper quartile) 76-77, 82-83
quadratic equations 9-16, 19, 25,
quadratic formula 14-16
quadratic graphs 9, 18
quadratic inequalities 18
Quartile Coefficient of Skewness 82-83
quartiles 76, 82-83

R

radians 51
radius 30-31
random events 86
random number tables 84
random sampling 84
random variable 94, 98-100, 104
range 74, 152-153
ranks 113
rationalising the denominator 7
really squishy sponge 100
rebounding balls 148
recurrence relation 41
reflections 33
regression 114-115
remainders 20-21
remainder theorem 20
residuals 115
response variable 114
resolving 122-125, 136-139, 142-145, 150-153
resultant force 136
resultant vectors 122-125, 128
right-angled triangles 51
right-angles 39
rigid bodies 126-127
rod 126
roots of equations 13, 15, 21, 53
rough surfaces 126, 132, 133, 144
running total 74, 76

S

sample 84-85, 107
sample mean 85, 107
sample space 86
sampling 84-85
sampling distribution 107
sausages 87
scalars 122
scatter diagram 110-112, 115
scitardauq gnisirotcaf 10
sequences 41-42
series 43-48
short trousers 75
significance level 48-51
simple random sampling 84
simplifying expressions 2-5
simultaneous equations 24-26, 28
sin 51
sine rules 51-52, 130
sketching quadratic graphs 9
skewness 82-83
slopes 125, 137-139, 144, 145
smooth pulleys 142
smooth surfaces 126, 127
solutions in an interval 56
solving triangles 52
Spearman's Rank Correlation Coefficient (SRCC) 113
speed 122, 123
splendid 17
spread 76, 78, 81
squared brackets 2
square roots 6-7, 32
stages 20
standard deviation 78-80, 83, 99, 104, 105
standard error 107
statics 126-135
stationary points 37-38
statistics 73, 85
stem and leaf diagrams 73
stretching 33
substitution 19, 25
subtracting integrals 67
sum of squares 78, 115
summarised data 80
summations 112
surds 7
surveys 87
sweeping statements 112, 115
S_{xx} 110, 112, 114
S_{xy} 110, 112, 114
symmetrical distribution 82, 104, 105

T

taking out common factors 3
tan 51
tangents to circles 31
tangents to curves 26, 36, 39
tasteless colour scheme 126
tension 126, 130, 142
terms 1, 46
the clever bit 96
thin bodies 126
thrust 130
time of flight 152-153
top to tail arrows 122-125, 128-129
train carriages 142
translating 33
transformations 33
trapezium rule 68-69
tree diagrams 88-90
trial 86
tricky things 87
trig functions 3
trigonometry 58, 123, 128-131, 150
turning effect 134-135
turning points 37, 40

U

uniform distribution 97
uniform mass 126, 134
unit vectors 122
upper class boundary 71, 75-76
upper fence 83
upper quartile (Q_3) 76-77, 82-83
u-shaped graphs 9, 32
"uvast" variables 116

V

variable acceleration 120-121
variables 1
variance 78-79, 97-101, 104-105, 107
vector motion 124-125
vectors 120-125, 146
velocity 116-122
velocity-time graphs 118
Venn diagram 86-88

W

waffle 124
weight 126, 130
WG Grace 77
wind 127, 153

Z

Z 104-108